Adaptive Diversification

MONOGRAPHS IN POPULATION BIOLOGY
EDITED BY SIMON A. LEVIN AND HENRY S. HORN

A complete series list follows the index

Adaptive Diversification

MICHAEL DOEBELI

PRINCETON UNIVERSITY PRESS

Princeton and Oxford

Copyright © 2011 by Princeton University Press
Published by Princeton University Press, 41 William Street,
Princeton, New Jersey 08540
In the United Kingdom: Princeton University Press,
6 Oxford Street, Woodstock,
Oxfordshire OX20 1TW
press.princeton.edu

Library of Congress Cataloging-in-Publication Data
Doebeli, Michael, 1961
Adaptive diversification / Michael Doebeli.
    p. cm. — (Monographs in population biology)
  Summary: "Adaptive biological diversification occurs when frequency-dependent
selection generates advantages for rare phenotypes and induces a split of an ancestral lineage
into multiple descendant lineages. Using adaptive dynamics theory, individual-based simulations,
and partial differential equation models, this book illustrates that adaptive diversification due to
frequency-dependent ecological interaction is a theoretically ubiquitous phenomenon"– Provided
by publisher.
  Includes bibliographical references and index.
  ISBN 978-0-691-12893-1 (hardback) — ISBN 978-0-691-12894-8 (paperback) 1. Adaptation
(Biology)—Mathematical models. 2. Biodiversity—Mathematical models. 3. Evolution
(Biology)—Mathematical models. I. Title.

  QH546.D64 2011
  578.4—dc22
                        2011006879

British Library Cataloging-in-Publication Data is available

This book has been composed in Times New Roman

Printed on acid-free paper. ∞

Printed in the United States of America

1   3   5   7   9   10   8   6   4   2

10   9   8   7   6   5   4   3   2   1

To Gabriela and Carlos, my beautiful children

"There are two camps. There always are.

That's the worst thing about democracy: there have

to be two opinions about every issue."

—Ross Macdonald, *Black Money*

Contents

Acknowledgments   xi

1. Introduction   1

2. Evolutionary Branching in a Classical Model for Sympatric
   Speciation   9

3. Adaptive Diversification Due to Resource Competition in Asexual
   Models   38
   3.1 Adaptive dynamics with symmetric competition kernels   50
   3.2 Adaptive dynamics with asymmetric competition kernels   64

4. Adaptive Diversification Due to Resource Competition in Sexual
   Models   74
   4.1 Evolutionary branching in sexual populations when assortative
        mating is based on the ecological trait (one-allele models)   82
   4.2 Evolution of assortative mating   90
   4.3 Evolutionary branching in sexual populations when assortative
        mating is not based on the ecological trait (two-allele models) 100
   4.4 A footnote on adaptive speciation due to sexual selection   110

5. Adaptive Diversification Due to Predator-Prey Interactions   113
   5.1 Adaptive diversification in classical predator-prey models   114
   5.2 An example of evolutionary branching in host-pathogen
        models   132

6. Adaptive Diversification Due to Cooperative Interactions   139
   6.1 Diversification in models for intraspecific cooperation   139
   6.2 Diversification in coevolutionary models of cooperation   148

7. More Examples: Adaptive Diversification in Dispersal Rates,
   the Evolution of Anisogamy, and the Evolution of Trophic
   Preference   163
   7.1 Diversification in dispersal rates   163

7.2 Diversification in gamete size: evolution of anisogamy   180
7.3 Diversification in trophic preference: evolution of
     complexity in ecosystems   189

 8. Cultural Evolution: Adaptive Diversification in Language and
     Religion   195
     8.1 Diversification of languages   197
     8.2 Diversification of religions   200

 9. Adaptive Diversification and Speciation as Pattern Formation in
     Partial Differential Equation Models   217
     9.1 Partial differential equation models for adaptive
          diversification due to resource competition   218
     9.2 Partial differential equation models for predator-prey
          interactions   236
     9.3 Partial differential equation models for adaptive
          diversification in spatially structured populations   242
     9.4 A general theory of diversification in partial differential
          equation models   258

10. Experimental Evolution of Adaptive Diversification in Microbes 262

Appendix: Basic Concepts in Adaptive Dynamics   279

Bibliography   306

Index   323

Acknowledgments

I am indebted to Steve Stearns, who took me under his wings when all I had was a PhD in Mathematics and no clue about evolution. Steve had a tremendous impact on evolutionary biology in Europe, and he had a tremendous impact on me. Along the way, I have met a number of scientists who I greatly admire, and who greatly influenced me in one way or another. Among them are Hans Metz, the visionary inventor of adaptive dynamics, Ulf Dieckmann, whose technical versatility always awed me, and Martin Ackermann, a great scientist, and a great friend. I would also like to thank Tim Killingback and Christoph Hauert for many interesting collaborations, and Karl Sigmund, Alan Hastings, and Peter Abrams for their continued support. Many thanks also go to all the people who passed through my research group and made my daily life interesting. I was particularly impressed with Maren Friesen, who, as an undergraduate, single-handedly established my experimental evolution lab, and with Slava Ispolatov, who seems to understand everything. Slava made essential contributions to Chapter 9. Alistair Blachford and Erik Hanschen were a great help with many technical aspects of producing this book. At the University of British Columbia I have a great group of colleagues, and I am particularly grateful to Dolph Schluter, Martin Barlow and Ed Perkins for their support. My ultimate thanks go to Nelly Pante for staying the course.

Adaptive Diversification

CHAPTER ONE

Introduction

Evolution occurs when organisms reproduce so that their offspring inherit certain characteristics, or traits. Variation in heritable traits, together with variation in reproductive success, generates evolutionary change in trait distributions. If the correlation between heritable variation and reproductive variation is (close to) zero, evolutionary change is neutral, and the trait distribution performs an evolutionary random walk. In contrast, evolution is adaptive if the correlation between heritable variation and reproductive variation is significantly different from zero.

Adaptive evolution is generally thought to be of central importance for the history of life on earth. The process of adaptation, whereby types that are better adapted to the prevalent circumstances leave more offspring than types that are less well adapted, is, for example, believed to have been the main driving force generating major evolutionary transitions (Szathmáry & Maynard Smith, 1995). By far the most widespread view of adaptation, both among experts and laymen, is that of an optimization process: Given a set of environmental conditions, the type that is best adapted to these conditions prevails. Determining the optimal type in a given situation, and understanding how genetic and developmental constraints impinge on the evolutionary trajectory toward such optimal types, have been among the main objectives in evolutionary theory.

One of the problems with viewing evolution as an optimization process is that this perspective leaves little room for diversity: the optimally adapted type has more offspring than all other types, and so eventually, all other types will go extinct, leaving the optimal type as the single type present. Of course, recurring mutations may constantly introduce genetic variation into a population, but optimization essentially generates uniformity. In particular, evolution of distinct ecological types out of a uniform ancestral lineage at the same physical location is precluded under the tenet of evolutionary optimization.

Yet understanding the evolution of diversity is one of the central and most fundamental problems in biology. To explain the evolution of diversity in the realm of the traditional optimization perspective, one needs to invoke geographical heterogeneity: if environmental conditions differ between different geographical locations, then different optimization problems must be solved, and hence different adaptations evolve in different locations. The process of diversification due to local adaptation to different environments is usually called ecological speciation (Schluter, 2000, 2009), but different local adaptations can also be generated by sexual selection (e.g., Lande, 1981). After their formation in separate geographical areas, different types may migrate to and coexist at the same location due to a plethora of genetic and ecological mechanisms, which have been the subject of intense study. However, physical separation, and hence an intrinsically nonbiological ingredient, is necessary to explain the emergence of diverse life forms if one views evolution primarily as an optimization process. Note that geographical isolation is also necessary for diversity to arise due to neutral evolution, but such a neutral theory of diversification has become less popular among evolutionary biologists (e.g., Hendry, 2009; Schluter, 2009), partly because it runs contrary to the generally accepted notion that diversity is paramount in nonneutral traits (i.e., in traits in which heritable variation and variation in reproductive success are significantly correlated).

Optimization theory has proved to be useful for gaining many evolutionary insights. However, it misses out on a class of ecological and evolutionary mechanisms that are intuitively appealing, and that opens up a whole new perspective on the problem of the evolution of diversity. These mechanisms operate whenever the relevant components of the environment determining selection pressures on a given focal type not only consist of abiotic, physical ingredients that may remain constant over evolutionary time, but also comprise other organisms that may be present in the environment. Whether these other organisms are individuals of the focal type's species, or part of other species with which the focal type interacts, it is often obvious that an individual's survival and fecundity generally depend on the ecological impact of other organisms. For example, if organisms with different traits eat different types of food, then whether a given trait confers a high food intake will depend on the traits of the other organisms currently present in the population (with food intake low if the other organisms have similar traits, and hence eat similar food). Moreover, the food intake of a given organism may change as the distribution of traits in the population changes. As a consequence, adaptation to constant conditions may rarely occur: as the population evolves, the biological environment changes, and hence the optimization problem changes as evolution unfolds.

The phenomenon that the evolving population is part of the changing environment determining the evolutionary trajectory is usually referred to as frequency-dependent selection. In the language of correlations used above, selection is frequency-dependent if the sign and magnitude of the correlations between heritable variation and reproductive variation change as a consequence of changes in the trait distribution that are themselves generated by such correlations. From an anthropocentric perspective, frequency dependence is of course very familiar, as it is obvious to us that in many of our enterprises, what is "good" for us depends very much on what everyone else is doing. In principle, applying such anthropocentric insights to evolutionary theory is deeply problematic, but it works in the context of frequency dependence, as it seems obvious that whether a trait or a behavioral strategy of an organism confers a fitness advantage may very much depend on the traits or strategies of other organisms.

In fact, from the perspective of mathematical modeling, the realm of frequency dependence in evolution is larger than the realm of situations in which selection is not frequency-dependent, because the absence of frequency dependence in a mathematical model of evolution essentially means that some parameters describing certain types of biological interactions are set to zero (or almost zero). Thus, in a suitable parameter space, frequency independence corresponds to the region around zero, while everything else corresponds to frequency dependence. In this way, frequency independence can be seen as a special case of frequency dependence, much like neutral evolution is a special case of adaptive evolution (with neutral evolution corresponding to certain correlations being near zero, see earlier). From a theoretical point of view, frequency-dependent selection should therefore be considered the norm, not the exception, for evolutionary processes.

Using the metaphor of a fitness landscape for evolutionary optimization problems, frequency dependence implies that the fitness landscape is changing as a consequence of evolutionary change. Thus, frequency dependence generates a kind of evolutionary feedback mechanism. In line with general dynamical systems theory, where feedback mechanisms can lead to many complicated scenarios, frequency dependence can lead to very interesting evolutionary dynamics. This is true in particular in models for the evolution of diversity. For example, it has long been known that frequency dependence can mediate the coexistence of different types without geographical isolation. This is most easily seen using the mathematical framework of evolutionary game theory, where frequency dependence can allow for coexistence between different strategies already in very simple games such as the Hawk-Dove game (Doebeli & Hauert, 2005). However, in evolutionary game theory, coexistence typically requires the a priori presence of different types. Moreover, game theory is limited in that it typically assumes payoff functions that are linear functions of the strategy distribution. These restrictions imply that traditional game theory cannot explain the origin of diversity out of uniform ancestral populations.

However, such explanations can be provided by an extension of game theory to nonlinear payoff functions that depend on continuously varying strategy traits. This extension is known as *adaptive dynamics*, a mathematical framework that has been developed by Hans Metz and others (Dercole & Rinaldi, 2008; Dieckmann et al., 2004; Dieckmann & Law, 1996; Geritz et al., 1998, 1997; Metz et al., 1996), and that has proved to be a very convenient and useful tool for studying many different aspects of long-term evolutionary dynamics of quantitative traits under frequency-dependent selection. In particular, using the concepts of evolutionarily singular points and evolutionary branching, adaptive dynamics has allowed to generally identify and classify adaptive processes that are conducive to evolutionary diversification. Evolutionary branching points are points in phenotype space that have, roughly speaking, two characteristic properties: they are attractors for the evolutionary dynamics, and they are fitness minima. Points with these properties can only exist if fitness landscapes change as the population evolves, that is, if selection is frequency-dependent. Evolutionary branching occurs if a fitness minimum is "catching up" with the evolving population: As long as the population is away from the evolutionary branching point, it sits on one side of the fitness minimum, but as it is moving up that side of the disruptive landscape, the landscape changes so that the population comes to lie closer and closer to the trough of the landscape until eventually it reaches the singular point, which coincides with the fitness minimum. At this point, the population experiences disruptive selection, and hence is prone to splitting into two diverging lineages.

If diversification occurs due to evolutionary branching, it is adaptive in the sense that when the population is sitting at the evolutionary branching point, being different from the population mean confers an adaptive advantage. Thus, the splitting of the population into different phenotypic branches is itself an adaptive process that is driven by frequency-dependent biological interactions. Such diversification due to frequency-dependent selection is called *adaptive diversification* (Dieckmann et al., 2004). In contrast to diversification scenarios unfolding due to geographical isolation, adaptive diversification requires ecological contact between the diverging lineages. Using traditional terminology, adaptive diversification can occur in sympatry or parapatry, but not in strict allopatry. Thus, adaptive diversification is in some sense the antithesis to allopatric speciation.

I am of course well aware that diversification under conditions of sympatry continues to have a rather tainted reputation in evolutionary biology. The prevalent skepticism is perhaps best reflected in treatments in which authors present a whole catalog of conditions that presumably need to be satisfied for "true" sympatric speciation to occur (Bolnick & Fitzpatrick, 2007; Coyne & Orr, 2004). Such lists often contain circular elements, for example, when one purported requirement is the presence of gene flow in sympatric speciation, even though it is clear that any form of speciation, whether sympatric or not, requires the eventual disruption of gene flow. To me, such discussions are reminiscent of medieval scholarly disputes about how many angels can dance on the head of a pin. In fact, I think that placing the geographic context at the center of the speciation problem detracts from the more important question about the actual biological mechanisms that lead to diversification and speciation. For example, Schluter (2009) provides an interesting discussion of speciation mechanisms, arguing that mutation order speciation, which occurs when different mutations are fixed in separate populations adapting to similar selection pressures, is generally less likely than ecological speciation, which subsumes all speciation processes that are driven by adaptation to divergent environmental conditions. Nevertheless, this discussion does not distinguish between different mechanisms for ecological speciation, and in particular does not mention the possibility that ecological speciation can be driven by frequency-dependent selection.

More generally, frequency-dependent selection does not really seem to be on the radar of mainstream evolutionary biology as a potential mechanism for generating major evolutionary patterns. For example, Estes & Arnold (2007) conducted an extensive study to test the feasibility of various population genetic models to explain the mechanisms generating evolutionary data series on many different time scales. Estes & Arnold (2007) had no qualms about using optimization models, but frequency-dependent selection was not even mentioned in their article, nor was it mentioned in a subsequent discussion of their work (Hendry, 2007).

Perhaps this book will serve the purpose of generating more awareness of frequency-dependent selection as a potentially powerful evolutionary mechanism, so that evolutionary biologists at least think twice before dismissing this perspective from their approach. After all, when studying a particular problem, the outcome of the analysis is often partly determined by the perspectives one has in mind at the outset. It is therefore important to keep an open mind, and this seems particularly true when studying the evolution of diversity, a field that seems to have been dominated by established doctrine for quite some time. That said, this book is of course itself biased toward demonstrating the potential role of frequency dependence for evolutionary diversification. However, the book is not about making any claims about the ubiquity of such adaptive diversification in real empirical systems (except to say that it does seem to occur in one particular microbial system, see Chapter 10), and it is not my intention to advocate adaptive diversification and speciation as the predominant mode of evolutionary diversification. Determining the mechanisms generating biological diversity is ultimately an empirical, not a theoretical, question. However, what I do indeed want to say with this book is that from a theoretical point of view, adaptive diversification, that is, diversification as a response to disruptive selection caused by frequency-dependent selection, is, after all, little more enigmatic than alternative explanations of diversification based on geographical isolation or neutral evolution. Thus, hopefully, this book will contribute to paving the way for a more parsimonious approach than currently often used when it comes to assessing the causes and mechanisms of diversification in natural systems.

One insight gained from modeling adaptive diversification is that convergence toward points in phenotype space at which selection turns disruptive does not require fine tuning of parameters. Instead, evolutionary branching is a structurally stable and robust dynamical process, and the conditions for its occurrence can be readily assessed using the theory of adaptive dynamics. Moreover, the basic insights about adaptive diversification obtained from studying adaptive dynamics and evolutionary branching can be corroborated using other methods, most notably stochastic, individual-based models and partial differential equation models, with which one can investigate the dynamics of pattern formation in polymorphic phenotype distributions. These frameworks confirm the basic results obtained from the theory of evolutionary branching and show that adaptive diversification into distinct phenotypic clusters is a generic outcome of frequency-dependent selection, and hence a theoretically plausible evolutionary process. It should be noted that this does not imply that this process should continually occur and that we should see it in operation at all times in a given lineage. On the one hand, diversification in the form of lineage splitting is but one possible evolutionary consequence of frequency-dependent selection, which can also cause other forms of increased phenotypic variation, such as sexual dimorphism or phenotypic plasticity (Rueffler et al., 2006b). On the other hand, once a lineage has diversified, the conditions for further evolutionary branching are often harder or impossible to satisfy, leading to a saturation of diversity much like that envisaged in adaptive radiations (Schluter, 2000).

But here I concentrate on the primary process of adaptive diversification, and the goal of this book is to make the theory and principles of adaptive diversification accessible to a wide audience by providing an overview of different types of models from a diverse range of ecological settings (including cultural evolution, see Chapter 8). In populations in which sexual reproduction leads to a reshuffling of phenotypes, the emergence of distinct lineages through adaptive diversification requires assortative mating mechanisms, which I will discuss fairly extensively in Chapter 4. In the presence of prezygotic isolating mechanisms due to assortative mating, adaptive diversification leads to adaptive speciation (Dieckmann et al., 2004), that is, to reproductive isolation between the emerging lineages. However, I want to emphasize that the main purpose of this book is not to address the various genetic mechanisms of assortment enabling adaptive speciation once the ecological conditions for diversification are satisfied, but to show that there are many different ecological scenarios that are conducive to diversification in the first place. Perhaps heretically, I view the evolution of assortative mating mechanisms as a secondary problem for diversification. The primary problem is to identify the circumstances that lead to selection for diversification, that is, to disruptive selection due to frequency-dependent ecological interactions. Once there is selection for diversification, evolution will seek a solution to the problem of allowing diversification to happen, and in particular to the problem of allowing assortative mating mechanisms to induce speciation in sexual populations. Although relying on more elaborate assumptions, this perspective is similar to the one traditionally taken when modeling evolution as an optimization problem: there the task is to find the evolutionary optimum given a certain set of selective circumstances, and once an optimum is found under the constraints assumed, one simply implies that the genetic machinery allows the organisms to achieve that optimum. Admittedly, the process of splitting is genetically much more complicated than converging to an optimal trait value. Nevertheless, in the perspective taken here splitting is a solution to a problem posed by frequency-dependent interactions, and just as evolution is likely to find an optimal solution, evolution is also likely to find a splitting solution, at least some of the time. The basic ecological question is: What are the ecological environments in which adaptive diversification becomes a solution?

This book is aimed at theoreticians, as well as theoretically inclined empirical researchers, starting at the graduate student level. Technical prerequisites for making the most of this book are a good understanding of calculus and of some basic concepts from dynamical systems theory. The appendix introduces some of the basic concepts from adaptive dynamics theory and hopefully provides a useful theoretical backdrop. Throughout the book, I have included a number of problems and exercises, which are set apart from the main text as Challenges. Many of these require a fair amount of work, and some of them, marked by a superscript *, are difficult and open ended, and may have the scope of a scientific publication. I hope that graduate students in particular will find these challenges useful, as only practice makes perfect.

Evolutionary Branching in a Classical Model for Sympatric Speciation

In 1966 John Maynard Smith published an article in which he countered the argument, made most forcibly by Ernst Mayr, that there is no mechanism for sympatric speciation consistent with the known facts of evolution. A mischievous John once told me during a workshop in the Swiss Alps how part of the motivation for writing this article was to "get on Mayr's nerves." Nevertheless, his model turned out to be the starting point for much subsequent work analyzing the conditions for speciation in the Levene type niche models (Levene, 1953) that were the basis for the Maynard Smith model. Much of this work, as reviewed, for example, in Fry (2003) and Kawecki (2004), concentrated on the genetic mechanisms for assortative mating and reproductive isolation, based on the assumption that the underlying niche ecology would generate disruptive selection. However, understanding the conditions under which disruptive selection arises in the first place is equally important, and indeed necessary for assessing whether diversification is a general outcome in the Maynard Smith model. In my opinion, formulating the model in the framework of adaptive dynamics offers the best perspective for understanding the ecological underpinnings of this model and how likely it generates disruptive selection and the maintenance of polymorphisms. Eva Kisdi and Stefan Geritz have provided a very nice explanation of this perspective (Geritz & Kisdi, 2000; Kisdi, 2001; Kisdi & Geritz, 1999), on which the exposition in this chapter is based. The purpose of this chapter is twofold. First, it shows that disruptive selection and polymorphism are scenarios that occur generically, that is, for a wide range of parameters, in a classical and widely used speciation model. Second, it provides an on-the-go introduction to some of the basic concepts of adaptive dynamics theory. A more detailed and technical treatment of elementary adaptive dynamics can be found in the Appendix.

The most famous scenario of biological diversification that has been studied using Levene type models is speciation through host race formation in phytophagous insects, and in particular in the apple maggot fly *Rhagoletis pomonella* (Berlocher & Feder, 2002; Bush, 1969; Johnson et al., 1996). However, conceptually such models potentially apply to many other scenarios of habitat specialization. In their simplest form, they work as follows (Levene, 1953; Maynard Smith, 1962, 1966). A population is assumed to occupy two habitats, in each of which individuals first undergo genotype-specific survival from zygotes to adult, after which genotype-unspecific density regulation brings the densities to cN and $(1 − c)N$. N is the constant total population size, and c and $1 − c$ are the relative contributions of the two habitats to the total population. All N surviving adults mate randomly, and each mating produces the same number of zygotes, which are dispersed randomly over the two habitats to form the next generation. Thus, generations are discrete, and viability in the two habitats is the only component of selection. The model assumes a diploid locus with three genotypes AA, Aa, and aa, whose relative probabilities of survival are $1 + K$, $1 + K$, and 1 in one habitat, and 1, 1, and $1 + k$ in the other habitat (with $K, k > 0$). That is, allele A is fully dominant in both habitats, and there is a trade-off between the viabilities in the two habitats. The assumption of dominance is not crucial and was made to avoid any kind of "concealed" heterozygote advantage that could potentially lead to polymorphism (Maynard Smith, 1966). This model is of course very simple, yet biologically intuitive, as it can be thought of as describing the simplest form of an environmental gradient or cline, along which a population experiences two different sets of external conditions (e.g., riverine and lacustrine). The main conclusions from this model in Maynard Smith (1966) were that maintenance of a polymorphism at the A-locus either requires strong selection, that is, large K and k, or fine tuning of K and k to the relative habitat sizes c and $1 − c$. The latter condition seems unlikely, but in fact turns out to be the key to understanding why the former condition is a natural outcome of gradual evolution. To show this, we need to formulate a related model in which allelic effects are continuous, and which therefore describes the evolutionary dynamics of a continuously varying, quantitative trait x.

To simplify the genetics, let's assume that the quantitative trait x is determined by a continuum of possible alleles at a haploid locus. This assumption is equivalent to asexual inheritance of the trait and would therefore trivialize questions related to reproductive isolation. However, the assumption is not crucial for understanding the ecological workings of the Maynard Smith model, and essentially the same outcomes as described below are obtained if one assumes a continuum of alleles with additive effects at a diploid locus (Kisdi, 2001; Kisdi & Geritz, 1999). Instead of assigning two habitat-specific fitness values to each of the two alleles A and a, we now need two functions describing the relative viability of a range of trait values x in each of the two habitats.

These viability functions should reflect a trade-off: trait values that have high viability in one habitat should have low viability in the other habitat, and vice versa. A natural way to define those functions is to assume that viability in habitat 1 is maximized for some trait value m_1 and decreases monotonically as the distance $|x - m_1|$ increases, whereas viability in habitat 2 is maximized for some other trait value m_2 and decreases monotonically as the distance $|x - m_2|$ increases. Following Kisdi & Geritz (1999), if d is half the distance between the two viability maxima, $2d = |m_1 - m_2|$, this can be achieved by choosing Gaussian viability functions of the form:

$$w_1(x) = \exp\left(-\frac{(x+d)^2}{2\sigma^2}\right) \quad \text{viability in habitat 1} \quad (2.1)$$

$$w_2(x) = \exp\left(-\frac{(x-d)^2}{2\sigma^2}\right) \quad \text{viability in habitat 2} \quad (2.2)$$

The parameter σ measures how fast viability declines with distance from the maximum, and for simplicity I assume that this parameter is the same in both habitats. Also for simplicity, I have assumed that the maximum relative viabilities are the same in both habitats. Obviously, these two assumptions should be relaxed in general, but relaxing them does not have a qualitative effect on the basic results described below. The habitat viabilities given by eqs. (2.1) and (2.2) are illustrated schematically in Figure 2.1. Note that it is not the absolute magnitude of the viability functions that matters, but their relative magnitude in the two habitats. Therefore, the scale on the y-axis in Figure 2.1 is arbitrary.

We now consider populations that are monomorphic for some trait value x and ask how such populations would change over evolutionary time. Clearly, if x is smaller than $-d$, the viability in both habitats increases as x increases. Therefore, in such populations selection will drive the trait x to larger values. Similarly, if a population is monomorphic for a trait x that is larger than d, selection will act to decrease x. So the question is, what happens if x lies between $-d$ and d? To answer this, we consider the fate of a mutant y appearing in the resident population that is monomorphic for x. Assuming discrete generations, the dynamics of the frequency q of the mutant y is given according to standard haploid population genetics (e.g., Otto & Day, 2007):

$$q_{t+1} = \left[c\frac{w_1(y)}{(1-q_t)w_1(x) + q_t w_1(y)} + (1-c)\frac{w_2(y)}{(1-q_t)w_2(x) + q_t w_2(y)}\right]q_t, \quad (2.3)$$

where c and $1 - c$ are the relative habitat sizes, and where q_t is the frequency of y in generation t. To assess invasion success of the mutant y in the resident x,

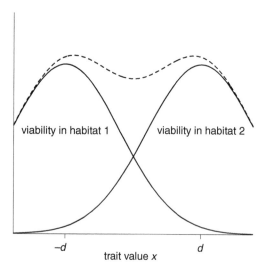

FIGURE 2.1. Example of viability functions in two habitats as a function of the trait value x, one having a maximum at trait value $x = -d$ and the other at trait value $x = d$. The dashed line represents the sum of the two viability functions.

we need to know its *invasion fitness*, that is, the expected number of offspring of the mutant when it is rare. The invasion fitness is denoted by $f(x, y)$ to emphasize that it depends on both the resident and the mutant trait values, and in the present situation it is given as the right hand side of eq. (2.3), divided by q_t and in the limit $q \to 0$:

$$f(x, y) = c\frac{w_1(y)}{w_1(x)} + (1 - c)\frac{w_2(y)}{w_2(x)}. \tag{2.4}$$

Note that $f(x, x) = 1$, which is the consistency condition that a mutant with the same trait value as the resident is neutral, and hence has the same growth rate as the resident, which is assumed to be in equilibrium. Note also that the invasion fitness function $f(x, y)$ is frequency-dependent, meaning that the invasion fitness of a mutant y depends on the current resident trait value x. To determine the evolutionary dynamics, we need to know, for any given resident x, the direction in which invasion fitness increases. That is, we need to determine the derivative of the invasion fitness with respect to the mutant trait and evaluated at the resident trait value. This quantity is called the *selection gradient*:

$$D(x) = \left.\frac{\partial f(x, y)}{\partial y}\right|_{y=x} = c\frac{w_1'(x)}{w_1(x)} + (1 - c)\frac{w_2'(x)}{w_2(x)}. \tag{2.5}$$

The adaptive dynamics of the trait x is then governed by the selection gradient. More precisely:

$$\frac{dx}{dt} = mD(x). \tag{2.6}$$

This is the *canonical equation* of adaptive dynamics for 1-dimensional trait spaces (Dieckmann & Law, 1996) (see Appendix for more details). The equation simply says that the rate of evolutionary change in the trait x, that is, the adaptive dynamics of the trait x, is proportional to the selection gradient. The constant of proportionality, that is, the parameter $m > 0$, is determined by the processes generating mutations in the trait x (see Appendix). In general, m depends on population size, because larger populations produce more mutations, and population size may in turn depend on the resident trait values (we will see many examples of this latter dependence in later chapters). Thus, in general, m is a function of x. However, in the simple models used here the population size is constant, and hence so is m, and by rescaling time in the adaptive dynamics (2.6) we can assume $m = 1$.

The next task is to find *singular points*, which are defined as the equilibrium points for the adaptive dynamical system given by eq. (2.6). Thus, singular points are solutions x^* to the equation

$$D(x^*) = 0. \tag{2.7}$$

Because $w_1'(-d) = 0$ and $w_2'(d) = 0$, as well as $w_2'(-d) > 0$ and $w_1'(d) < 0$, we have $D(-d) > 0$ and $D(d) < 0$. Therefore, there is at least one solution of $D(x^*) = 0$, that is, one singular point in the trait interval $[-d, d]$. In fact, with the viability functions w_1 and w_2 given by eqs. (2.1) and (2.2), it is easy to see that

$$D(x) = \frac{d - 2cd - x}{\sigma^2}. \tag{2.8}$$

It follows that there is exactly one singular point $x^* \in [-d, d]$, which is located at

$$x^* = d(1 - 2c). \tag{2.9}$$

One of the great virtues of the framework of adaptive dynamics is that it clearly highlights the difference between two types of stability that can be associated with singular points. The first type, *convergence stability*, concerns the stability of the singular point as an equilibrium point of the adaptive dynamical system given by the differential equation (2.6). Here the question is, if the current trait value is different from x^*, does it evolve toward x^*, eventually converging to the singular point? This question can be answered using basic calculus tools: an equilibrium x^* of the differential equation (2.6) is (locally) stable if and only if the derivative of the function D at x^* is negative, that is, if and only if

$$\frac{dD}{dx}(x)\bigg|_{x=x^*} < 0. \tag{2.10}$$

In the present case, it is clear from (2.8) that this condition is always satisfied, since $\frac{dD}{dx} = -1/\sigma^2$ is a negative constant. Moreover, because x^* is the only solution of $D(x^*) = 0$, we have $D(x) > 0$ for all $x < x^*$ and $D(x) < 0$ for all $x > x^*$. Therefore, whenever the resident value x is smaller than x^*, selection favors larger trait values, and whenever the resident value x is larger than x^*, selection favors smaller trait values. Thus, the singular point x^* is a global attractor of the differential equation (2.6), and the adaptive dynamics of the trait x always converges to the singular point.

But what happens evolutionarily after convergence to the singular point? This question is addressed by considering *evolutionary stability*, which is the second type of stability associated with singular points. By definition of D as the selection gradient, the first derivative of the invasion fitness function vanishes at the singular point:

$$\frac{\partial f}{\partial y}(x^*, y)\bigg|_{y=x^*} = D(x^*) = 0. \tag{2.11}$$

In general, the singular point is therefore either a maximum or a minimum of the invasion fitness function. More precisely, the singular point is *evolutionarily stable* if

$$\frac{\partial^2 f}{\partial y^2}(x^*, y)\bigg|_{y=x^*} < 0, \tag{2.12}$$

and the singular point is *evolutionarily unstable* if

$$\frac{\partial^2 f}{\partial y^2}(x^*, y)\bigg|_{y=x^*} > 0. \tag{2.13}$$

In the latter case, the singular point is a fitness minimum. Distinguishing between convergence stability and evolutionary stability is of fundamental importance, because the condition for convergence stability and the condition for evolutionary stability are in general not identical. In terms of the invasion fitness functions $f(x, y)$, the condition for convergence stability, eq. (2.10), is

$$\frac{dD}{dx}(x)\bigg|_{x=x^*} = \left[\frac{\partial^2 f(x, y)}{\partial x \partial y} + \frac{\partial^2 f(x, y)}{\partial y^2}\right]\bigg|_{y=x=x^*} < 0, \qquad (2.14)$$

whereas as the condition for evolutionary stability is given by (2.12). The two conditions differ by the mixed derivative $\partial^2 f(x, y)/\partial x \partial y$, evaluated at the singular point $y = x = x^*$. In particular, if this term is negative, it is possible that a singular point is convergent stable, but evolutionarily unstable, that is, that both inequalities (2.10) and (2.13) are satisfied. Such singular points are called *evolutionary branching points* for reasons that will be explained shortly.

Evolutionary branching points can easily occur in the Maynard Smith model. We already know that the singular point is always convergent stable, and the evolutionary stability is determined by

$$\frac{\partial^2 f}{\partial y^2}(x^*, y)\bigg|_{y=x^*} = c\frac{w_1''(x^*)}{w_1(x^*)} + (1 - c)\frac{w_2''(x^*)}{w_2(x^*)}. \qquad (2.15)$$

Thus, if the singular point lies for example in an area of trait space where both habitat viability functions w_1 and w_2 are convex (i.e., have positive second derivatives), then it is an evolutionary branching point, that is, convergent stable, but evolutionarily unstable. For instance, if $c = 0.5$, so that $x^* = d(1 - 2c) = 0$ lies exactly in the middle between the two habitat maxima $m_1 = -d$ and $m_2 = d$, then $w_1''(x^*) = w_2''(x^*) > 0$, that is, x^* is an evolutionary branching point, if and only if $d > \sigma$. Note that this is equivalent to requiring that the sum $w_1(x) + w_2(x)$ of the two habitat viability functions has a minimum at the midpoint $x = 0$, see Figure 2.1. Thus, in this case the convergent stable singular point is a fitness minimum whenever the two habitat maxima are distant enough compared to how fast within habitat viability decreases with phenotypic distance from the maximum.

Another way to understand the condition for evolutionary branching is to consider the trade-off between the viabilities in the two habitats, that is, by considering the graph of the curve $(w_1(x), w_2(x))$ for different trait values x around the singular point. This trade-off is illustrated schematically in Figure 2.2. It is easy to see that the condition for evolutionary branching, $d > \sigma$, is equivalent to the condition that the trade-off in habitat viabilities is convex around the singular point, that is, that the second derivative of the trade-off curve is positive. Compared to the generalist at the singular point, becoming more specialized for one of the habitats generates an increase in viability in that habitat, at the cost of a decrease in viability in the other habitat. If the trade-off is convex, then on balance it pays to become more specialized for either habitat,

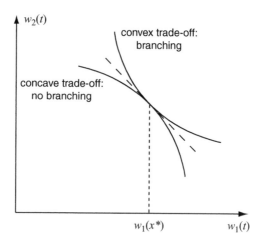

FIGURE 2.2. Schematic representation of the trade-off between habitat viabilities $w_1(x)$ and $w_2(x)$ around the singular point x^*. For convex trade-offs, the generalist strategy x^* is an evolutionary branching point, whereas for concave trade-offs, the generalist strategy is an ESS. Linear trade-offs (more precisely, trade-off curves for which the second derivative at x^* is 0) delineate the boundary between the two regimes.

whereas if the trade-off is concave, the generalist is on balance better off than more specialized types. The role of the curvature of viability trade-offs has been investigated in detail by Kisdi (2001) for the models considered here, and qualitatively similar results can be expected to hold in many other settings. For example, Claessen et al. (2007) showed that convex trade-offs are necessary for evolutionary branching in a consumer-resource model with two different types of resources, and Box 10.1 in Chapter 10 describes another example where convex trade-off between preferences for different resources are a necessary condition for evolutionary diversification.

Challenge: Show that under the assumptions used above, the trade-off curve $(w_1(x), w_2(x))$ is convex at the singular point x^* if and only if $d > \sigma$.

The key to understanding how a fitness minimum can be an attractor for the evolutionary dynamics lies in the frequency dependence of the invasion fitness function. Assuming that a singular point x^* is an attractor, and that the current resident x is some distance away from the singular point, the invasion fitness must have a positive slope in the direction of x^*. Thus, to first order the invasion fitness is a line through x with positive slope if $x < x^*$, and a line with negative slope if $x > x^*$, as illustrated in Figure 2.3. As the trait x approaches the singular point, this slope becomes shallower, until the invasion fitness vanishes to first order, that is, has slope 0, at $x = x^*$. This of

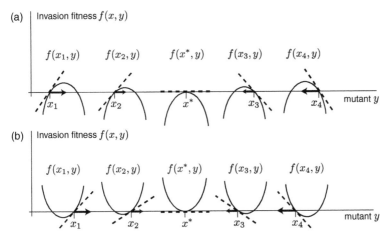

FIGURE 2.3. The two generic scenarios for approaching an evolutionary attractor in 1-dimensional trait spaces. The various continuous line segments show the invasion fitness of mutants y for different resident trait values x_i. The dashed lines indicate the selection gradients, that is, the slopes of the invasion fitness function at the resident values (i.e., when $y = x_i$). When this slope is negative, lower trait values than the current resident are favored, and vice versa when the selection gradient is positive. In both panels, the selection gradients generate convergence to the singular point x^*. In panel (a), this convergence is generated by invasion fitness functions with a negative curvature, so that the invasion fitness function $f(x^*, y)$ with the singular value x^* as the resident has a maximum at x^* (note that the selection gradient must vanish at the singular point x^*). Conversely, in panel (b) convergence to the singular point is generated by invasion fitness functions with a positive curvature, so that the invasion fitness function $f(x^*, y)$ has a minimum at x^*, and hence even though x^* is convergent stable, it can be invaded by all nearby mutants.

course already shows the action of frequency dependence: The invasion fitness changes as x converges toward x^*.

But what about the second order of the invasion fitness function? To second order, the invasion fitness function is a parabola, which is either open to below (concave) or open to above (convex). If it is open to below, then the current trait value x must lie on the side of the parabola that is farther from x^* in order for the slope of the parabola to be positive in the direction of x^*, as is illustrated in panel (a) of Figure 2.3. In this case, as the trait x converges to the singular point, the invasion fitness function becomes a parabola with maximum at x^*. Therefore, the convergent stable singular point is also evolutionarily stable in this case.

However, if to second order the fitness function is a parabola that is open to above, evolution can still drive the trait x toward x^* if x lies on the side of the parabola that is closer to x^*, as is illustrated in panel (b) of Figure 2.3. In this case, as the trait x converges to the singular point, the invasion fitness function becomes a parabola with minimum at x^*. Therefore, the convergent stable singular point is a fitness minimum, that is, an evolutionary branching point. Note that as long as the current population trait value x is different from x^*, there is directional selection toward x^*, despite the fact that the invasion fitness function is a parabola with positive curvature (and hence with a minimum). This is because as long as $x \neq x^*$, the trait x always lies to the side of the fitness minimum of the invasion fitness. However, as the trait x converges to x^*, the invasion fitness function keeps changing in such a way that when the trait x reaches x^*, the fitness minimum comes to lie exactly at x^* as well. Thus, frequency dependence, that is, the dependence of the invasion fitness function on the current trait value, generates a situation in which the fitness minimum catches up with the current trait value exactly as the trait value reaches the singular point.

It is very important to realize that the two scenarios depicted in Figure 2.3 are essentially the only possible scenarios for convergence to evolutionary attractors in one-dimensional trait spaces. Other convergence scenarios can only occur under special circumstances, in which the invasion fitness function vanishes not just to first, but also to second order at the singular point. Therefore, with frequency-dependent invasion fitness functions, convergence to a fitness minimum is in some sense just as likely as convergence to an evolutionarily stable singular point. For example, in the continuous Maynard Smith model with $c = 1/2$ discussed here, evolutionary branching points occur whenever $d > \sigma$, that is, roughly speaking for one half of parameter space. In particular, no fine tuning of parameters is necessary for evolutionary convergence to fitness minima. This is important for understanding the evolution of diversity, because evolutionary branching points are potential starting points for adaptive diversification, as we will see later.

Another way to detect the existence of evolutionary branching points is by means of *pairwise invasibility plots*, as lucidly explained in Geritz et al. (1998). In a pairwise invasibility plot, the x-axis represents resident trait values while the y-axis represent mutant trait values. For each coordinate pair (x, y), the plot indicates whether the mutant y can invade the resident x or not. Thus, when invasion dynamics occur in discrete time as in the Maynard Smith model discussed in this chapter (see eq. (2.4)), what is shown in a pairwise invasibility plot is the sign of $\ln f(x, y)$: if $\ln f(x, y) < 0$, then the long term expected number of offspring of the mutant is <1, and hence the mutant cannot invade, whereas if $\ln f(x, y) > 0$, the long term expected number of offspring of the mutant is >1, hence the mutant has a positive probability of invading the resident x (Dieckmann & Law, 1996). Similarly, when invasion dynamics occur in continuous time (as in many models discussed in this book), a pairwise invasibility plot shows the sign of $f(x, y)$ itself: if $f(x, y) < 0$, then the long-term growth rate of the mutant is <0, and hence the mutant cannot invade, whereas the mutant can invade if its long-term growth rate > 0, that is, if $f(x, y) > 0$. In either case, a pairwise invasibility plot thus consists of a number of different regions in the (x, y)-plane corresponding to different invasion dynamics. These regions are separated by curves defined by $f(x, y) = 1$ if invasion dynamics is discrete, and by $f(x, y) = 0$ if invasion dynamics is continuous. For example, in every pairwise invasion plot the diagonal (x, x) is part of the boundary separating the $+$ and $-$ regions, because when the mutant value y is the same as the resident value x, the growth rate of the mutant must be the same as that of the resident, which is assumed to neither grow nor decline in the long term (we have already seen that $f(x, x) = 1$ in the model studied in this chapter).

Pairwise invasibility plots are very useful because they allow one to detect the evolutionary dynamics graphically. For any given resident x, one simply needs to check the vertical axis through x in order to detect which mutants y can invade the resident and which ones cannot. Those that can may become the new resident if one assumes that invasion implies fixation, that is, that invasion of a mutant implies subsequent extinction of the resident. This may seem like a very restrictive assumption, but it turns out that it is not: as long as the mutant is close to the resident, the invasion implies fixation assumption may only be violated at special points, which are exactly the singular points of the adaptive dynamics (Geritz, 2005; Geritz et al., 1998). Away from these points, invasion of nearby mutants implies that the mutants become the new residents, which allows one to follow the evolutionary dynamics through a series of trait substitutions, as shown in Figure 2.4. Such a series is represented by arrows in the pairwise invasion plot, the first one of which starts at the "ancestral" resident, then points to the next invading mutant, which becomes the new resident, etc. The series of arrows ends at convergence stable singular points. In general, singular points are given in a pairwise invasibility plot as intersection points of the diagonal (x, x) with another part of the curve defined by $f(x, x) = 1$ (or by $f(x, x) = 0$ in continuous time). The evolutionary stability of singular point is then easily determined: if the vertical line through a singular point lies in an area of negative growth rates of possible mutant values, then no nearby mutant can invade, and the singular point is evolutionarily stable. Such a scenario is shown in Figure 2.4(a). On the other hand, if the vertical line through a singular point lies in an area of positive growth rates of possible mutant values, then all nearby mutants can invade, and the singular point is evolutionarily unstable. In that case, if the singular point is also convergence stable, it is an evolutionary

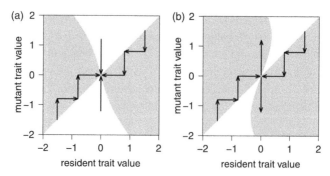

FIGURE 2.4. Pairwise invasibility plots for the invasion fitness function (2.4) corresponding to the two generic scenarios depicted in Figure 2.3. Gray areas indicate regions in (x, y)-space where $f(x, y) > 1$, and white areas indicate regions where $f(x, y) < 1$. Panel (a) illustrates convergence to the trait value 0 through trait substitutions, starting from either low or high trait values. The singular point 0 is evolutionarily stable, because the vertical line through this point is contained in the white region, indicating that all mutants have a growth rate that is smaller than 1 in the resident with trait value 0. In panel (b), the trait value 0 again attracts evolutionary trajectories starting from either low or high trait values, but in this case the singular point is evolutionarily unstable, because the vertical line through this point is contained in the gray region, indicating that all (nearby) mutants have a growth rate that is larger than 1 in the resident with trait value 0. Thus, in panel (b) the singular point 0 is an evolutionary branching point. This is reflected in the parameter values used: in both panels, $\sigma = 1$ and $c = 0.5$, the latter ensuring that 0 is a singular point; in (a), $d = 0.75 < \sigma$, and in (b), $d = 1.5 > \sigma$, which is the condition for evolutionary branching.

branching point. Such a scenario is shown in Figure 2.4(b). It is apparent from Figure 2.4 that pairwise invasibility plots are very useful for determining qualitative features of one-dimensional adaptive dynamics, and Geritz et al. (1998) have shown how such plots can be used for a much more detailed analysis than the one given here. Pairwise invasibility plots also reveal certain features of two-dimensional adaptive dynamics, as we will see later.

Convergence stable singular points are the attractors of adaptive dynamics in a given dimension, but not necessarily the endpoint of the evolutionary process. If a convergence stable singular point is evolutionarily stable, then by definition of evolutionary stability, no nearby mutant has a positive growth rate, and hence no nearby mutant can invade (at least not based on the deterministic dynamics deriving from the assumption of infinite population size that we used here all along). Thus, if only sufficiently small mutations are allowed, the population will simply remain at the singular point, which therefore indeed represents the endpoint of the evolutionary process in this case.

However, if the singular point is a fitness minimum, *every* nearby mutant has a positive growth rate and hence can invade. Moreover, two nearby mutants on either side of the singular point can coexist. Mathematically speaking, convergence stability together with evolutionary instability guarantee that two phenotypes $x_1 = x^* - \epsilon$ and $x_2 = x^* + \delta$ with $\epsilon, \delta > 0$ satisfy $f(x_1, x_2) > 0$ and $f(x_2, x_1) > 0$ if ϵ and δ are small, where f is the invasion fitness function (2.4). The pertaining mathematical argument is explained in the Appendix. Thus, two nearby mutants on either side of x^* can each invade the other when rare, which implies coexistence. Therefore, two such mutants satisfy the requirements for polymorphism in the original Maynard Smith model for discrete types (strictly speaking, I am now referring to a haploid version of the Maynard Smith model). It is interesting to note that two coexisting phenotypes x_1 and x_2 as shown earlier are necessarily close to each other and hence have similar viabilities in both habitats. Therefore, such mutants would have small and similar values of K and k in the corresponding classical Maynard Smith model. According to the classical results, coexistence of these phenotypes thus requires fine-tuning of the habitat viabilities. However, it is a crucial insight that this fine-tuning is provided by the evolutionary dynamics itself: the fine-tuning necessary for coexistence of nearby phenotypes is a consequence of convergence to the evolutionary branching point x^*. After convergence, coexistence of nearby phenotypes on either side of x^* is guaranteed by the conditions for convergence stability and evolutionary instability (see Appendix).

Given coexistence of the phenotypes $x_1 = x^* - \epsilon$ and $x_2 = x^* + \delta$, we can now ask again how the population state changes if we allow for (small) mutations. Thus, we now view the two coexisting phenotypes x_1 and x_2 as two resident "strains," and we try to determine the fate of rare mutants occurring in either of the resident strains. In general, the dynamics of the frequency q of a mutant type y in a resident population (x_1, x_2) is given by

$$q_{t+1} = q_t \left[c \frac{w_1(y)}{p_1 w_1(x_1) + p_2 w_1(x_2) + q_t w_1(y)} \right]$$
$$+ q_t \left[(1 - c) \frac{w_2(y)}{p_1 w_2(x_1) + p_2 w_2(x_2) + q_t w_2(y)} \right], \qquad (2.16)$$

where p_1 and p_2 are the frequencies of the resident types x_1 and x_2. If the mutant y is rare, so that $q_t \approx 0$, we can neglect the terms $q_t w_1(y)$ and $q_t w_2(y)$ in the denominators on the right-hand side of this equation, and we can assume that the resident population is at equilibrium. In fact, it can be shown that a system of coexisting resident strains x_1 and x_2 converges to a unique equilibrium at which $p_1 = p^*$ and $p_2 = 1 - p^*$, where p^* is the equilibrium frequency of strain x_1. This resident equilibrium p^* is itself a function of the two resident phenotypes x_1 and x_2 and can be calculated analytically (see Challenge later). Therefore, the invasion fitness of rare mutants y in the resident (x_1, x_2) becomes:

$$f(x_1, x_2, y) = c \frac{w_1(y)}{p^* w_1(x_1) + (1 - p^*) w_1(x_2)}$$

$$+ (1 - c) \frac{w_2(y)}{p^* w_2(x_1) + (1 - p^*) w_2(x_2)}. \qquad (2.17)$$

Inserting the expression for p^* as a function of x_1 and x_2 into this expression for the invasion fitness functions and using elementary algebraic manipulations, it is easy to see that the invasion fitness function is actually independent of the parameter c. More precisely:

$$f(x_1, x_2, y) = \frac{w_1(y) \, (w_2(x_2) - w_2(x_1)) + w_2(y) \, (w_1(x_1) - w_1(x_2))}{w_1(x_1) w_2(x_2) - w_1(x_2) w_2(x_1)}. \qquad (2.18)$$

This may be surprising at first but reflects the fact that any habitat asymmetry in the invasion fitness is exactly cancelled by the effect of habitat asymmetry on the equilibrium frequencies of resident strains, which are in turn determinants of the frequency-dependent invasion fitness (through their appearance in the denominators of eq. (2.17)).

Challenge: Use (2.3) to calculate the equilibrium frequencies p^* and $1 - p^*$ of two coexisting strains, then prove the equality (2.18) by substituting the expression for p^* into the invasion fitness function (2.17).

The selection gradients in the two resident strains x_1 and x_2 are again given by the derivative of the invasion fitness functions with respect to mutant trait values and evaluate it at the resident strains:

$$D_1(x_1, x_2) = \left. \frac{\partial f(x_1, x_2, y)}{\partial y} \right|_{y=x_1}$$

$$= c \frac{w_1'(x_1)}{p^* w_1(x_1) + (1 - p^*) w_1(x_2)}$$

$$+ (1 - c) \frac{w_2'(x_1)}{p^* w_2(x_1) + (1 - p^*) w_2(x_2)} \qquad (2.19)$$

$$D_2(x_1, x_2) = \left.\frac{\partial f(x_1, x_2, y)}{\partial y}\right|_{y=x_2}$$

$$= c\frac{w_1'(x_2)}{p^* w_1(x_1) + (1 - p^*)w_1(x_2)}$$

$$+ (1 - c)\frac{w_2'(x_2)}{p^* w_2(x_1) + (1 - p^*)w_2(x_2)}. \tag{2.20}$$

Thus, the two-dimensional adaptive dynamics of the population of two coexisting strains become

$$\frac{dx_1}{dt} = D_1(x_1, x_2) \tag{2.21}$$

$$\frac{dx_2}{dt} = D_2(x_1, x_2) \tag{2.22}$$

Here I have again omitted the details of the mutational process that generates new mutations in each strain. Taking the mutational process into account would rescale the time in the two equations above, but would otherwise not have a qualitative effect on the adaptive dynamics (at least as long as we assume that mutations occur independently in the two resident strains; see the Appendix). This system thus describes the adaptive dynamics of the population after convergence to the evolutionary branching point and establishment of an initial polymorphism consisting of two phenotypes on either side of the branching point. Note that the adaptive dynamics given by eqs. (2.21) and (2.22) is independent of the relative habitat productivity c, because the invasion fitness function (2.17), which determines the selection gradients and hence the adaptive dynamics, is independent of c.

For the habitat viability functions w_1 and w_2 given by eqs. (2.1) and (2.2), one can explicitly calculate the equilibrium p^* as a function of x_1 and x_2 (see Challenge above), which upon inserting into eqs. (2.19) and (2.20) yields the following expressions for the selection gradients:

$$D_1(x_1, x_2) = \frac{-x_1}{\sigma^2} + \frac{d\left(\exp\left(\frac{2x_1 d}{\sigma^2}\right) + \exp\left(\frac{2x_2 d}{\sigma^2}\right)\right)}{\sigma^2\left(\exp\left(\frac{2x_1 d}{\sigma^2}\right) - \exp\left(\frac{2x_2 d}{\sigma^2}\right)\right)}$$

$$- \frac{2d\exp\left(\frac{(2d - x_1 + x_2)(x_1 + x_2)}{2\sigma^2}\right)}{\sigma^2\left(\exp\left(\frac{2x_1 d}{\sigma^2}\right) - \exp\left(\frac{2x_2 d}{\sigma^2}\right)\right)} \tag{2.23}$$

$$D_2(x_1, x_2) = \frac{-x_2}{\sigma^2} + \frac{d\left(\exp\left(\frac{2x_1 d}{\sigma^2}\right) + \exp\left(\frac{2x_2 d}{\sigma^2}\right)\right)}{\sigma^2\left(\exp\left(\frac{2x_2 d}{\sigma^2}\right) - \exp\left(\frac{2x_1 d}{\sigma^2}\right)\right)}$$

$$- \frac{2d\exp\left(\frac{(2d - x_2 + x_1)(x_1 + x_2)}{2\sigma^2}\right)}{\sigma^2\left(\exp\left(\frac{2x_2 d}{\sigma^2}\right) - \exp\left(\frac{2x_1 d}{\sigma^2}\right)\right)}. \qquad (2.24)$$

The approach to understanding the two-dimensional adaptive dynamical system given by these selection gradients is in principle the same as in the one-dimensional case considered before: We first look for equilibrium points of the adaptive dynamics, and then analyze convergence stability as well as evolutionary stability of these equilibria. The first thing to note is that as long as x_1 and x_2 are close to x^* and $x_1 < x^* < x_2$, the direction of adaptive change in both residents is away from x^*, that is, $dx_1/dt < 0$ and $dx_2/dt > 0$. There-fore, initially the adaptive dynamics is diverging from the singular point x^*. This is not hard to see from (2.23) and (2.24), and in fact, initial evolutionary divergence from a branching point in one-dimensional phenotype spaces does not depend on the particular form of the selection gradients. Instead, initial divergence is a necessary consequence of the requirements for an evolutionary branching point, that is, from convergence stability and evolutionary instability of x^*. This is explained in the Appendix. Therefore, the existence of the evo-lutionary branching point x^* guarantees evolutionary diversification, at least in the haploid asexual case considered here: Evolutionary convergence to x^* is followed by the establishment of two adaptively diverging branches.

To investigate the dynamics after this initial divergence, we first need to find singular points (x_1^*, x_2^*) in two-dimensional phenotype space, that is, equilibrium points of the adaptive dynamics given by the system of differ-ential equations (2.21) and (2.22). Such points are given by the conditions $D_1(x_1^*, x_2^*) = D_2(x_1^*, x_2^*) = 0$, that is, by the condition that both selection gra-dients vanish at (x_1^*, x_2^*). Convergence stability of singular points is tested using a classical tool from the theory of differential equations: the Jacobian matrix. This matrix is obtained by taking derivatives of the selection gradients with respect to all arguments and evaluating at the singular point. Thus, the Jacobian matrix of the adaptive dynamical system given by (2.21) and (2.22) at the singular point is given by

$$J(x_1^*, x_2^*) = \begin{pmatrix} \left.\dfrac{\partial D_1(x_1, x_2)}{\partial x_1}\right|_{x_1 = x_1^*, x_2 = x_2^*} & \left.\dfrac{\partial D_1(x_1, x_2)}{\partial x_2}\right|_{x_1 = x_1^*, x_2 = x_2^*} \\ \left.\dfrac{\partial D_2(x_1, x_2)}{\partial x_1}\right|_{x_1 = x_1^*, x_2 = x_2^*} & \left.\dfrac{\partial D_2(x_1, x_2)}{\partial x_2}\right|_{x_1 = x_1^*, x_2 = x_2^*} \end{pmatrix} \qquad (2.25)$$

According to general theory for differential equations (e.g., Edelstein-Keshet, 1988), the Jacobian matrix defines a system of linear differential equations that approximate the dynamics of the full system (2.21) and (2.22) in the vicinity of the equilibrium (x_1^*, x_2^*). In particular, the Jacobian matrix determines the local stability of the singular point, which is convergent stable if both eigenvalues of the Jacobian matrix have negative real parts. In that case, all trajectories of the adaptive dynamics that are started close enough to the singular point will converge to the singular point. (Note that this generalizes the notion of convergence stability in the one-dimensional system (2.6) studied earlier, in which the Jacobian matrix is one-dimensional, that is, simply a number. That number is $dD/dx(x^*)$, which must be negative for convergence stability of the singular point x^*.)

On the other hand, evolutionary stability of a singular point (x_1^*, x_2^*) is checked by evaluating the second derivatives of the invasion fitness function $f(x_1^*, x_2^*, y)$ with respect to y at the values $y = x_1^*$ and $y = x_2^*$. This determines whether, in a resident population consisting of the two strains x_1^* and x_2^*, the invasion fitness function has, as a function of mutant trait values, a fitness minimum or a fitness maximum at the values x_1^* and x_2^*.

In the two-habitat model studied here, it is in general not possible to derive analytical expressions for the singular points (x_1^*, x_2^*), and hence for the Jacobian matrices $J(x_1^*, x_2^*)$ or for the second derivatives of the invasion fitness functions. However, numerically, one can easily study many different examples of the adaptive dynamics given by the selection gradients (2.23) and (2.24), as well as the corresponding questions of convergence and evolutionary stability. Based on such investigations I make the following conjectures.

- The adaptive dynamical system given by the selection gradients (2.23) and (2.24) has a unique convergent stable singular point (x_1^*, x_2^*), which is symmetrical, that is, $x_2^* = -x_1^*$. The phenotypic values of the singular point satisfy $-d < x_1^* < 0$, and hence $0 < x_2^* = -x_1^* < d$; therefore, the two phenotypic branches lie symmetrically in between the two phenotypes d and $-d$ that are optimal in the two habitats. In fact, at the singular point the two phenotypic branches are typically close, but not identical to the respective habitat optima. Thus, the two emerging phenotypic branches can be considered as specialists in their respective habitat.
- The convergent stable singular point $(x_1^*, -x_1^*)$ of the adaptive dynamics given by the selection gradients (2.23) and (2.24) is evolutionarily stable. That is, the invasion fitness function $f(x_1, x_2, y)$ given by eq. (2.18)

satisfies the two inequalities

$$\left.\frac{\partial^2 f(x_1^*, -x_1^*, y)}{\partial y^2}\right|_{y=x_1^*} < 0 \tag{2.26}$$

$$\left.\frac{\partial^2 f(x_1^*, -x_1^*, y)}{\partial y^2}\right|_{y=-x_1^*} < 0 \tag{2.27}$$

(Note again that these claims are independent of the relative habitat size c, because the invasion fitness function (2.18) is independent of c.)

*Challenge**: Prove the above conjecture.

Assuming that this conjecture is true, we can amalgamate the information to obtain a classification of the possible evolutionary scenarios in the continuous Maynard Smith model introduced in this chapter. First, if the phenotypic distance between the two habitat optima is relatively small compared to the range of phenotypes having a high viability within each habitat, the trait x converges to an evolutionarily stable singular point representing a generalist strategy, which is evolutionarily stable and hence represents the endpoint of the evolutionary process. This makes sense intuitively, for under these conditions it is possible to attain a high viability in both habitats by having a phenotype that lies somewhere in between the two habitat optima. Since we assume the two habitat viability functions to only differ in their optimum but otherwise have the same functional form, this generalist lies exactly in the middle between the optima if the two habitats are equally productive.

Second, if the phenotypic distance between the two habitat optima is relatively large compared to the range of phenotypes having a high viability within each habitat, the trait x first converges to an evolutionary branching point representing a generalist strategy that is a minimum for the invasion fitness function. After convergence to this generalist strategy, the population splits into two coexisting phenotypic branches. These branches diverge evolutionarily from each other, and the system converges toward an evolutionarily stable singular point representing two coexisting habitat specialists. Thus, in this case evolutionary branching leads to *adaptive diversification*. Interestingly, the resulting coexisting strains lie symmetrically with respect to the intermediate generalist strategy $x = 0$ even if the habitats have different productivities, that is, even if the parameter $c \neq 1/2$. This is because habitat asymmetries not only generate differential contributions of the two habitats to invasion fitness, but also lead to unequal frequencies of coexisting resident strains, and it turns out that these two effects exactly cancel each other in the invasion fitness function (2.18), so that the two-dimensional adaptive dynamics is independent of c. Thus, while evolutionary branching points occurring with asymmetric habitat productivity are always asymmetric with respect to the two habitat optima (i.e., are never exactly at the midpoint between these optima), the two-dimensional adaptive dynamics ensuing after evolutionary branching results in symmetric diversification with respect to habitat optima even when habitat productivity is asymmetric. (As we will see later, it is of course possible to devise alternative models in which differential habitat productivity does indeed result in evolutionary asymmetry.)

Because the two coexisting specialists resulting from evolutionary branching have very different phenotypes, their viabilities in the two habitats differ substantially. Therefore, coexistence of these phenotypes is mediated by a large difference in the parameters K and k in the classical Maynard Smith model, where these parameters define the trade-off between the viabilities in the two habitats. According to classical theory, a large difference between K and k is a robust way of achieving coexistence in the Maynard Smith model, in the sense that no fine-tuning of the parameters is required for polymorphism, which is reflected in the fact that in the model with continuous phenotypes, coexistence is not only possible at the singular point $(x_1^*, -x_1^*)$, but typically for a large region of phenotype pairs (x_1, x_2) surrounding the singular point. This can be seen with the help of pairwise invasibility plots, as follows. Pairwise invasibility plots, such as shown in Figure 2.4, tell us which mutants y can invade which residents x. By taking the mirror image of such a plot along the diagonal, that is, by switching the x- and y-axes, the plot tells us which mutants x can invade which residents y. Now coexistence of two types x_1 and x_2 requires that each type can invade the other when rare (note that this is also the condition for polymorphism in the classical Maynard Smith model). Therefore, superposition of a pairwise invasibility plot with its mirror image along the diagonal reveals the regions of coexistence, that is, those regions in (x_1, x_2)-space in which x_1 can invade x_2 when x_1 is rare (i.e., the mutant) and x_2 is common (i.e., the resident), and vice versa. This is illustrated in Figure 2.5 for a case in which the one-dimensional adaptive dynamics of the continuous Maynard Smith model has an evolutionary branching point, and the two-dimensional adaptive dynamics ensuing after evolutionary branching results in coexistence of two evolutionarily stable types (x_1^*, x_2^*). As can be seen from Figure 2.5, the singular point (x_1^*, x_2^*) lies in the interior of a large region of coexistence. This shows that after evolutionary branching, the adaptive dynamics takes the system to a region of robust coexistence in the original Maynard Smith model.

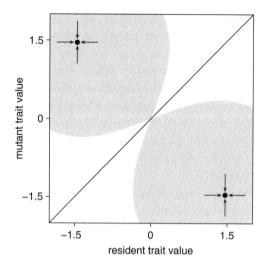

FIGURE 2.5. Superposition of the pairwise invasibility plot from Figure 2.4b with its mirror image along the diagonal. The gray region indicates coexistence of resident and mutant trait values: for resident-mutant pairs in this region, the mutant trait can invade a population that is monomorphic for the resident trait, and the resident trait can invade a population that is monomorphic for the mutant trait. Within the gray region, the two-dimensional adaptive dynamics given by (2.21) and (2.22) converges to the singular points indicated in the diagrams and having approximate coordinates (−1.48, 1.48) and (1.48, −1.48), respectively. Thus, the coordinates of the singular points lie very close to the respective maxima of the viability functions in the two habitats (which are at −1.5 and 1.5). Both singular points are convergent stable, with the initial conditions determining which singular point the adaptive dynamics approaches. Moreover, both singular points are evolutionarily stable in the sense that no mutants can invade either phenotypic branch if the two resident branches have the trait values given by the singular points.

We can now see that putting the classical Maynard Smith model into the context of continuously varying phenotypes using the framework of adaptive dynamics provides a very useful perspective for understanding the conditions for polymorphism in this classical model. In the classical model, one set of conditions, valid when differences in habitat viability are small, requires fine-tuning of parameters, while the other set of conditions requires large differences in habitat viabilities. In the corresponding adaptive dynamics model, convergence to an evolutionary branching point provides the fine-tuning necessary for polymorphism when viability differences are small, and subsequent adaptive divergence of coexisting phenotypic branches takes the polymorphic population to regions in phenotype spaces where viability differences are large and coexistence therefore robust. Since existence of evolutionary branching points is a generic phenomenon occurring for a large region in parameter space, adaptive diversification into coexisting specialists is a robust and general phenomenon, at least in the haploid (asexual) version of the continuous Maynard Smith model considered thus far.

Frequency dependence is the driving force behind adaptive diversification in the continuous Maynard Smith model (and in all the other models that will be discussed in this book). In the Maynard Smith model, frequency dependence occurs because habitat viabilities determine relative fitness in each habitat. More precisely, the within habitat fitness of a given phenotype is given by the phenotype's viability in that habitat divided by the mean viability of the population in that habitat. The mean viability in turn depends on the phenotypic composition of the population, that is, on the frequencies of the various phenotypes present.

Frequency dependence based solely on relative fitness in an otherwise constant environment is conceptually the simplest scenario to produce evolutionary branching. More complicated scenarios occur when there is an explicit feedback from the phenotypes to the environment, which happens for example when different phenotypes generate different ecological dynamics. In this case, the population dynamics generated by a resident phenotype can impinge on the invasion fitness of a mutant. In the Maynard Smith model, this could occur with explicit density-dependent population regulation in each habitat. In this scenario, the parameter c in the classical model, describing the relative contribution of the two habitats to the total population, becomes a dynamic variable that is determined by the population dynamics in each habitat, which is in turn determined by the habitat viability. Thus, the contribution of each habitat now depends on the viabilities, which is usually referred to as hard selection. Under hard selection, the invasion fitness is not only determined by the relative viabilities of residents and mutants, but also by the equilibrium population density of the resident population. Meszéna et al. (1997) analyzed a hard selection version of the continuous Maynard Smith model in which the number of adults recruited in each patch is not a constant. Due to the additional complication of hard selection, their model is harder to analyze mathematically, but allows for a greater variety of evolutionary scenarios. For example, the adaptive dynamics does not always converge to a generalist singular point, and instead a single specialist may be convergent and evolutionarily stable. This can happen because if the resident is a specialist in one habitat, then a mutant that is slightly worse in that habitat, but slightly better in the other habitat, may not be able to make up for the disadvantage in the first habitat because recruitment from the second habitat is too low (since the mutant, being only slightly different from the resident would still be much better adapted to the first habitat than to the second). Thus, hard selection hinders the evolution of generalists, and hence evolutionary diversification emanating from generalist singular points. Nevertheless, Meszéna et al. (1997) have shown that in their hard selection model, evolutionary branching is still a generic outcome that occurs for large regions in parameter space. See Brown & Pavlovic (1992) for another example of diversification in Levene models with hard selection and migration.

> *Challenge*: Investigate diversification in the habitat model of Brown & Pavlovic (1992) using adaptive dynamics (see also Chapter 8 in Vincent & Brown (2005)).

Further interesting complications can arise when frequency-dependence is due to selection in more than two habitats. This case has been considered by Geritz et al. (1998), who showed that in the case of three habitats, evolutionary branching from a generalist singular point can lead to the establishment of a convergent stable dimorphic population in which one of the strains is evolutionarily unstable. As a consequence, secondary evolutionary branching occurs in that strain, leading to a convergent and evolutionarily stable trimorphic population in which the three coexisting strains correspond to specialists in the three habitats.

> *Challenge*: Generalize the model introduced here to $n > 3$ habitats and investigate patterns of evolutionary branching.

Eva Kisdi and Stefan Geritz have studied the continuous Maynard Smith model under the assumption of continuously varying alleles at a diploid locus in great detail (Geritz & Kisdi, 2000; Kisdi & Geritz, 1999). In this case, the adaptive dynamics also always first converges to a generalist singular point, after which the population may or may not undergo evolutionary branching. However, when evolutionary branching occurs the ensuing adaptive dynamics of polymorphic populations can be more complicated than in the haploid case. Polymorphism in the haploid case as described earlier corresponds to symmetric polymorphic singular points in the diploid model at which both homozygotes are habitat specialists, and the heterozygote is an intermediate generalist that has a lower average viability than both homozygotes. This case was the focus of interest in the classical Maynard Smith model (Maynard Smith, 1966), and Kisdi & Geritz (1999) have shown how it can arise through evolutionary branching in continuous traits. However, in diploid models evolutionary branching can also lead to asymmetric polymorphic singular points, at which one of the homozygotes is a specialist in one habitat, the heterozygote is specialist in the other habitat, and the second homozygote lies either to the left or to the right of both habitat maxima and hence has low viability in both habitats. Moreover, it can happen that symmetric and asymmetric polymorphic singular points are simultaneously convergent stable, in which case the trajectory of the adaptive dynamics depends on the initial polymorphic conditions established after convergence to the evolutionary branching point.

Regardless of whether the adaptive dynamics converges to a symmetric or to an asymmetric polymorphism after evolutionary branching, it must be kept in mind that in sexual populations, the evolution of a polymorphism itself does not correspond to speciation as long as mating is random. To obtain speciation, some form of assortative mating mechanism would have to be established, whereby individuals with a given ecological phenotype (given habitat viabilities) preferentially mate with similar phenotypes. Geritz & Kisdi (2000) have studied this problem, showing that in all polymorphic scenarios, the populations are likely to go through a phase in which heterozygotes are at a disadvantage. This phase can be either permanent (e.g., when there are no asymmetric singular points) or transient (under certain conditions when there are asymmetric singular points). Maynard Smith (1966) has already noted that heterozygote inferiority generates selection for assortative mating, and the question of the evolution of various types of assortative mating in populations undergoing evolutionary branching will be investigated in a different modeling context in Chapter 4 of this book. Here we note that Geritz & Kisdi (2000) have shown that evolutionary branching in the diploid, continuous Maynard Smith model generally favors the evolution of reproductive isolation, even if assortative mating is determined by an ecologically neutral locus, in which case a linkage disequilibrium between the assortative mating locus and the ecological locus is necessary for reproductive isolation. Interestingly, even when there are convergent stable asymmetric singular points in the diploid model with random mating, the evolution of assortative mating always restores the outcome of the haploid model and leads to a symmetric state with both homozygotes being habitat specialists.

Other potentially important mechanisms for reproductive isolation in the Maynard Smith model are reduced migration between the habitats and habitat choice (Kawecki, 2004). In general, reduced migration and habitat choice should be favored because they increase the probability that phenotypes stay in the habitat for which they were selected. The role of migration and habitat choice for speciation in the Maynard Smith model has been studied quite extensively (see Kawecki (2004) for a review), and the same is true for assortative mating mechanisms in general (Fry, 2003; Kawecki, 2004). The majority of these studies assumed discrete allelic effects at the ecological locus determining habitat viabilities and were therefore based on the seemingly tenuous assumptions for the maintenance of polymorphism in the classical Maynard Smith model. The purpose of this chapter was to show that polymorphism should be viewed as a generic and robust feature of the Maynard Smith model, as it naturally arises through evolutionary branching in the adaptive dynamics of continuous traits.

This demonstrates the usefulness of both the adaptive dynamics framework, and of the classical niche models introduced by Levene (1953), for studying the problem of evolutionary diversification as a natural consequence of gradual convergence to a fitness minimum in phenotype space. However, as Rueffler et al. (2006b) have pointed out, evolutionary branching is not the only possible consequence of convergence to regimes of disruptive selection, and instead other forms of diversity may arise under these conditions. Most notably, various forms of phenotypic plasticity have traditionally been regarded as a mechanism for phenotypic diversification. For example, in the Levene models discussed in this chapter, one could envisage a scenario in which a single genotype produces different phenotypes with certain probabilities, each adapted to one of the habitats. Exactly this possibility has been studied by Leimar (2005), who used Levene models to compare the likelihood of evolutionary branching and phenotypic plasticity as alternative consequences of disruptive selection. Box 2.1 explains some of the main results of Leimar (2005), which make it clear that it is important to keep in mind that adaptive diversification in the form of evolutionary branching is only one of a number of possible evolutionary outcomes of disruptive selection due to frequency-dependent interactions (Rueffler et al., 2006b). A further possibility, which consists of widening resource utilization, is discussed in the next chapter (Box 3.1). Nevertheless, in this book I concentrate on scenarios of genetic diversification due to evolutionary branching, rather than on scenarios resulting in genetically monomorphic populations consisting of phenotypically plastic individuals.

BOX 2.1

EVOLUTIONARY BRANCHING VERSUS PHENOTYPIC PLASTICITY IN LEVENE AND LOTTERY MODELS

Leimar (2005) extended Levene models with two habitat patches specified by two viability functions w_1 and w_2 given by (2.1) and (2.2) and with a single phenotypic dimension x to models with three-dimensional phenotypes (z_1, z_2, q). In any given generation and habitat, such a phenotype expresses trait value z_1 with probability q and trait value z_2 with probability $1 - q$. Thus, in this model phenotypic plasticity means that different trait values are expressed with different probabilities (and not that trait z_1 is expressed in habitat 1 and trait z_2 is expressed in habitat 2). In each of the two habitats, a phenotype (z_1, z_2, q) is still assigned two viability values $\tilde{w}_1$ and $\tilde{w}_2$, respectively, but these functions are now given by

$$\tilde{w}_1(z_1, z_2, q) = qw_1(z_1) + (1 - q)w_1(z_2) \quad \text{viability in habitat 1} \quad \text{(B2.1)}$$

$$\tilde{w}_2(z_1, z_2, q) = qw_2(z_1) + (1 - q)w_2(z_2) \quad \text{viability in habitat 2} \quad \text{(B2.2)}$$

Note that a three-dimensional phenotype (z_1, z_2, q) is the same as a one-dimensional phenotype x if $z_1 = z_2 = x$, irrespective of the value of q. In particular, if x^* is a branching point for the adaptive dynamics of the one-dimensional trait, then a population that is monomorphic for the three-dimensional phenotype (x^*, x^*, q) can be expected to be under disruptive selection. However, genetic diversification due to evolutionary branching is now not the only option anymore. Rather, instead of undergoing a split into two clusters consisting of genotypes (z_1, z_1, q_1) and (z_2, z_2, q_2) with $z_1 \neq z_2$, the population could stay monomorphic, but move in three-dimensional trait space to a point (z_1, z_2, p) with $z_1 \neq z_2$ (note that with $z_1 \neq z_2$, the viability in both habitats depends on p). This would correspond to a population that is genetically monomorphic but consists of individuals that are phenotypically plastic and express the two trait values z_1 and z_2 with probabilities p and $1 - p$.

To investigate the evolution of plasticity, one has to consider adaptive dynamics in the three-dimensional trait space (z_1, z_2, q). In principle, such an analysis is again based on the invasion fitness function and proceeds in analogy to the case of one-dimensional trait spaces, as explained in the Appendix (see also Dieckmann & Law, 1996; Leimar, 2005). However, both convergence and evolutionary stability of singular points are technically more complicated concepts in higher dimensions, and the evolutionary dynamics near singular points have not been fully classified to date. Nevertheless, following Leimar (2005) interesting insights can be gained regarding the evolution of randomized strategies (z_1, z_2, q).

Let $f(z_1, z_2, q, z_1', z_2', q')$ be the invasion fitness (i.e., the long-term growth rate) of a rare mutant type (z_1', z_2', q') in a resident population that is monomorphic for (z_1, z_2, q). The the selection gradient is a vector

$$D(z_1, z_2, q) = \begin{pmatrix} \left.\dfrac{\partial f}{\partial z_1'}\right|_{z_1'=z_1, z_2'=z_2, q'=q} \\[2mm] \left.\dfrac{\partial f}{\partial z_2'}\right|_{z_1'=z_1, z_2'=z_2, q'=q} \\[2mm] \left.\dfrac{\partial f}{\partial q'}\right|_{z_1'=z_1, z_2'=z_2, q'=q} \end{pmatrix} = \begin{pmatrix} D_1(z_1, z_2, q) \\ D_2(z_1, z_2, q) \\ D_3(z_1, z_2, q) \end{pmatrix}. \quad \text{(B2.3)}$$

BOX 2.1 (*continued*)

The adaptive dynamics of the trait (z_1, z_2, q) is then given as follows (see Appendix):

$$
\begin{pmatrix}
\dfrac{dz_1}{dt} \\[2mm]
\dfrac{dz_2}{dt} \\[2mm]
\dfrac{dq}{dt}
\end{pmatrix}
= m(z_1, z_2, q) G(z_1, z_2, q) D(z_1, z_2, q). \tag{B2.4}
$$

Here $m(z_1, z_2, q)$ is a scalar function describing the rate of production of new mutations, and $G(z_1, z_2, q)$ is a 3×3-matrix describing the covariances of mutational effects on the three traits (Dieckmann & Law, 1996). Unless the covariance matrix G has special properties (more precisely, unless this matrix is singular, i.e., has a nonzero eigenvector), singular points of the adaptive dynamics (B2.4) are solutions (z_1^*, z_2^*, q^*) of $D(z_1^*, z_2^*, q^*) = 0$ (i.e., all partial derivatives defining D are 0). Convergence stability of a singular point is then determined by the Jacobian matrix J of first derivatives of the selection gradient D, evaluated at the singular point, whereas evolutionary stability is determined by the Hessian matrix H of second derivatives of the invasion fitness functions f_i $(i = 1, \ldots, 3)$, also evaluated at the singular point. The Jacobian matrix at the singular point is

$$
J =
\begin{pmatrix}
\left. \dfrac{\partial D_1}{\partial z_1} \right|_{z_1 = z_1^*, z_2 = z_2^*, q = q^*} & \cdots & \left. \dfrac{\partial D_1}{\partial q} \right|_{z_1 = z_1^*, z_2 = z_2^*, q = q^*} \\[2mm]
\vdots & \ddots & \vdots \\[2mm]
\left. \dfrac{\partial D_3}{\partial z_1} \right|_{z_1 = z_1^*, z_2 = z_2^*, q = q^*} & \cdots & \left. \dfrac{\partial D_3}{\partial q} \right|_{z_1 = z_1^*, z_2 = z_2^*, q = q^*}
\end{pmatrix}, \tag{B2.5}
$$

whereas the Hessian matrix at the singular point is

$$
H =
\begin{pmatrix}
\left. \dfrac{\partial^2 f}{\partial z_1^2} \right|_{z_1 = z_1^*, z_2 = z_2^*, q = q^*} & \cdots & \left. \dfrac{\partial_2 f}{\partial z_1 \partial q} \right|_{z_1 = z_1^*, z_2 = z_2^*, q = q^*} \\[2mm]
\vdots & \ddots & \vdots \\[2mm]
\left. \dfrac{\partial_2 f}{\partial q \partial z_1} \right|_{z_1 = z_1^*, z_2 = z_2^*, q = q^*} & \cdots & \left. \dfrac{\partial^2 f}{\partial q^2} \right|_{z_1 = z_1^*, z_2 = z_2^*, q = q^*}
\end{pmatrix} \tag{B2.6}
$$

The singular point (z_1^*, z_2^*, q^*) is called strongly convergent stable if the Jacobian matrix is negative definite in the sense that all eigenvalues of its symmetric part (with entries $(J_{ij} + J_{ji})/2$) are strictly negative. In that case, the singular point is asymptotically stable for any biologically realistic covariance matrix G (mathematically, for any covariance matrix that is positive definite). The singular point is called evolutionarily stable if the Hessian matrix H is negative definite.

In analogy to the one-dimensional case (eq. (2.4)), the invasion fitness f is determined by the habitat viabilities $\tilde{w}_1$ and $\tilde{w}_2$, given by eqs. (B2.1) and (B2.2), as

$$f(z_1, z_2, q, z_1', z_2', q') = c \frac{\tilde{w}_1(z_1', z_2', q')}{\tilde{w}_1(z_1, z_2, q)} + (1 - c) \frac{\tilde{w}_2(z_1', z_2', q')}{\tilde{w}_2(z_1, z_2, q)}. \qquad \text{(B2.7)}$$

For simplicity, we assume $c = 1/2$ in what follows. It is convenient to introduce the following change of coordinates:

$$\zeta = q z_1 + (1 - q) z_2 \qquad \text{(B2.8)}$$

$$\eta = \frac{z_2 - z_1}{2} \qquad \text{(B2.9)}$$

$$\rho = 1 - 2q \qquad \text{(B2.10)}$$

Now ζ is the mean of the two primary traits z_1 and z_2 that are expressed plastically, η measures the phenotypic plasticity, and ρ parametrizes the probability q. Using these new variables and basic rules of calculus, one can then derive the following facts (Leimar, 2005):

- Singular points of the three-dimensional adaptive dynamics (B2.4) are of the form $(x^*, 0, \rho)$, where x^* is the singular trait value from the one-dimensional adaptive dynamics (2.6), and ρ is any number in the interval $[-1, 1]$. In particular, there is a whole line of singular points for the adaptive dynamics (B2.4).

- The Jacobian J at an equilibrium point $(x^*, 0, \rho)$ has only two nonzero entries. The first one measures convergence stability of x^* in the ζ-direction and is always <0, which is equivalent to the result that x^* is always convergent stable in the one-dimensional case. The second one, denoted by B, measures convergence stability in the η-direction and may be positive or negative. This implies that the phenotype subspace defined by $\eta = 0$ may or may not be convergent stable. If it is not, that is, if $B > 0$, then there is scope for the evolution of phenotypic plasticity, that is, for convergence to points in phenotype space with $\eta \neq 0$.

- The Hessian H at an equilibrium point $(x^*, 0, \rho)$ has only two nonzero entries. The first one, denoted by A, measures evolutionary stability of x^* in the ζ-direction and is equivalent to the second derivative of the

BOX 2.1 (*continued*)

invasion fitness function at x^* in the one-dimensional case (eq. (2.14)). Thus, depending on whether $A < 0$ or $A > 0$, the singular point $(x^*, 0, \rho)$ is evolutionarily stable or unstable for mutations affecting only the trait ζ. The other nonzero term of the Hessian is equal to B (i.e., the same as the second nonzero term in the Jacobian).

In fact, it is easy to see that in the simple Levene model studied here, we always have $A = B$ (Leimar, 2005). If $A = B < 0$, both the Jacobian J and the Hessian H are negative definite, and hence the singular point is both strongly convergent stable and evolutionarily stable. However, if $A = B > 0$, the singular point is not only evolutionarily unstable, but also convergent unstable in the η-direction. This means that in this case, either evolutionary branching or the evolution of phenotypic plasticity ($\eta \neq 0$) can occur. Leimar (2005) argued that which one of these two scenarios occurs in any given situation with $A = B > 0$ will depend on the covariance matrix G and on mutational chance events (i.e., on which mutations occur first, the ones leading to genetic diversification in the primary trait or the ones leading to nonzero η).

In more general models, A and B may have different magnitudes but may still be both positive. In this case, $A > B$ would indicate a higher propensity toward genetic diversification, whereas $B > A$ would indicate a higher propensity toward the evolution of phenotypic plasticity. For example, the model studied here is a special case of a class of Levene models studied in Leimar (2005) that incorporate different degrees of migration between the two habitats. The present model corresponds to a migration probability of $m = 1/2$. Leimar (2005) showed that for the general case $m < 1/2$, one generally has $B < A$, so that it is possible to have evolutionary instability $A > 0$ but convergence stability in the η-direction $B < 0$. Such scenarios are conducive to genetic diversification, but not to phenotypic plasticity. In general $B < A$ implies that the propensity toward genetic diversification is larger than the propensity for plasticity to evolve.

One can extend the above considerations from Levene type models incorporating spatial heterogeneity to lottery models incorporating temporal fitness fluctuations. For example, assume that in each generation, individuals encounter either habitat 1 or habitat 2 with equal probability, and that in each generation, a fraction b of the adult population is replaced by offspring that are produced according to the viability functions w_1 and w_2 (eqs. (2.1) and (2.2)) in the case of a one-dimensional phenotype space, and $\tilde{w}_1$ and $\tilde{w}_2$ (eqs. (B2.1) and (B2.2)) in the case of a three-dimensional phenotype space. One can show (Leimar, 2005) that the invasion fitness functions then become

$$f(x, y) = \left(1 - b + b\frac{w_1(y)}{w_1(x)}\right)^{1/2} \left(1 - b + b\frac{w_2(y)}{w_2(x)}\right)^{1/2} \qquad \text{(B2.11)}$$

in the case of a one-dimensional trait space (cf. eq. (2.4)), and

$$f(x, y) = \left(1 - b + b\frac{\tilde{w}_1(z_1', z_2', q')}{\tilde{w}_1(z_1, z_2, q)}\right)^{1/2} \left(1 - b + b\frac{\tilde{w}_2(z_1', z_2', q')}{\tilde{w}_2(z_1, z_2, q)}\right)^{1/2} \qquad \text{(B2.12)}$$

in the case of a three-dimensional trait space (cf. eq. (B2.7)). A similar analysis as earlier shows that just as in the case of spatial heterogeneity, the midpoint x^* between the two optima of w_1 and w_2 is always a convergent stable singular point for the one-dimensional adaptive dynamics, and the three-dimensional adaptive dynamics have a corresponding line of singular points $(x^*, 0, q)$, $q \in [0, 1]$. Using the coordinates ζ, η and ρ form above, this line is convergent stable in the ζ direction. The Jacobian and Hessian matrices J and H have the same structure as before, but in contrast to the case of spatial heterogeneity, with temporal heterogeneity one always has $A < B$ (Leimar, 2005). Thus, evolutionary instability in the ζ-direction is always less pronounced than convergence instability in the η-direction, and hence the propensity for genetic diversification is always smaller than the propensity for the evolution of phenotypic plasticity. In particular, it can now happen that $A < 0 < B$, so that plasticity evolves even in the absence of evolutionary instability in the ζ-direction.

These examples show that under certain conditions, evolution of phenotypic plasticity in a genetically homogenous population is an alternative to genetic diversification due to evolutionary branching. Moreover, based on the simple models explained here and studied in detail in Leimar (2005), it appears that phenotypic plasticity is more likely to replace evolutionary branching as a diversifying mechanism when heterogeneity in fitness is temporal rather than spatial. Temporal fitness fluctuations can also be generated intrinsically if fitness depends on population density, and if population densities fluctuate. Svanback et al. (2009) have investigated the evolution of phenotypic plasticity as an alternative to evolutionary branching in predator-prey models in which predation on two alternative prey types can lead to complicated prey population dynamics. In these models, the existence of alternative prey types can drive diversification in the predator, either in the form of phenotypic plasticity or of evolutionary branching. In accordance with Leimar (2005), the results show that phenotypic plasticity is a more likely outcome than evolutionary branching if temporal fluctuations in prey population densities are large.

CHAPTER THREE

Adaptive Diversification Due to Resource Competition in Asexual Models

The idea that competition for limiting resources can drive the evolution of diversity was already present in Darwin's work. The following quotes illustrate Darwin's contention that natural selection can favor rare types that are sufficiently different from common types, and that such selection for being different could eventually lead to the formation of new species:

> Consequently, I cannot doubt that in the course of many thousands of generations, the most distinct varieties of any one species [...] would always have the best chance of succeeding and increasing in numbers, and thus in supplanting the less distinct varieties; and varieties, when rendered very distinct from each other, take the rank of species. (Darwin, 1859, 155)

> Natural selection, also, leads to divergence of character; for more living beings can be supported on the same area the more they diverge in structure, habits, and constitution [...]. Therefore, during the modifications of the descendants of any one species, and during the incessant struggle of all species to increase in numbers, the more diversified these descendants become, the better will be their chance of succeeding in the battle of life. Thus the small differences distinguishing varieties of the same species, will steadily tend to increase till they come to equal the greater differences between species of the same genus, or even of distinct genera. (Darwin, 1859, 169)

These passages reflect the astounding fact of Darwin's realization that frequency-dependent ecological interactions can give rise to disruptive selection and adaptive divergence. More recently, Rosenzweig (1978) has coined the term "competitive speciation," which essentially encompasses the same basic idea: when the majority phenotype of a population is competing for one type of resource, selection may favor minority phenotypes that consume different types of resources, which could result in phenotypic differentiation and divergence. The idea of divergence due to competition is also the basis for the well-known concept of ecological character displacement, although here the focus is not so much on the origin of diversity arising in a single species, but rather on the evolutionary dynamics of existing diversity between different and already established species. Ecological character displacement embodies the possibility that competition between species can drive divergence in characters determining resource use, and a famous example of this process is provided by Darwin's finches, in which divergence in beak morphology led to consumption of different seed types in coexisting species (Grant & Grant, 2006). Ecological character displacement is thought to play an important role in adaptive radiations (Schluter, 2000), although the potential role of competition in driving divergence between existing species is typically emphasized much more than its potential role in generating new species themselves. This bias is reflected in the fact that the theoretical literature on ecological character displacement has quite a long tradition (Abrams, 1987a, 1987b; Doebeli, 1996a; MacArthur & Levins, 1967; Roughgarden, 1979; Slatkin, 1980; Taper & Case, 1985), whereas corresponding models for competitive speciation have received relatively little attention until recently. Models showing that competition can generate diversity and speciation have been analyzed for example in Seger (1985), Christiansen (1991), and Doebeli (1996b). However, in my opinion only the advent of adaptive dynamics has provided the perspective allowing us to assess the generality of adaptive diversification due to competitive interactions. The previous chapter gave an example of the usefulness of this perspective when analyzing competition generated by two discrete resource niches. In this chapter, we turn our attention to models with continuously distributed resources and explicit population dynamics, as traditionally assumed in models for ecological character displacement driven by frequency-dependent competition.

Throughout this chapter, we will consider a single (one-dimensional) trait x that can vary continuously and that determines resource uptake and competitive interactions. In general, the trait x determines the per capita birth and death rates of its carriers, which in turn determine the population dynamics of individuals with trait x. Per capita birth and death rates can be used both in deterministic models assuming infinite population size, and in individual-based stochastic simulations of finite populations. Both these approaches will be used in this chapter, and we start out with the deterministic dynamics of a monomorphic population, that is, a population in which all individuals have the same trait value x. I assume that the per capita growth rate is independent of the population density and is given by a function $b(x) > 0$, which may, for example, be a constant, or a unimodal function with some intermediate optimum. To incorporate density dependence, I assume that the per capita death rate is proportional to the current population density N. The constant of proportionality is given by a function $c(x) > 0$, that may again, for example, be constant, or have an intermediate minimum corresponding to the trait value with the smallest death rate for a given population density N. Note that $c(x)$ is a property of individual organisms that measures how much the death rate of an individual with phenotype x increases due to competition from a population of unit density. Some phenotypes will have a low competition tolerance per unit density, corresponding to a high $c(x)$, for example, because they experience high starvation stress, whereas other phenotypes will have a high tolerance, corresponding to low $c(x)$.

Assuming infinite population size (but of course finite population density), as is customary in many ecological models, the dynamics of the population monomorphic for x is then given by

$$\frac{dN}{dt} = N(b(x) - c(x)N). \tag{3.1}$$

This is of course the logistic equation for population dynamics. Note that more generally, the per capita death rate could be a linear function $d(x) + c(x)N$ of the current population density. In this case, the density independent death term $d(x)$ can be incorporated into the function $b(x)$ (which might thus become negative). By assuming $b(x) > 0$ for all x in eq. (3.1), I implicitly assume that all monomorphic populations are viable in the sense that $dN/dt > 0$ for small N. As a consequence, the population dynamics given by eq. (3.1) converges, from any initial population density $N_0 > 0$, to the equilibrium $K(x) = b(x)/c(x)$. The population dynamic equilibrium $K(x)$ is the carrying capacity of a population monomorphic for x. Note that using $K(x)$ instead of $c(x)$, eq. (3.1) could be written in the more familiar logistic form

$$\frac{dN}{dt} = N\left(b(x) - \frac{b(x)}{K(x)}N\right) = b(x)N\left(1 - \frac{N}{K(x)}\right). \tag{3.2}$$

As an equilibrium population density, the carrying capacity $K(x)$ is a property of the population, more precisely of a population that is monomorphic for the phenotype x. However, by definition $K(x) = b(x)/c(x)$ is also a property of individuals with phenotype x, because $b(x)$ and $c(x)$ are individual properties. Therefore, as a parameter in the above population dynamical equation, $K(x)$ can just as well be interpreted as a property of individuals with trait x, determining the rate of increase of individual death rates as a function of the effective density experienced. It has recently been argued by Rueffler et al. (2006a) that the carrying capacity should not be viewed as an individual property in evolutionary models. This may be true if one starts out with the carrying capacity as a parameter in the model, in which case that parameter represents a population property. However, in the approach used here the carrying capacity is an emergent property of the individual-level traits $b(x)$ and $c(x)$. Therefore, $K(x)$ can be viewed as a trait of individuals with phenotype x.

It is important to note that when working with the carrying capacity function $K(x)$, one can either simply assume a certain shape of this function based on what seems biologically reasonable, or one can attempt to derive this function from underlying assumptions about the distribution of different resource types and how different consumer phenotypes x utilize the different resource types. Such attempts could for example be based on the classical approach of MacArthur & Levins (1964, 1967). Following Roughgarden (1979), most of the theoretical literature on evolutionary models based on the ecological dynamics given by eq. (3.1) adheres to the first option and makes heuristic assumptions about $K(x)$, rather than deriving it from underlying consumer-resource dynamics. However, the latter approach is feasible as well, as we showed in Ackermann & Doebeli (2004). Somewhat unfortunately, the carrying capacity functions derived from underlying consumer-resource dynamics are in general not the same as those commonly used in the theoretical literature (including my own papers). Box 3.1 explains how one can derive the carrying capacity function using MacArthur & Levins' (1967) consumer-resource model. This derivation is of course also based on underlying assumptions, most notably about the carrying capacities of different types of resources in the absence of consumers. Whether it is preferable to make such assumptions at the level of the resource or at the level of the consumer seems to be a moot point. For simplicity, for the most part in this chapter I will take the traditional approach of making a priori assumptions about the shape of the consumer carrying capacity $K(x)$ (or, equivalently, about the shape of the functions $b(x)$ and $c(x)$), rather than deriving that function from underlying assumptions about the distribution of different resource types.

Following the adaptive dynamics approach, we will determine the evolutionary dynamics of the trait x by considering the fate of rare mutants y in resident populations that are monomorphic for a given trait value x. For this, we assume that the resident population is at its population dynamical equilibrium, $K(x)$. Because the mutant y is assumed to be rare, the mutant's population dynamics is only affected by the density of the resident, which in turn is unaffected by the mutant's invasion attempt, and hence remains at $K(x)$. As a consequence, the effective density that the mutant experiences during the invasion attempt is determined by $K(x)$. However, because the mutant's phenotype is different from the resident phenotype, the effective density is not equal to $K(x)$. Rather, the effective density is proportional to $K(x)$, with the constant of proportionality a function $\alpha(x, y)$ of the trait values of the resident and the mutant. The function $\alpha(x, y)$ is called the competition kernel and describes the strength of competition that phenotype x exerts on phenotype y. In most models in the existing literature, this function is assumed to be symmetric in x and y, which means that for any given phenotypes x and y, the competitive effect of x on y is the same as the competitive effect of y on x. However, this symmetry assumption may not be satisfied in real systems, and it is important to consider asymmetric competition kernels as well. For example, asymmetric competition kernels are more realistic if the trait x is related to body size, and if larger body size confers an inherent advantage in interference competition, in which case one would expect $\alpha(x, y) > \alpha(y, x)$ for $x < y$. Below we will consider the effects of both symmetric and asymmetric competition kernels.

A canonical biological example that is sometimes invoked to rationalize the competition models discussed here consists of thinking about birds and seeds: the trait x is beak size and determines the preferred size of seeds, $K(x)$ describes the abundance of seeds preferred by birds with different beak sizes, and the competition kernel α incorporates the fact that birds with similarly sized beaks compete more strongly, because they prefer similar types of seeds. However, there are of course many other biological scenarios that one could envisage being described, at least conceptually, by logistic competition models. To put results obtained from these models in perspective, it is interesting to note that logistic competition models are thought to be mathematically representative of a large class of models (Durinx et al., 2008).

In the models considered here, the competition kernel is the source of frequency-dependence in the competitive interactions, because the effective density that an individual with a given focal phenotype experiences depends not only on the total density of all the phenotypes that are present in the population, but also on the frequency distribution of these phenotypes, that is, on their competitive impact on the focal phenotype, which is described by the competition kernel α. In particular, the effective density that a rare mutant y experiences in a resident x that is at equilibrium $K(x)$ is given by

$$N_{\text{eff}} = \alpha(x, y)K(x). \tag{3.3}$$

For example, if $\alpha(x, y)$ decreases with phenotypic distance $|x - y|$, as is commonly assumed, then the effective density decreases with increasing phenotypic difference between mutant and resident. This could for example be the case if the trait x under consideration is a proxy for resource preference. In that case, mutants that have resource preferences that are very different from the preference of the resident may experience a lower effective density, and hence a lower competitive impact, than mutants with similar resource preference as the resident. (It is important to note, though, that larger differences in resource preference may not always lead to lower competitive impacts; see Box 3.1.) Viewed from the perspective of the mutant, this means that the effective density experienced by the mutant changes as the resident phenotype changes, which is another way of saying that competition is frequency-dependent.

To determine the dynamics of the population density N_{mut} of the mutant y during the invasion attempt, that is, during the phase when the mutant is rare, we use the birth and death rates of the mutant, $b(y)$ and $c(y)$, together with the effective population density given by eq. (3.2) to obtain

$$\frac{dN_{mut}}{dt} = N_{mut}\left(b(y) - c(y)N_{\text{eff}}\right) = N_{mut}\left(b(y) - c(y)\alpha(x, y)K(x)\right)$$

$$= N_{mut}\left(b(y) - \frac{b(y)\alpha(x, y)K(x)}{K(y)}\right). \tag{3.4}$$

In particular, the net per capita birth rate of the mutant y in the resident x, that is, the invasion fitness $f(x, y)$, is given by

$$f(x, y) = b(y) - c(y)N_{\text{eff}} = b(y) - c(y)\alpha(x, y)K(x)$$

$$= b(y) - \frac{b(y)\alpha(x, y)K(x)}{K(y)}. \tag{3.5}$$

Following Dieckmann & Law (1996) (see also the Appendix), the canonical equation of the adaptive dynamics of the trait x is then given by

$$\frac{dx}{dt} = m(x)D(x), \tag{3.6}$$

where

$$D(x) = \left.\frac{\partial f(x, y)}{\partial y}\right|_{y=x}$$

$$= b'(x) - \left(c'(x)\alpha(x, x)K(x) + c(x)\left.\frac{\partial\alpha(x, y)}{\partial y}\right|_{y=x}K(x)\right) \tag{3.7}$$

is the selection gradient. The quantity $m(x) > 0$ in eq. (3.7) influences the speed of the evolutionary dynamics and is determined by the mutational process. More precisely, $m(x)$ depends on both the rate at which new mutations occur and on the distribution of effects that mutations have on the trait x.

Even if the distribution of mutational effects is independent of x, the rate at which new mutations occur depends on the current population size, which itself depends on the current resident trait x. Therefore, the rate $m(x)$ is in general a function of the trait x. As long as evolution occurs in essentially monomorphic populations along the one-dimensional trait axis x, the dependence of m on x is not essential, as m effectively only scales time. However, if the population were polymorphic, for example, due to one or more bouts of evolutionary branching, the different coexisting phenotypic branches would have, in general, different population sizes, and hence new mutations would occur at different rates in the different branches. This rate difference can, in principle, affect the evolutionary dynamics of polymorphic populations (as we will see for example in Chapter 5). First, however, we want to study singular points of the one-dimensional adaptive dynamics given by eq. (3.6), for which it suffices to study the selection gradient $D(x)$.

Before we proceed, we make a number of simplifying assumptions. First, we can assume without loss of generality that $\alpha(x, x) = 1$, that is, that the competitive impact of individuals of the same phenotype on each other is scaled to unity. Second, we are mostly interested in the disruptive effects of frequency-dependent competition generated by the competition kernel α, but we also want to incorporate some stabilizing force on the trait x in order to avoid regions of extreme trait values that would be biologically unrealistic. Stabilizing selection can be introduced by assuming that the birth rate $b(x)$ becomes very small for extreme x-values, or that the death rate $c(x)$ becomes very large for extreme x-values. In the existing literature, it is mainly assumed either that the birth rate $b(x)$ is a unimodal function of x and the death rate $c(x)$ is independent of x (e.g., Kisdi, 1999), or that the birth rate $b(x)$ is independent of x and the carrying capacity $K(x)$ is a unimodal function of x (e.g., Dieckmann & Doebeli, 1999). Both of these assumptions imply a stabilizing selective force for the trait value maximizing either the birth rate or the carrying capacity, respectively. Anticipating the basic result of this chapter, frequency-dependent competition can generate evolutionary branching under both of these assumptions. Rather than repeating all the results for the two different scenarios for the stabilizing component of selection, we will concentrate on the case where the carrying capacity $K(x)$ is a unimodal function of x and the birth rate $b(x) \equiv b$ is independent of x. Note that one could of course also consider more complicated scenarios in which both $b(x)$ and $c(x)$ (respectively $K(x)$) vary with x (for an example, see Box 3.1). In such cases, the frequency-independent component of fitness may become more complicated, for example, exhibiting multiple local optima.

BOX 3.1
AN EXPLICIT DERIVATION OF CARRYING CAPACITIES AND COMPETITION
COEFFICIENTS FROM UNDERLYING RESOURCE DYNAMICS

Based on the consumer-resource models introduced by MacArthur & Levins (1967), Ackermann & Doebeli (2004) have derived explicit expressions for the carrying capacity and the competition coefficients in consumer populations in which individuals are characterized by how they utilize the available resources. It is assumed that different resource types are characterized by a continuous quantity z (e.g., seed size). If $F(z)$ denotes the density of resources of type z, $F(z)$ grows logistically to carrying capacity $S(z)$:

$$\frac{dF(z)}{dt} = r \cdot F(z) \cdot \left(1 - \frac{F(z)}{S(z)}\right). \tag{B3.1}$$

The carrying capacity $S(z)$ is of Gaussian form

$$S(z) = S_0 \cdot \exp\left(\frac{-z^2}{2\sigma_S^2}\right). \tag{B3.2}$$

The distribution of the resource is assumed to be spatially homogenous at all times, so that each consumer individual always has access to the whole range of resource types at densities corresponding to the momentary steady-state resource distribution.

A consumer is characterized by how it utilizes the resources $F(z)$ along the z-axis, that is, by its utilization function. This function describes the relative effort a consumer invests in harvesting resources of type z. If all resource types would be equally common (i.e., if the resource distribution was flat), then the utilization curve would describe the distribution of resources consumed by an individual. If, as in our model, the resource distribution is not flat, then the distribution of resources actually consumed by an individual depends both on the effort spent on the different resource types (given by the utilization curve) and the momentary density of the different resource types.

The utilization curve is of Gaussian form and is determined by two phenotypic properties of the consumer: the position of the maximum, x, and the standard deviation, y. The consumer (x, y) utilizes resources $F(z)$ at position $z = x$ most intensively, and the intensity of utilization declines with increasing distance from $z = x$ at a rate that is determined by y. Thus, whereas x determines the preferred position along the resource axis, y determines the degree of specialization, with small y corresponding to specialists and large y corresponding to generalists. x thus denotes the niche position, while y is a measure for the

BOX 3.1 (*continued*)

niche width. Utilization $a_{x,y}(z)$ of resources of type z by consumer phenotype (x, y) is described by the function

$$a_{x,y}(z) = \frac{\exp[-c \cdot y]}{\sqrt{2\pi} y} \exp\left(\frac{-(z - x)^2}{2y^2}\right),$$ (B3.3)

where c is a measure of the costs or benefits of larger niche widths y. The integral $\int_z a_{x,y}(z)dz$ is a measure of the total effort devoted to resource consumption. If $c = 0$, then $\int_z a_{x,y}(z)dz = 1$ for all phenotypes (x, y). If $c > 0$, then $\int_z a_{x,y}(z)dz$ decreases as y increases, indicating a cost of being a generalist. If $c < 0$, $\int_z a_{x,y}(z)dz$ increases as y increases, indicating a cost of specialization. Both scenarios are biologically plausible. Note that even without costs to generalists (i.e., $c = 0$), widening the utilization curve leads to a decrease in niche depth, that is, in the maximum of the function $a_{x,y}(z)$.

The presence of consumer (x, y) changes the dynamics of the resources to

$$\frac{dF(z)}{dt} = r \cdot F(z) \cdot \left(1 - \frac{F(z)}{S(z)}\right) - F(z) \cdot a_{x,y}(z) \cdot N_{x,y}$$ (B3.4)

Here $N_{x,y}$ is the density of consumers with phenotype (x, y), so that $a_{x,y}(z)N_{x,y}$ is the rate at which resources of type z are consumed. The consumer density $N_{x,y}$ itself changes over time due to resource consumption, and according to the original model in MacArthur & Levins (1967), the dynamics of $N_{x,y}$ is

$$\frac{dN_{x,y}}{dt} = R \cdot N_{x,y} \cdot \left(b \cdot \int_z a_{x,y}(z)F(z)dz - m\right).$$ (B3.5)

Here b represents the net energy per food item acquired, so that the total energy acquired per individual is b times the total amount of resources acquired, $\int_z a_{x,y}(z)F(z)$. We assume that all resource types contribute equally to energy gain, so that $b(z) = b$ is a constant. The total amount of energy available for reproduction is the total gain minus m, the per individual costs for maintenance. Finally, R is the number of individuals that are produced per unit of energy available for reproduction.

We first bring the consumer equation (B3.4) into logistic form by assuming a time scale separation between the dynamics of the resources and that of the consumer, so that the resource is always assumed to be in its equilibrium state. Thus, for a given consumer density $N_{x,y}$, one first finds the nonzero solution $\hat{F}(z)$ of the resource equation (B3.4), and then substitutes this solution into eq. (B3.5), which yields a logistic equation for $N_{x,y}$ of the form

$$\frac{dN_{x,y}}{dt} = R_{x,y} \cdot N_{x,y} \cdot \left(1 - \frac{N_{x,y}}{K_{x,y}}\right), \tag{B3.6}$$

(see Appendix 1 in Ackermann & Doebeli (2004) for details).

This logistic equation of population growth allows one to interpret $R_{x,y}$ as the intrinsic growth rate, and $K_{x,y}$ as the carrying capacity of a consumer population that is monomorphic for phenotype (x, y).

General analytical expressions for $R_{x,y}$ and $K_{x,y}$ are given in Appendix 1 of Ackermann & Doebeli (2004). For example, for $r = 1$ and $b = 1$ in eqs. (B3.1) and (B3.5), one gets

$$K_{x,y} = y\sqrt{2\pi}\sqrt{2\sigma_S^2 + y^2} \exp\left[\frac{x^2 + 4c\sigma_S^2 y + 2cy^3}{2\sigma_S^2 + y^2}\right]$$

$$\times \left(\frac{-m}{\sigma_S S_0} + \frac{\exp\left[-cy - \frac{x^2}{2\sigma_S^2 + 2y^2}\right]}{\sqrt{\sigma_S^2 + y^2}}\right). \tag{B3.7}$$

This expression calculates the carrying capacity, that is, the equilibrium density, of a consumer with phenotype (x, y), given the resource parameters σ_S and S_0, the consumer parameter m, and the cost parameter c. It is important to note that, as a function of the niche position x, the consumer carrying capacity $K_{x,y}$ does not necessarily have a maximum at the value $x = 0$ corresponding to the maximum of the carrying capacity of the resource. In other words, a consumer phenotype that most prefers the resource with the highest carrying capacity does not necessarily attain the highest equilibrium density.

The corresponding expression for $R_{x,y}$ is

$$R_{x,y} = R\left[\frac{\sigma_S S_0 \exp\left[-cy - \frac{x^2}{2\sigma_S^2 + 2y^2}\right]}{\sqrt{\sigma_S^2 + y^2}} - m\right]. \tag{B3.8}$$

Note that the expressions for $K_{x,y}$ and $R_{x,y}$ given here differ slightly from eqs. (7) and (A5) given in Ackermann & Doebeli (2004). Seiji Kumagai pointed out to us that the original formulas contain some errors, and that the correct formulas are the ones given here. As a consequence, m needs to be replaced by $m/2$ in eq. (14) in Ackermann & Doebeli (2004), and for the numerical simulations reported in Ackermann & Doebeli (2004), one needs the additional information that $r = 144$ and $\lambda = \exp(R_0)$. With these corrections, the results reported in Ackermann & Doebeli (2004) remain valid.

BOX 3.1 (*continued*)

To determine the competition coefficients in the consumer population, Ackermann & Doebeli (2004) derived the dynamics of consumer type (x, y) in the presence of a competitor (u, v) and showed that

$$\frac{dN_{x,y}}{dt} = R_{x,y} \cdot N_{x,y} \cdot \left(1 - \frac{N_{x,y} + \beta(x, y, u, v) \cdot N_{u,v}}{K_{x,y}}\right), \qquad \text{(B3.9)}$$

where $\beta(x, y, u, v)$ is the relative competitive impact of a consumer individual of type (u, v) on consumers of type (x, y) and is given by

$$\beta(x, y, u, v)$$

$$= \exp\left[c \cdot (y - v)\right]$$

$$\times \exp\left[\frac{-2\sigma_S^4(u - x)^2 + v^2 x^2 y^2 + \sigma_S^2 y^2 \left(-3u^2 + 2ux + x^2\right) - u^2 y^4}{2\left(2\sigma_S^2 + y^2\right)\left(v^2 y^2 + \sigma_S^2 \left(v^2 + y^2\right)\right)}\right]$$

$$\times \frac{y\sqrt{2\sigma_S^2 + y^2}}{\sqrt{v^2 y^2 + \sigma_S^2 \left(v^2 + y^2\right)}}. \qquad \text{(B3.10)}$$

In contrast to many competition models studied in the literature (as well as in this chapter), these competition coefficients are in general not symmetric in the two types (x, y) and (u, v) (i.e., $\beta(x, y, u, v) \neq \beta(u, v, x, y)$ in general). In particular, the competitive effect exerted by one consumer on another depends not only on the distance between the consumer's niches, but also on the niche positions of the two competitors relative to the resource. Moreover, it is possible that the competitive impact increases with increasing distance between niches, which is related to the fact that carrying capacities may increase with increasing niche distance to the resource optimum (Ackermann & Doebeli, 2004).

Is interesting to note that the competition coefficients β reduce to the Gaussian competition coefficients given by eq. (3.13) under the following conditions: $c = 0$ (no costs or benefits for widening the niche), all competitors have a fixed niche width y_0 of the utilization function, and the carrying capacity $S(z)$ of the resource is constant and independent of z. Then

$$\beta(x, y_0, u, y_0) = \exp\left[\frac{-(u - x)^2}{4y_0^2}\right]. \qquad \text{(B3.11)}$$

This Gaussian symmetry of competitive impacts only occurs under the assumption of constant resource carrying capacities, which in turn imply constant

consumer carrying capacities $K(x, y_0)$ as a function the position of the preferred resources x (eq. (B3.7) with $S(z)$ constant, i.e., with $\sigma_S \to \infty$). Thus,
while models with both symmetric Gaussian competition coefficients and a
nonconstant consumer carrying capacity as a function of niche position x
may be biologically plausible, such models cannot be derived from the basic
consumer-resource model of MacArthur & Levins (1967).

■

The above assumptions amount to $b'(x) = 0$ and $c(x) = b/K(x)$, hence
$c'(x) = -\frac{bK'(x)}{K(x)^2}$. With $\alpha(x, x) = 1$ and $m(x) = 1$ in eq. (3.7), the adaptive
dynamics of the trait x becomes

$$\frac{dx}{dt} = D(x) = \frac{bK'(x)}{K(x)} - b \left. \frac{\partial \alpha(x, y)}{\partial y} \right|_{y=x}. \tag{3.8}$$

To investigate the adaptive dynamics given by eq. (3.8), we now turn our
attention back to the competition kernel. Just as with the carrying capacity
function $K(x)$, it is in principle possible to derive the shape of the competition
kernel from underlying assumptions about the distribution of resource types
and how different consumer types utilize the different resource types. This
approach is outlined in Box 3.1. Again somewhat unfortunately, it does not
yield the competition kernels commonly used in the literature. For example,
Roughgarden (1979) has derived competition kernels based on the assumption
of uniform resource distributions in which all resource types have the same
carrying capacities. However, this latter assumption would imply that all consumer types have the same carrying capacity as well, independent of consumer
resource preference. That is, it would imply that the carrying capacity function
$K(x)$ is a constant. But the competition kernels derived in Roughgarden (1979)
are most often used in conjunction with nonconstant $K(x)$ (typically, unimodal
$K(x)$), which constitutes a basic inconsistency if one views these functions as
derived from underlying consumer-resource dynamics. In view of the complicated nature of the competition kernels derived from consumer-resource
dynamics (Box 3.1), the most pragmatic approach is to work with competition kernels that reflect biologically reasonable assumptions about the strength
of competition between different phenotypes, without any explicit claims as to
how these functions are derived from underlying consumer-resource models.
As with the carrying capacity function, this is the approach I will take for the
most part in this chapter.

3.1 ADAPTIVE DYNAMICS WITH SYMMETRIC
COMPETITION KERNELS

To model symmetric competition, we assume that the competition kernel is a function of the distance $|x - y|$ only. It is then also reasonable to assume that as a function of $|x - y|$, the competition kernel has a maximum at 0, that is, that competitive impacts decrease with phenotypic distance. If we also assume that $\alpha(x, y)$ is differentiable in y at $y = x$ (so that $\left.\frac{\partial \alpha(x,y)}{\partial y}\right|_{y=x} = 0$), the adaptive dynamics given by eq. (3.6) yields

$$\frac{dx}{dt} = D(x) = b\frac{K'(x)}{K(x)}. \tag{3.9}$$

This means that in monomorphic populations, the evolution of the trait x is determined solely by the stabilizing component of selection, that is, by the gradient $K'(x)$ of the carrying capacity function. Under the given assumptions, this is not surprising and can easily be understood mathematically as follows. For any given resident x, a mutant y that is different from x has an advantage due to a reduction in competitive impact felt from the resident population, that is, because $\alpha(x, y) < \alpha(x, x) = 1$ for any $y \neq x$. However, because $\left.\frac{\partial \alpha(x,y)}{\partial y}\right|_{y=x} = 0$ by assumption, this advantage is only of second order. In contrast, the advantage of moving up the gradient of the stabilizing component is described by the first order derivative $K'(x)$ and therefore determines the evolutionary dynamics. In particular, assuming a unimodal carrying capacity function, the adaptive dynamics will initially always converge to the maximum of $K(x)$. This can be made explicit using the concepts of singular points and convergence stability. A singular point x^* of the adaptive dynamics (3.6) must satisfy $K'(x^*) = 0$, and hence (differentiable) unimodality of $K(x)$ implies the existence of a unique singular point at the trait value maximizing the carrying capacity. Convergence stability of this singular point is determined by the quantity $dD/dx(x^*)$. Because x^* is a maximum of $K(x)$, we have

$$\left.\frac{dD}{dx}\right|_{x=x^*} = \left[b\frac{K''(x)}{K(x)} - b\frac{K'(x)^2}{K(x)^2}\right]\bigg|_{x=x^*} = b\frac{K''(x^*)}{K(x^*)} < 0, \tag{3.10}$$

and hence the singular point is always convergence stable. By definition, selection vanishes to first order at the singular point. Therefore, after convergence to the singular point, second order effects of selection come into play, which is why the competition kernel, and hence frequency dependence, starts to be important.

As explained in Chapter 2 and in the Appendix, a convergent stable singular point of one-dimensional adaptive dynamics is either a minimum or a maximum for the invasion fitness function given by eq. (3.5). In a resident population sitting at the singular point x^*, a newly arising mutant $y \neq x^*$ has an advantage due to the frequency-dependent nature of competition: because $\alpha(x^*, y) < \alpha(x^*, x^*) = 1$ for any $y \neq x^*$, the effective density that a mutant y experiences, $\alpha(x^*, y)K(x^*)$, is smaller than the effective density felt by a resident individual, $K(x^*)$. However, a mutant y also has a disadvantage, because $y \neq x^*$ implies $K(y) < K(x^*)$ (as the singular point is the maximum of K). Whether x^* can be invaded by all nearby mutants or none is determined by the relative magnitude of these two effects, which is in turn determined by the curvatures of the competition kernel and of the carrying capacity. More precisely, the second derivative of the fitness function $f(x^*, y)$ with respect to y is

$$\frac{\partial^2 f(x^*, y)}{\partial y^2}\bigg|_{y=x^*} = \frac{\partial^2}{\partial y^2}\left[-b\frac{\alpha(x^*, y)K(x^*)}{K(y)}\right]\bigg|_{y=x^*}$$

$$= b\left[\frac{K''(x^*)}{K(x^*)} - \frac{\partial^2 \alpha(x^*, y)}{\partial y^2}\bigg|_{y=x^*}\right]. \tag{3.11}$$

Note that in this expression, the curvature of the carrying capacity at the singular point, $K''(x^*)$, is scaled by the carrying capacity itself, $K(x^*)$. Intuitively, this is because for any given curvature $K''(x^*)$, that is, for any given decrease in carrying capacity per unit distance from the singular point, the relative disadvantage due to moving away from the maximum carrying capacity is larger if the carrying capacity is already small. It follows from eq. (3.11) that the maximum of the carrying capacity function, x^*, is a fitness minimum, and therefore an evolutionary branching point, if

$$\frac{\partial^2 \alpha(x^*, y)}{\partial y^2}\bigg|_{y=x^*} < \frac{K''(x^*)}{K(x^*)}. \tag{3.12}$$

In this inequality, the curvature given by the second derivative on the left-hand side is a measure of the strength of frequency dependence, that is, of the advantage of being different from the resident strain. In contrast, the right-hand side reflects the force of stabilizing selection, that is, the disadvantage of being different from the maximum of the carrying capacity function. Since the left-hand side is independent of the right-hand side, condition (3.12) reinforces the point, already made in the previous chapter (see Fig. 2.3), that evolutionary branching is a generic phenomenon: for any given unimodal carrying capacity function, evolutionary branching occurs whenever the rare type advantage due to frequency dependence is strong enough. Note that if evolutionary branching does not occur, then the convergent stable singular point is also evolutionarily stable, and therefore represents the final state of the evolutionary process.

We now apply this theory to some concrete examples. The most commonly used symmetric competition kernel is the Gaussian function

$$\alpha(x, y) = \exp\left[\frac{-(x-y)^2}{2\sigma_\alpha^2}\right]. \tag{3.13}$$

Here the parameter σ_α is a measure for the strength of frequency dependence: the smaller σ_α, the faster competitive impacts decrease with increasing phenotypic distance. Similarly, the most commonly used carrying capacity function is also of Gaussian form:

$$K(x) = K_0 \exp\left[\frac{-(x-x_0)^2}{2\sigma_K^2}\right]. \tag{3.14}$$

Here the parameter K_0 is the maximal carrying capacity, and the parameter x_0 is the phenotype with the maximal carrying capacity, for which it is usually assumed, without loss of generality, that $x_0 = 0$. By what was said above, this is also the unique singular point x^* of the adaptive dynamics (3.9). The parameter σ_K measures how fast the carrying capacity drops off with increasing distance from the maximum.

With these functions, inequality (3.12) becomes

$$\frac{-1}{\sigma_\alpha^2} < \frac{-1}{\sigma_K^2}, \tag{3.15}$$

which is equivalent to

$$\sigma_\alpha < \sigma_K. \tag{3.16}$$

This is a familiar result that has appeared in various incarnations in the literature (e.g., Roughgarden, 1979 (Chapter 24); Dieckmann & Doebeli, 1999). Intuitively and somewhat imprecisely speaking, it simply says that evolutionary branching should occur if, with increasing phenotypic distance from the singular point, competitive impacts from the singular resident decrease faster than the carrying capacity. Corresponding results can be derived for other functional forms of the competition kernel and the carrying capacity (e.g., Baptestini et al., 2009).

If condition (3.16) is satisfied, so that the maximum of the carrying capacity is an evolutionary branching point, it follows from general adaptive dynamics theory, as explained in the Appendix, that phenotypes x_1 and x_2 on either side of the singular point can coexist, at least if they are sufficiently close to the singular point. Thus, after convergence to the singular point, the resident population will become dimorphic. Moreover, selection will act to increase the phenotypic distance between the two coexisting branches. This can be seen by considering the invasion fitness of rare mutants y in a dimorphic resident population (x_1, x_2). This invasion fitness is determined by the equilibrium population densities of the two coexisting resident branches, as well as by the competitive impact that the resident phenotypes have on the mutant. Let N_1 and N_2 denote the population densities of the two coexisting resident phenotypes. Extrapolating from the population dynamics of a monomorphic resident, eq. (3.2), the equilibrium densities of coexisting resident strains are determined by the following ecological dynamics:

$$\frac{dN_1}{dt} = bN_1 \left(1 - \frac{N_1 + \alpha(x_2, x_1)N_2}{K(x_1)} \right) \tag{3.17}$$

$$\frac{dN_2}{dt} = bN_2 \left(1 - \frac{N_2 + \alpha(x_1, x_2)N_1}{K(x_2)} \right). \tag{3.18}$$

For example, if $x_2 = -x_1$, that is, if the two resident phenotypes have the same distance from the singular point (recall that we assumed that the maximum of the carrying capacity is at 0), the equilibrium densities resulting from the ecological dynamics are

$$N_1^* = N_2^* = \frac{\exp\left(x_1^2 \left(\frac{2}{\sigma_\alpha^2} - \frac{1}{2\sigma_K^2} \right) \right)}{1 + \exp\left(\frac{2x_1^2}{\sigma_\alpha^2} \right)}. \tag{3.19}$$

The effective density experienced by a rare mutant y in the resident (x_1, x_2) is a weighted sum of the equilibrium densities N_1^* and N_2^* of the resident strains, with weights the competitive impacts of the resident phenotypes on the mutant, as determined by the competition kernel α. Thus, in analogy to the invasion fitness in monomorphic populations, eq. (3.5), the invasion fitness of a mutant y into the resident (x_1, x_2) is:

$$f(x_1, x_2, y) = b \left(1 - \frac{\alpha(x_1, y)N_1^* + \alpha(x_2, y)N_2^*}{K(y)} \right). \tag{3.20}$$

The selection gradients in the two resident strains are given the derivative of the invasion fitness function with respect to the mutant trait y and evaluated at the resident value. Thus, the selection gradient in the resident x_1 is

$$D_1(x_1, x_2) = \left. \frac{\partial f(x_1, x_2, y)}{\partial y} \right|_{y=x_1}$$

$$= -b \left(\frac{\left. \frac{\partial \alpha(x_2, y)}{\partial y} \right|_{y=x_1} N_2^*}{K(x_1)} - \frac{K'(x_1)(N_1^* + \alpha(x_2, x_1)N_2^*)}{K(x_1)^2} \right),$$

$$(3.21)$$

and the selection gradient in the resident x_2 is

$$D_2(x_1, x_2) = \left. \frac{\partial f(x_1, x_2, y)}{\partial y} \right|_{y=x_2}$$

$$= -b \left(\frac{\left. \frac{\partial \alpha(x_1, y)}{\partial y} \right|_{y=x_2} N_1^*}{K(x_2)} - \frac{K'(x_2)(N_2^* + \alpha(x_1, x_2)N_1^*)}{K(x_2)^2} \right).$$

$$(3.22)$$

These selection gradients yields the two-dimensional adaptive dynamical system

$$\frac{dx_1}{dt} = m_1(x_1, x_2)D_1(x_1, x_2) \qquad (3.23)$$

$$\frac{dx_2}{dt} = m_2(x_1, x_2)D_2(x_1, x_2). \qquad (3.24)$$

Here the functions $m_1(x_1, x_2)$ and $m_2(x_1, x_2)$ describe the mutational process in the two resident strains (see Appendix). Even if the per capita rate of mutation and the distribution of mutational effects does not depend on the phenotype, the total rate of mutations will, in general, depend on the population densities of the two resident strains, that is, on N_1^* and N_2^*, and hence on the resident trait values themselves. Nevertheless, as long as population densities are positive, it is clear that singular points of the two-dimensional adaptive dynamics, that is, equilibrium points of the dynamical system (3.23) and (3.24), are given by solutions x_1^* and x_2^* of the system of equations

$$D_1(x_1^*, x_2^*) = 0 \qquad (3.25)$$

$$D_2(x_1^*, x_2^*) = 0. \qquad (3.26)$$

With symmetric Gaussian competition kernel and Gaussian carrying capacity, it can be shown analytically that this system has unique solutions

$$x_1^* = -x_2^* = \sqrt{\ln\left(\frac{2\sigma_K^2}{\sigma_\alpha^2} - 1\right)\frac{\sigma_\alpha^2}{2}}. \qquad (3.27)$$

Note that for these strategies to be well defined, we need $\sigma_\alpha < \sigma_K$, that is, we need to assume that the one-dimensional adaptive dynamics (3.9) exhibits an evolutionary branching point. To determine whether the singular coalition (x_1^*, x_2^*) given by (3.27) is an attractor for the two-dimensional adaptive dynamics, that is, whether the singular coalition is convergent stable, one needs to evaluate the Jacobian matrix

$$J(x_1^*, x_2^*) =$$

$$\left(\begin{array}{cc} \left.\dfrac{\partial[m_1(x_1, x_2)D_1(x_1, x_2)]}{\partial x_1}\right|_{x_1=x_1^*, x_2=x_2^*} & \left.\dfrac{\partial[m_1(x_1, x_2)D_1(x_1, x_2)]}{\partial x_2}\right|_{x_1=x_1^*, x_2=x_2^*} \\[3mm] \left.\dfrac{\partial[m_2(x_1, x_2)D_2(x_1, x_2)]}{\partial x_1}\right|_{x_1=x_1^*, x_2=x_2^*} & \left.\dfrac{\partial[m_2(x_1, x_2)D_2(x_1, x_2)]}{\partial x_2}\right|_{x_1=x_1^*, x_2=x_2^*} \end{array} \right).$$

$$(3.28)$$

It can be shown (see *Challenge* below) that the real parts of the eigenvalues of this Jacobian matrix are always negative, that is, that the singular coalition (x_1^*, x_2^*) is always convergent stable, and hence after evolutionary branching, the two phenotypic branches converge to x_1^* and x_2^*, respectively. After convergence, the obvious question is again whether the singular strategies are evolutionarily stable or unstable. Note that because of the symmetry assumptions made (the competition kernel is symmetric, and the carrying capacity function is symmetric around the maximum), the two singular strategies either both represent (local) maxima for the invasion fitness function (3.20), or they both represent fitness minima of this function. Again it can be shown analytically that whenever $\sigma_\alpha < \sigma_K$ (i.e., when frequency-dependent competition generates evolutionary branching in the first place), both singular strategies are fitness minima, and hence evolutionarily unstable.

Challenge: Prove the existence of the singular coalition (x_1^*, x_2^*) given in (3.27), its convergence stability, as well as the evolutionary instability of x_1^* and x_2^*.

So what happens next? As a thought experiment, we can fix one of the resident strains in the singular coalition, say x_2^*, and only consider mutants in the other resident, x_1^*. For the same reasons that apply to evolutionary branching points in monomorphic populations, strategies u_1 and u_2 on either side of and close to x_1^* can invade the resident. Moreover, such strategies can coexist, thereby driving the resident x_1^* to extinction. Similar considerations apply to the singular strategy x_2^*, so that we can now envisage a situation in which the evolutionary instability of the singular strategies x_1^* and x_2^* leads to coexistence of four strategies (u_1, u_2, u_3, u_4), where u_1 and u_2 lie close to and on either side of x_1^*, and u_3 and u_4 lie close to and on either side of x_2^*. These four coexisting strategies form a new polymorphic resident population, in which we can define a four-dimensional adaptive dynamical system by considering the invasion fitness of rare mutants in each of the four resident branches. The invasion fitness $f(u_1, u_2, u_3, u_4, y)$ of rare mutants y is given by a straightforward extension of eq. (3.20) to four resident strains, and the four selection gradients that define the four-dimensional adaptive dynamics are

$$D_i(u_1, \ldots, u_4) = \left. \frac{\partial f(u_1, \ldots, u_4, y)}{\partial y} \right|_{y=u_i} \tag{3.29}$$

for $i = 1, \ldots, 4$. To find singular points of the four-dimensional adaptive dynamics, one has to find solutions $u_1^*, \ldots, u_4^*$ of the four equations $D_i(u_1^*, \ldots, u_4^*) = 0$ for $i = 1, \ldots, 4$, and one then needs to investigate the convergence and evolutionary stability of these singular strategies. Unfortunately, this problem appears to be analytically intractable. However, for any given values of $\sigma_\alpha < \sigma_K$, it is easy to find numerical solutions of the four-dimensional adaptive dynamics. This is illustrated in Figure 3.1, which shows the equilibria of the adaptive dynamics after a number of successive bouts of evolutionary branching. For any given number of resident strains (one at the outset, two after the first branching event, four after the second branching event, and so on), the adaptive dynamics was solved numerically, starting from resident strains straddling the previous branching points, and run to equilibrium, that is, to the singular coalition corresponding to the given number of resident strains. Figure 3.1 shows the invasion fitness at these successive singular coalitions for parameter values that ensure the existence of an evolutionary branching point in the basic one-dimensional adaptive dynamics (3.9). In the example shown, singular strategies emerging after successive bouts of evolutionary branching are always fitness minima, that is, evolutionarily unstable. However, as the resident population becomes more polymorphic (i.e., as the number of branches increases), the fitness minima at the singular strategies become less pronounced. Nevertheless, I conjecture that the pattern of diversification illustrated in Figure 3.1 is typical for Gaussian competition kernels and carrying capacity functions and will, in principle, lead to an ever-increasing number of coexisting branches. More precisely, I predict that with such ecological functions, and for any $\sigma_\alpha < \sigma_K$, the series of branching events leading to a doubling

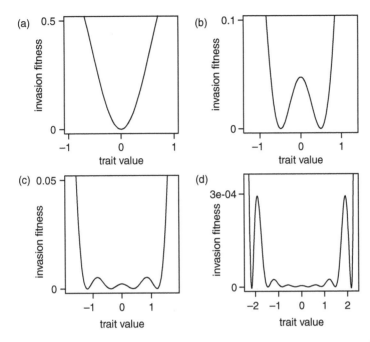

FIGURE 3.1. Invasion fitness functions at singular strategies emerging after successive bouts of evolutionary branching with Gaussian carrying capacity and competition kernel. (a) For the one-dimensional adaptive dynamics, the maximum of the carrying capacity at 0 is an evolutionary branching point, as indicated by the positive curvature of the invasion fitness function when the resident is monomorphic for trait value 0. (b) Singular coalition of two strategies at approximately −0.49 and 0.49 for the two-dimensional adaptive dynamics obtained from (a) after evolutionary branching. At both singular values, the invasion fitness function has again a minimum. (c) Singular coalition of four strategies at approximately −1.2, −0.38, 0.38, and 1.2 for the four-dimensional adaptive dynamics obtained from (b) after (simultaneous) evolutionary branching in both phenotypic branches resulting from the two-dimensional adaptive dynamics. At all four singular values, the invasion fitness function has a minimum. (d) Singular coalition of eight strategies at approximately −2.16, −1.47, −0.86, −0.28, 0.28, 0.86, 1.47, and 2.16 for the eight-dimensional adaptive dynamics obtained from (c) after (simultaneous) evolutionary branching in all four phenotypic branches resulting from the four-dimensional adaptive dynamics. At all eight singular values, the invasion fitness function has a minimum. Notice that the invasion fitness function becomes more shallow after each bout of evolutionary branching. Parameter values were $\sigma_K = 1$, $\sigma_\alpha = 0.5$, $K_0 = 1$, and $r = 1$.

in the number of branches in the resident population would go on ad infinitum in a strictly deterministic model (i.e., in a model that can pick up the selective signal of even very weak disruptive invasion fitness profiles). Thus, successive bouts of evolutionary branching would in principle lead to an infinite number of coexisting strains. Moreover, the abundance of the coexisting strains would match the carrying capacity curve in the sense that strains near the maximum of $K(x)$ would have the highest abundance, and strains farthest away from the maximum would have the lowest abundance.

Challenge[*]: Prove the conjecture that with Gaussian competition kernels and carrying capacity functions, the adaptive dynamics will exhibit an infinite sequence of evolutionary branching events as described in Figure 3.1 if the initial branching condition $\sigma_\alpha < \sigma_K$ is satisfied, and that the pattern of abundance in the infinitely many coexisting branches resulting from this process is unimodal.

If the evolutionary dynamics illustrated in Figure 3.1 were run for many successive branching events, the evolving population would eventually consist of very many coexisting strains with short phenotypic distances between neighboring strains and with a unimodal pattern of abundance along the phenotypic axis. In phenotype space, such a population would be represented by a single polymorphic cluster, rather than by a number of discrete and separate clusters representing differentiated subpopulations. Thus, in this model there is in some sense too much diversification due to evolutionary branching for the evolutionary process to result in distinct phenotypic clusters. However, this does not appear to be a robust phenomenon, and instead is a result of the particular choice of Gaussian functions for the competition kernel and the carrying capacity, and due to the deterministic nature of the adaptive dynamics illustrated in Figure 3.1. As we will see shortly, using different ecological functions can drastically reduce the number of evolutionary branching events in deterministic models, and using Gaussian ecological functions in stochastic individual-based models for finite populations typically also leads to only a small number of successive branching events.

For example, in the deterministic derivation given before one could replace the Gaussian carrying capacity by the quadratic function

$$K(x) = 1 - ax^2, \tag{3.30}$$

where a is now the parameter measuring how fast the carrying capacity decreases from the maximum at $x = 0$. Clearly, we now have to assume that possible phenotypes are restricted to the interval $(-1/\sqrt{a}, 1/\sqrt{a})$, but we will see a difference in the evolutionary dynamics already in that region of phenotype space. Since this carrying capacity is also unimodal, its maximum is a convergent stable singular point by (3.10). Assuming the Gaussian competition kernel (3.13), inequality (3.12) shows that this singular point is an evolutionary

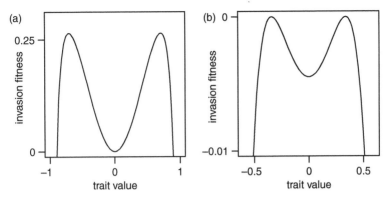

FIGURE 3.2. Invasion fitness functions at singular strategies emerging before and after evolutionary branching. (a) For the one-dimensional adaptive dynamics, the maximum of the carrying capacity function at 0 is an evolutionary branching point, as indicated by the positive curvature at 0 of the invasion fitness function when the resident is monomorphic for trait value 0. (b) Singular coalition of two strategies at approximately -0.34 and 0.34 for the two-dimensional adaptive dynamics obtained from (a) after evolutionary branching. At both singular values, the invasion fitness function has a maximum, and hence the singular coalition consisting of two phenotypic branches is evolutionarily stable. Parameter values were $a = 1$, $\sigma_\alpha = 0.5$, and $r = 1$.

branching point if

$$\sigma_\alpha < \frac{1}{\sqrt{2a}} \qquad (3.31)$$

Thus, again, evolutionary branching occurs if the frequency-dependent component of competition is strong enough, that is, if the advantage due to being different from the resident at the singular point increases fast enough with increasing phenotypic distance. If condition (3.31) holds, the two-dimensional adaptive dynamics ensuing after convergence to the branching point can be investigated as outlined before. It is not possible in general to give explicit expressions for the singular coalition of the two-dimensional dynamics, but it is easy to give examples where the singular coalition is convergent stable, and the two coexisting singular strategies in the coalition are both evolutionarily stable (see Figure 3.2). A simple numerical analysis shows that this is generally the case as long as σ_α is not too small, that is, as long as frequency dependence is not too strong. For example, for $a = 0.125$ in eq. (3.30), evolutionary branching occurs for $\sigma_\alpha < 2$, and the corresponding two-dimensional singular coalition is evolutionarily stable for $\sigma_\alpha \gtrsim 1.25$. Even if the two-dimensional singular coalition is evolutionarily unstable, so that both singular strategies can undergo another round of evolutionary branching, the ensuing four-dimensional adaptive dynamics can have a singular coalition in which all strains are evolutionarily stable (prove this).

It follows from this that even if frequency dependence is strong enough to induce diversification, it does not necessarily lead to an infinite series of subsequent branching events. In fact, such infinite branching appears to be a special property of models with Gaussian competition kernels and carrying capacities that is related to particular mathematical properties of Gaussian functions. These properties tend to make models base on Gaussian functions structurally unstable, as is explained in more detail in Chapter 9. Indeed, for many other choices of symmetric competition kernels and carrying capacities, infinite branching will not occur, and instead diversification will come to halt after a finite number of branching events. This is, for example, also the case for the model introduced in Box 3.1, in which the carrying capacity and the competition kernel are derived from underlying consumer resource dynamics (see e.g., Fig. 4 in Ackermann & Doebeli (2004)).

But even when both the competition kernel and the carrying capacity are Gaussian, we have seen in Figure 3.1 that after only a few branching events, the fitness minima at the singular strategies tend to become very shallow, and hence selection for diversification becomes very weak. Deterministically, even weak disruptive selection generates diversification, but the situation could be different in models for finite populations, in which stochastic effects can wash out the effects of selection. It is fairly straightforward to construct individual-based stochastic models that correspond to the deterministic models described so far. To do this, we use the so-called Gillespie algorithm (Erban et al., 2007; Gillespie, 1976, 1977; Pineda-Krch, 2010) that was developed for simulating systems of interacting particles in which certain (chemical) reactions take place at certain rates. By treating individuals as particles, and birth and death events as reactions, one can adopt these algorithms for simulating population biological stochastic processes. Recall that in the deterministic Gaussian model, the per capita birth rate of phenotype x is b and its per capita death rate is $bN_{\text{eff}}/K(x)$, where N_{eff} is the effective density experienced by phenotype x. In the corresponding individual-based model, these quantities are interpreted as rates, or probabilities per unit time, with which an individual of phenotype x gives birth or dies. If at a given point in time t there are N individuals in the population with phenotypes $x_1, \ldots, x_N$, then the effective density experienced by the ith individual is the sum of the competitive impacts of all other individuals:

$$N_{\text{eff}}(i) = \sum_{j \neq i} \alpha(x_j, x_i). \tag{3.32}$$

(In practice, it does not matter whether the ith individual itself is included in this summation.) Thus, the death rate d_i of individual i at time t is

$$d_i = bN_{\text{eff}}(i)/K(x_i), \tag{3.33}$$

whereas the birth rate is the same for all individuals, $b_i = b$ for all i. We define the total birth rate at time t to be $B = \sum_{i=1}^{N} b_i = Nb$, and the total death rate as $D = \sum_{i=1}^{N} d_i$. The Gillespie algorithm for the individual-based model is then implemented as follows. At any given time t, all individual birth and death rates b_i and d_i, $i = 1, \ldots, N$, as well as the total birth and death rates B and D are calculated. Then the type of event that occurs next, birth or death, is chosen with probabilities proportional to the total rates for these events, B and D. Specifically, after generating a uniform random number v in the interval $(0, 1)$, a birth event occurs if $v < B/(B + D)$, and a death event occurs otherwise. If a death event occurs, the individual to die is chosen with probabilities proportional to the individual death rates d_i. More precisely, individual i is chosen for the death event, and hence removed from the population, with probability d_i/D. Similarly, if a birth event occurs, individual i is chosen to give birth with probability b_i/B. With probability $(1 - \mu)$, the phenotype of the newborn individual is the same as the chosen individual's phenotype x_i. Here μ is the per capita rate at which mutations occur, and with probability μ, the phenotype of the newborn individual is drawn from a distribution describing the mutational effects. For example, the mutated phenotype could be drawn from a uniform distribution or from a normal distribution with mean the parental phenotype x_i and a prescribed variance σ_μ, which is what we will use here.

It may sometimes be desirable to make the genotype-phenotype map used for determining the offspring phenotype more explicit. For example, one could assume that the above mutational procedure does not directly yield the phenotype of the offspring, but rather its genotype, which is then used to determine the offspring's phenotype by drawing a new random number from a distribution that is determined by the offspring genotype. This would necessitate keeping track of both the genotype and the phenotype of all individuals, and would correspond to incorporating phenotypic variance in addition to genetic variance for the production of offspring, which would decrease heritability and hence effectively weaken selective signals. However, as long as the phenotypic variance is sufficiently small, this would not have a qualitative effect on the evolutionary dynamics. We therefore make the simplifying assumption that the phenotypic variance is 0. Equivalently, the models can simply be thought of as tracking phenotype rather than genotype distributions.

Performing one birth or death event in the manner described above completes one computational step in the individual-based model, which advances the system from time t to time $t + \Delta t$ in real time. To make the translation from discrete computational steps to continuous real time, Δt is drawn from an exponential probability distribution with mean $1/E$, where E is the total event rate $E = B + D$. Thus, if the total event rate E is high, the time lapse Δt between one event and the next is small, and vice versa if the total event rate is low. Starting from some initial population containing N_0 individuals with phenotypes $x_1^0, \ldots, x_{N_0}^0$ at time 0, iteration of the computational steps described above generates the stochastic evolutionary dynamics of a finite population in continuous time.

In fact, this stochastic model is a natural generalization of the deterministic adaptive dynamics to finite populations. This has been made precise in a seminal paper by Dieckmann & Law (1996), who showed that the canonical equation of adaptive dynamics, of which (3.6) and (3.23), (3.24) are examples and which is given in full generality in the Appendix, can be derived from underlying individual-based stochastic models under the assumptions of rare mutations, small mutational effects, and infinite population sizes (also see Champagnat et al. (2006)). In view of this, results from the stochastic model should be more generally valid, but this gain in generality has a price: The stochastic models are less tractable analytically than their deterministic counterparts.

Figure 3.3 shows two examples of the dynamics of the individual-based model for a Gaussian competition kernel and a Gaussian carrying capacity, for which we saw earlier that the corresponding deterministic model is expected to yield an infinite sequence of branching events (see Figure 3.1). Starting from a fairly homogenous population away from the singular point at the maximum of the carrying capacity, the finite population also first converges to the singular point and then undergoes evolutionary branching into two diverging phenotypic clusters. Depending on parameter values, these clusters may converge to a singular coalition that is essentially already stable (Figure 3.3(a)), or that again undergoes branching (Figure 3.3(b)). However, even if there are multiple branchings, due to the stochasticity in finite populations these subsequent branchings may not be symmetric, and hence may not occur simultaneously in all coexisting phenotypic clusters. Moreover, in contrast to the deterministic model multiple branchings will eventually saturate at a small number of coexisting phenotypic clusters. Extinction of single branches may occur due to stochasticity (see Figure 3.3(a)), and at any point time only a few clusters will coexist. The reason is that even though in the corresponding deterministic model all branches in a singular coalition would be under disruptive selection, this deterministic component of selection is washed out in finite populations, because the assumption of strictly monomorphic resident strains is not satisfied

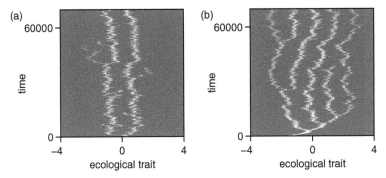

FIGURE 3.3. Evolutionary dynamics of individual-based competition models. The plots show frequency distributions of the population over time, with brighter areas indicating higher frequencies, for two examples in which the number of phenotypic branches remains small due to finite population size. Parameter values: (a) $\sigma_K = 2$, $\sigma_\alpha = 1.25$, $K_0 = 200$, $r = 1$, $\mu = 0.5$, $\sigma_\mu = 0.02$, and the population was initialized with a Gaussian distribution around trait value 0.8 with a small variance; (b) $\sigma_K = 2$, $\sigma_\alpha = 0.75$, $K_0 = 200$, $r = 1$, $\mu = 1$, $\sigma_\mu = 0.01$, and the population was initialized with a Gaussian distribution around trait value -1.5 with a small variance.

anymore. Instead, due to stochasticity the singular clusters consist of phenotypic clumps that are spread around the singular values, which flattens the fitness profile and prevents further diversification.

The fact that demographic stochasticity can blur the effects of selection may not only affect secondary evolutionary branching events, and hence the number of coexisting phenotypic branches in polymorphic populations, but also whether primary diversification through evolutionary branching occurs in monomorphic populations. For example, Claessen et al. (2007) studied the effects of demographic stochasticity on the likelihood of evolutionary branching in asexual consumer-resource models with two types of resources (see Claessen et al. (2008) for corresponding results in models for sexual populations). In individual-based simulations of deterministic models exhibiting evolutionary branching points, these authors found that evolutionary branching becomes less likely in small populations for two reasons. First, incipient diverging clusters can go extinct due to small population size, and second, the waiting time to branching increases because the disruptive selection regime at the branching point becomes less sharp due to stochastic fluctuations. Figure 3.4 illustrates that similar effects can be observed in the competition models studied here. However, the simulations indicate that robust evolutionary branching can occur even in very small populations if disruptive selection at the branching point is not too weak (see Figure 3.4(c)).

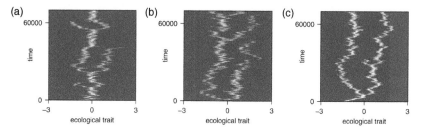

FIGURE 3.4. Effect of population size on evolutionary diversification. In all panels, the population size at the branching point is roughly equal to K_0. In panel (a), $K_0 = 300$, and disruptive selection is weak ($\sigma_K = 2$, $\sigma_\alpha = 1.75$). As a consequence, waiting times for diverging branches to emerge are long, and incipient phenotypic clusters tend to be short-lived. Panel (b) is the same as (a), except that the population size was doubled ($K_0 = 600$). This substantially decreases both the waiting times to branching and the tendency for the emerging branches to go extinct. Panel (c) is the same as (a) except that the strength of disruptive selection at the branching point was increased ($\sigma_\alpha = 1.5$). As a consequence, waiting times to branching decrease, and coexistence of different phenotypic clusters becomes more robust. Other parameter values were: $r = 1$, $\mu = 1$, $\sigma_\mu = 0.01$, and the population was initialized with a Gaussian distribution around trait value -1.5 with a small variance.

Overall, while the number of phenotypic clusters that evolve through adaptive diversification depends on the parameter values chosen for the ecological functions, the general pattern emerging from numerical simulations of the individual-based model is clear: when $\sigma_\alpha < \sigma_K$ and starting from nearly monomorphic populations, frequency-dependent competition often generates adaptive diversification into a small number of phenotypic clusters even with Gaussian carrying capacities and competition kernels. Similar results hold for other competition kernels and carrying capacities (e.g., for the model described in Box 3.1, Ackermann & Doebeli (2004)). Of course, in the models considered so far reproduction was asexual, and diversification is a strictly phenotypic phenomenon. In the next chapter we will investigate how diversification due to frequency-dependent competition can lead to adaptive speciation in sexual populations, but first we will finish the present chapter by taking a closer look at phenotypic diversification when competition is asymmetric.

3.2   ADAPTIVE DYNAMICS WITH ASYMMETRIC COMPETITION KERNELS

Asymmetric competition kernels describe scenarios in which having a larger (or smaller) trait value confers an intrinsic competitive advantage, as, for example, when larger plants take away more light energy from smaller plants than vice versa. In this case, if two individuals have ecological trait values

$x > y$, the competitive impact of x on y is larger than the competitive impact of y on x, that is, $\alpha(x, y) > \alpha(y, x)$, and hence the competition kernel is not symmetric. Asymmetric competition kernels have been used in various forms (e.g., Doebeli & Dieckmann, 2000; Kisdi, 1999; Law et al., 1997; Rummel & Roughgarden, 1985; Taper & Case, 1992; Vincent & Brown, 2005). In Box 3.1 we have already encountered asymmetric competition kernels, but here I adopt a more classical approach by using asymmetric competition kernels of the form:

$$\alpha(x, y) = \exp\left[\frac{\sigma_\alpha^2 \beta^2}{2}\right] \exp\left[\frac{-(x - y + \sigma_\alpha^2 \beta)^2}{2\sigma_\alpha^2}\right]. \tag{3.34}$$

These functions have the formal advantage of allowing a smooth transition from symmetric to increasingly asymmetric competition scenarios. If $\beta = 0$ in (3.34), the competition kernel has the familiar Gaussian symmetric shape, and competition is increasingly asymmetric with increasing $\beta > 0$. The asymmetric competition kernel (3.34) has been derived by Roughgarden (1979) from underlying resource dynamics, but as in the symmetric case, this derivation is only valid if the resource distribution is homogenous, and hence if the carrying capacity function K is a constant. However, the competition kernels (3.34) have typically been used with nonconstant $K(x)$ and should therefore be viewed as reflecting biologically reasonable a priori assumptions about the effects of asymmetric competition. We note that one of the assumptions reflected in (3.34) is that even though large phenotypes generally have a stronger impact on small phenotypes than vice versa, very distant phenotypes always tend to have a very small impact on each other.

For the functions $b(x)$ and $c(x)$ determining the per capita birth and death rates of individuals with phenotype x we make the same assumptions as in the symmetric case, that is, we assume that $b(x) \equiv b$ is a constant, and that $c(x) = b/K(x)$, where $K(x)$ is the carrying capacity function. This means that, just as in the symmetric case (eq. (3.5)), the invasion fitness of a rare mutant y into a resident x at its population dynamic equilibrium $K(x)$ is

$$f(x, y) = b - \frac{b\alpha(x, y)K(x)}{K(y)}. \tag{3.35}$$

Because the coefficient $\exp\left[\frac{\sigma_\alpha^2 \beta}{2}\right]$ in (3.34) ensures that $\alpha(x, x) = 1$ for all x, the adaptive dynamics under asymmetric competition is then again given by

$$\frac{dx}{dt} = D(x) = \left.\frac{\partial f(x, y)}{\partial y}\right|_{y=x} = \frac{bK'(x)}{K(x)} - b\left.\frac{\partial \alpha(x, y)}{\partial y}\right|_{y=x}. \tag{3.36}$$

However, because competition is asymmetric $\frac{\partial \alpha(x,y)}{\partial y}\Big|_{y=x} = 0$ does not hold anymore, and the ensuing analysis is different. To proceed, we make the assumption that the carrying capacity $K(x)$ is of Gaussian form with a maximum at $x = 0$, as in eq. (3.14). It then follows that

$$D(x) = -b\left(\frac{x}{\sigma_K^2} - \beta\right). \tag{3.37}$$

Therefore, the adaptive dynamics has a singular point x^* with $D(x^*) = 0$ at

$$x^* = \beta\sigma_K^2. \tag{3.38}$$

The singular point is always convergent stable because

$$\frac{dD}{dx}\Big|_{x=x^*} = -\frac{b}{\sigma_K^2}, \tag{3.39}$$

hence monomorphic resident populations will converge to x^*. Note that, not surprisingly, $x^* > 0$, that is, the singular point for asymmetric competition is not at the maximum of the carrying capacity function (and hence different from the singular point for symmetric competition), because by assumption, asymmetric competition confers an intrinsic advantage to larger phenotypes. Interestingly, in the asymmetric case the condition for the singular point to be an evolutionary branching point is exactly the same as in the symmetric case, namely

$$\frac{\partial^2 f(x^*, y)}{\partial y^2}\Big|_{y=x^*} = b\left(\frac{1}{\sigma_\alpha^2} - \frac{1}{\sigma_K^2}\right) > 0. \tag{3.40}$$

Thus, evolutionary branching again occurs if $\sigma_\alpha < \sigma_K$, that is, if frequency dependence induced by competitive interactions is strong enough. If the condition is satisfied, the two-dimensional adaptive dynamics of two phenotypic branches on either side of the singular point and diverging from each other can be investigated using the invasion fitness of a rare mutant y into a resident population containing two coexisting strains, as outlined in the previous section for the symmetric case. In general, the two-dimensional system converges toward a singular coalition containing two strategies, which may or may not be evolutionarily stable. Figure 3.5 shows the evolutionary dynamics of an individual-based model for which the corresponding deterministic model shows evolutionary branching into two evolutionarily stable phenotypic clusters. In contrast to the symmetric case, the coexisting strategies are of course

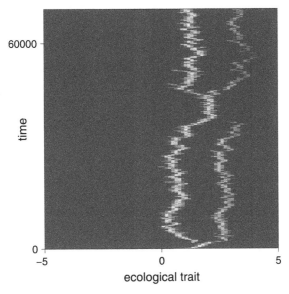

FIGURE 3.5. Evolutionary dynamics of an individual-based asymmetric competition model whose deterministic analogue shows evolutionary branching and subsequent convergence to a convergent and evolutionarily stable coalition of two coexisting phenotypic branches. The plot shows the phenotypic frequency distribution of the population over time, with brighter areas indicating higher frequencies. Note that after ca. 35,000 time units, the higher phenotypic branch goes extinct due to stochasticity, after which the lower branch again evolves to the branching point and subsequently undergoes diversification. Parameter values were $\sigma_K = 2$, $\sigma_\alpha = 1.5$, $\beta = 0.5$, $K_0 = 400$, $r = 1$, $\mu = 1$, $\sigma_\mu = 0.01$, and the population was initialized with a Gaussian distribution around trait value 0.8 with a small variance.

not symmetric anymore with respect to the maximum of the carrying capacity, and as a consequence the two singular strategies have different population densities.

Analytical determination of the two-dimensional singular strategies and their evolutionary stability is in general not possible, but for given choices of parameter values, these questions can be addressed numerically. As we have seen in the symmetric case, the deterministic model might lead to unrealistic predictions regarding multiple branching, and it is therefore advisable to study the possibility of multiple branching events using individual-based models of finite populations. This is done in exactly the same way as described above for the symmetric case, except that the symmetric competition kernel is replaced by the asymmetric kernel (3.34).

Figure 3.6 shows interesting examples of repeated branching due to asymmetric competition in individual-based models. In these examples, diversification continues through subsequent branchings after the first

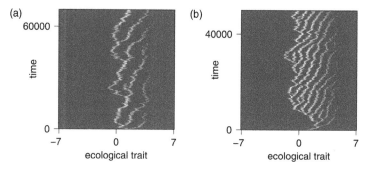

FIGURE 3.6. Evolutionary dynamics of individual-based asymmetric competition models with repeated diversification and extinction. The plots show the phenotypic frequency distribution of the population over time, with brighter areas indicating higher frequencies. In both plots, diversification events always occur in the lowest phenotypic branch, and directional evolution due to asymmetric competition always leads to extinction of the highest phenotypic branch. The examples differ in the resulting average number of coexisting phenotypic clusters, with weaker frequency dependence and weaker asymmetry leading to fewer branches. Parameter values: (a) $\sigma_K = 2, \sigma_\alpha = 1.25, \beta = 0.5, K_0 = 400, r = 1, \mu = 1,$ $\sigma_\mu = 0.01$, and the population was initialized with a Gaussian distribution around trait value 1.5 with a small variance; (b) $\sigma_K = 2, \sigma_\alpha = 0.5, \beta = 1, K_0 = 400, r = 1, \mu = 1, \sigma_\mu = 0.01$, and the population was initialized with a Gaussian distribution around trait value 1.5 with a small variance.

branching event. However, these secondary diversification events exclusively occur in the lowest phenotypic branch, whereas the other coexisting branches simply evolve to higher values over time. Moreover, the currently highest phenotypic branch always goes extinct after some time, because directional evolution to ever higher phenotypic values, driven by frequency-dependent competition, lowers its carrying capacity to values at which extinction due to stochastic events is inevitable. (Note that extinction due to stochastic events in finite populations can also happen if the individual-based model exhibits convergence to an evolutionary equilibrium, rather than indefinite directional evolution. See Figure 3.5 for an example.) Such directional evolution toward extinction is an example of the phenomenon of evolutionary suicide (Gyllenberg & Parvinen, 2001; Gyllenberg et al., 2002). Eventually, the rate of evolutionary extinction due to low carrying capacities in the currently highest branch and the rate of diversification fueled by branching in the currently lowest branch reach a balance, and despite continued branching events, the diversity of the system equilibrates at a finite number of coexisting phenotypic strains. The systems shown on Figures 3.6(a) and 3.6(b) differ in the number of branches maintained, but in both figures, the resulting pattern of evolutionary dynamics is reminiscent of taxon cycles (Maynard Smith & Brown, 1986; Roughgarden & Pacala, 1989; Taper & Case, 1992).

It is interesting to note that asymmetric competition is, in theory, not necessary to produce taxon cycles, as is shown by the models for repeated branching and extinction presented in Ito & Dieckmann (2007). These authors studied models for symmetric competition, in which individuals also carried a secondary trait that did not influence competitive interactions and instead was under directional selection. The mechanism for repeated adaptive radiations described in Ito & Dieckmann (2007) operates in cases where the symmetric competitive interactions alone would lead to repeated branching events and to permanent coexistence of multiple phenotypic clusters as described in this chapter. If during the evolutionary progression toward increased diversity there is also directional selection on a secondary trait, stochastic effects lead to some of the current clusters having higher fitness than other clusters. As a consequence, the lower-fitness clusters are driven to extinction by the higher-fitness clusters, leaving the system forever in a transitory state in which maximal diversity has not been achieved, and hence in which some clusters are always prone to be the origin of new bouts of diversification. The resulting evolutionary pattern resembles a growing tree in which new branches repeatedly appear from one of the currently existing branches, while the other currently existing branches go extinct, and which thus describes taxon cycles in the form of alternating adaptive radiations and mass extinctions.

Evolutionary branching in niche position due to frequency-dependent competition is the central theme of this chapter. However, as we saw in Box 2.1, there are alternative evolutionary scenarios for phenotypic diversification. In the context of resource competition, one such alternative is that individuals diversify their diet by evolving a wider niche. Based on the ecological model described in Box 3.1, Ackermann & Doebeli (2004) investigated the adaptive dynamics of the consumer trait (x, y) to test when widening the niche width y is an evolutionary alternative to evolutionary branching in the niche position x. To derive analytical results, the invasion fitness of a rare mutant (u, v) into a resident population that is monomorphic for (x, y) can be calculated using the expressions for the carrying capacity and the competition coefficients given in Box 3.1. Ackermann & Doebeli (2004) showed that there is always a unique convergence stable singular point (x^*, y^*), where x^* is the maximum of the resource carrying capacity (this is also true when x^* does not maximize the consumer carrying capacity!). The singular value of the niche width y^* is always evolutionarily stable in the y-direction. The singular value of the width position x^* may or may not be evolutionarily stable in the x-direction. If there are no costs or benefits to being a generalist (i.e., if $c = 0$ in eq. (B3.3)), the singular value of the niche width, y^*, is exactly such that the the second derivative of the invasion fitness in the x-direction vanishes at (x^*, y^*). That is, when $c = 0$, the niche width evolves to undo any disruptive selection of second order on niche position. Even though technically speaking, disruptivity only vanishes to second order, it is indeed the case that with $c = 0$, the singular point (x^*, y^*) is always evolutionarily stable in the x-direction (this can be shown both by considering higher order derivatives of the invasion fitness function, and individual-based models; see Ackermann & Doebeli (2004). Thus, evolving a wider niche is indeed an alternative to evolutionary branching in niche position. However, this is a "knife-edge" result in the sense that even with very small costs to being a generalist, the niche width does not evolve to large enough values to offset disruptive selection on niche position, so that the singular point (x^*, y^*) once again becomes an evolutionary branching point in the x-direction. In contrast, if there are benefits to being a generalist, the niche always widens more than enough to make (x^*, y^*) evolutionarily stable.

Challenge: Repeat the analysis given before for the asymmetric competition kernels used in Kisdi (1999).

*Challenge**: Repeat the salient results of this chapter under the assumption that the birth rate $b(x)$ is a unimodal function of x, whereas the function $c(x) \equiv c$ in eq. (3.1) is independent of x. Also, investigate the possibility of evolutionary branching in more complicated scenarios for the frequency-independent fitness components, in which both $b(x)$ and $c(x)$ (respectively $K(x)$) vary with x.

*Challenge**: Evolutionary branching due to competition with nonequilibrium ecological dynamics. So far, we have only considered invasion scenarios in which the resident was at a stable population dynamic equilibrium. However, it is one of the strengths of the adaptive dynamics framework that it can also be applied when the resident exhibits nonequilibrium ecological dynamics, such as cyclic or chaotic dynamics, although in such cases analytical treatments are typically much harder. In fact, it was such nonequilibrium scenarios that played a crucial role in the development of the concept of invasion fitness, as exemplified by the seminal paper of Metz et al. (1992). Consider for example discrete time models of the form

$$N_{t+1} = F(N_t) = N_t w(N_t), \tag{3.41}$$

in which N_t is the population size at time t, and w is the density-dependent per capita number of offspring. Positive equilibria N^* in such models are given as solution of $w(N^*) = 1$, and their stability is determined by the derivative of the next generation function F at the equilibrium.

If $|dF/dN(N^*)| < 1$, N^* is (locally) stable, and it is well known that as $dF/dN(N^*)$ passes through -1 and becomes more negative, the system exhibits the period-doubling route to chaos (May, 1976), examples of which will be discussed in more detail in Chapter 7. With

$$w(N) = \frac{\lambda}{1 + \left(\frac{N}{a}\right)^b}, \tag{3.42}$$

the unique positive equilibrium is $N^* = a(\lambda - 1)^{1/b}$, and the relevant derivative is $dF/dN(N^*) = 1 - b(\lambda - 1)/\lambda$. Therefore, for any given $\lambda > 1$, the system can exhibit chaotic dynamics if the parameter b is large enough. (Note that because the growth parameter λ roughly determines the amplitude of the fluctuations in population size, with values of λ close to 1 leading to very small fluctuations, this implies that this model can exhibit chaotic dynamics even if the amplitude of the resulting population fluctuations is small.) One can extend the discrete-time model (3.41) to frequency-dependent competition determined by a continuous character x by assuming, just as before, that the competitive impact between individuals with phenotypes x and y is described by a competition kernel $\alpha(x, y)$. Moreover, one can introduce a stabilizing component of selection by assuming that the quantity a is a unimodal function $a(x)$ of x, for example, of Gaussian form. Note that, similar to the carrying capacity $K(x)$ in the continuous-time logistic equation (3.2), the parameter $a(x)$ can be viewed as an individual-level property describing how an increase in unit density affects the survival of individuals with phenotype x.

One can then describe the population dynamics of two competing phenotypes x and y with densities $N(x)_t$ and $N(y)_t$ at time t by

$$N(x)_{t+1} = N(x)_t \frac{\lambda}{1 + \left(\frac{N(x)_t + \alpha(y,x)N(y)_t}{a(x)}\right)^b} \tag{3.43}$$

$$N(y)_{t+1} = N(y)_t \frac{\lambda}{1 + \left(\frac{N(y)_t + \alpha(x,y)N(x)_t}{a(y)}\right)^b}. \tag{3.44}$$

This dynamical system can be used to determine the invasion success of rare mutants y into a monomorphic resident x just as in the other models discussed in this chapter, except that now the population dynamics of the resident does not necessarily exhibit a stable equilibrium, and instead may be very complicated. Specifically, let $\{N_t(x), t = 1, \ldots\}$ be the time series of the resident population density. Then at any given time t the per capita reproductive output of a rare mutant y is

$$\frac{\lambda}{1 + \left(\frac{\alpha(x,y)N(x)_t}{a(y)}\right)^b} . \tag{3.45}$$

Over T generations $t = 1, \ldots, T$, the average reproductive output per generation of a mutant individual is given by the geometric mean

$$\sqrt[T]{\prod_{t=1}^{T} \frac{\lambda}{1 + \left(\frac{\alpha(x,y)N(x)_t}{a(y)}\right)^b}} . \tag{3.46}$$

Finally, following Metz et al. (1992) the long term growth rate of a rare mutant y in a resident x, that is, the invasion fitness of y in x, is given by the limit

$$f(x, y) = \ln\left(\lim_{T \to \infty} \sqrt[T]{\prod_{t=1}^{T} \frac{\lambda}{1 + \left(\frac{\alpha(x,y)N(x)_t}{a(y)}\right)^b}}\right)$$

$$= \lim_{T \to \infty} \frac{1}{T} \sum_{t=1}^{T} \ln\left(\frac{\lambda}{1 + \left(\frac{\alpha(x,y)N(x)_t}{a(y)}\right)^b}\right). \tag{3.47}$$

The adaptive dynamics of the trait x can in principle be derived based on this invasion fitness. (Note that the adaptive dynamics unfolds in continuous time, even though the underlying ecological dynamics is formulated in discrete time. In the derivation of the adaptive dynamics from the invasion fitness, the transition from discrete to continuous time occurs because ecological dynamics is assumed to occur on a much faster time scale than the evolutionary dynamics.)

Project 1: Assume that the competition kernel α is symmetric (and smooth), and that competitive impacts decrease with increasing phenotypic distance, just as we did in Section 3.1. Assume further that the function $a(x)$ has a unique maximum, so that the (possibly unstable) equilibrium density $N^*(x)$ has a unique maximum at the same trait value. Prove that, independent of the population dynamics generated by the discrete-time model (i.e., independent of the demographic parameters λ and b), this maximum is the unique singular point of the adaptive dynamics, and that this singular point is always convergent stable. Furthermore, prove that, independent of the population dynamics generated by the discrete-time model, the condition for evolutionary branching is essentially the same as in the continuous-time model with stable equilibrium population dynamics discussed in Section 3.1. In other words, the singular point is an evolutionary branching point exactly if condition (3.12), with $K(x^*)$ and $K''(x^*)$ replaced by $a(x^*)$ and $a''(x^*)$, is satisfied.

Project 2: Based on the deterministic model just described, and assuming specific functional forms for the competition kernel and the function $a(x)$ (for example, both these functions could have Gaussian form), construct individual-based models for the evolutionary dynamics of the trait x. In these individual-based model there is no time scale separation between ecological and evolutionary dynamics, and hence the individual-based models are now set in discrete time. There are various ways to implement these discrete-time dynamics, but the simplest way is to assume that in each generation, individuals first undergo viability selection, and then reproduce. Specifically, in each generation each individual survives with probability $1/\left(1 + [N_{\text{eff},t}/a(x)]^b\right)$, where x is the individual's phenotype, and the effective density N_{eff} experienced by the individual is calculated based on all individuals present at the start of the generation in a way that is analogous to expression (3.32). All surviving individuals then give birth to a number of offspring that is drawn from a Poisson distribution with mean λ (of course, other distributions could be used). Each offspring inherits the parental phenotype subject to mutations, as in the individual-based model described in Section 3.1. All the parents are then removed from the population, and the offspring make up the population at the start of the next generation. This defines a stochastic model for the ecoevolutionary dynamics of the trait x. Use this model to investigate evolutionary diversification in populations with nonequilibrium population dynamics. How does complex population dynamics affect the likelihood of evolutionary branching and the occurrence of multiple (sequential) branching events?

Adaptive Diversification Due to Resource Competition in Sexual Models

The previous chapter explained that adaptive diversification due to resource competition is a general and robust phenomenon in models for long-term evolution in clonally reproducing populations. Once such populations have converged to an evolutionary branching point, nothing hinders their splitting into diverging phenotypic clusters. Evolution in sexual populations should in general be qualitatively similar to clonal evolution during directional phases of adaptive dynamics, such as convergence to a singular point. Whether sexual or clonal, as long as selection gradients are nonzero, populations will move in the direction dictated by the selection gradient. However, as sexual populations converge to what would be a branching point in clonal models, splitting obviously becomes a problem, because mating between different marginal phenotypes generally creates intermediate phenotypes. Through segregation and recombination, sexual reproduction can prevent the establishment of diverging phenotypic clusters in randomly mating populations.

In some sense, the solution is obvious: to allow for a phenotypic split, mating needs to be assortative with respect to the ecological trait that is under disruptive selection. For example, a hypothetical situation in which individuals only mate with other individuals that have exactly the same ecological phenotype would be akin to asexual reproduction (at least on the phenotypic level), and hence evolutionary branching should be possible much like in the asexual models discussed in the previous chapter. Thus, the question of evolutionary branching in sexual populations, that is, of adaptive speciation (Dieckmann et al., 2004), is intimately tied to questions about the evolution of assortative mating. We can already note here that if evolutionary branching occurs in sexual populations due to the presence of assortative mating mechanisms, the diverging phenotypic clusters will show prezygotic reproductive isolation at least to some extent, and hence they can be viewed as representing incipient species.

Various mechanisms of assortative mating have been investigated in models for adaptive, sympatric speciation. The simplest assortment mechanisms that can lead to speciation are those in which the ecological trait itself dictates assortative mating through some form of pleiotropy. Such "direct selection" mechanisms (Kirkpatrick & Ravigne, 2002) could occur if the ecological trait not only determines fitness, but also the location or the timing of reproduction, for example, because the ecological trait is related to habitat choice. In such cases, no additional assortative mating traits need to be considered. However, the majority of models for sympatric speciation are based on two separate sets of loci: E-loci code for the (ecological) trait under disruptive selection, and A-loci code for the various types of assortative mating traits (Dieckmann & Doebeli, 2004). In the simplest case with two sets of loci, the A-loci only determine the degree of assortative mating with respect to the phenotypes determined by the E-loci. In this case, speciation can occur if the alleles at the A-loci inducing random mating are substituted in all individuals of the population by the alleles inducing stronger assortment. In particular, the same type of alleles at the A-loci must be substituted in all individuals for speciation to occur, which is why this type of assortative mating is classically called a "one-allele" mechanism (only allele substitutions of one type or in one direction are necessary). In the more complicated "two-allele" scenarios, assortative mating is not based directly on the ecological trait, but instead on some a priori ecologically neutral "marker" trait. In such scenarios, a splitting in the ecological trait requires a concomitant splitting in the marker trait, and an association between the different clusters of the marker trait and the different clusters of the ecological trait. A splitting in the marker trait corresponds to the substitution of different A-loci alleles in the different marker clusters, hence the name "two-allele mechanism."

In the classical Maynard Smith niche model introduced in Chapter 2, one-allele mechanisms have been discussed first in Maynard Smith (1966) and, more extensively, in Dickinson & Antonovics (1973). Starting with the influential papers by Udovic (1980) and Felsenstein (1981), two-allele mechanisms in the Maynard Smith model have received a lot of attention (see also Diehl & Bush, 1989; Fry, 2003; Johnson et al., 1996; Kawecki, 1996, 1997, 2004). In particular, Felsenstein's 1981 paper was seminal because it pointed out that recombination between E-loci and A-loci makes it difficult to establish the association between genetic clusters in the ecological trait and genetic clusters in the assortative mating traits that are required for speciation under two-allele mechanisms. Thus, the paper supported skepticism toward the theoretical feasibility of sympatric speciation.

Perhaps because frequency-dependent selection acts on traits determining performance in discrete ecological niches in the Maynard Smith models, most studies of the genetics of speciation based on these models involve only a small number of E- and A-loci with a small number of alleles at each locus (but see Fry, 2003; Geritz & Kisdi, 2000; Kisdi & Geritz, 1999). In contrast, frequency-dependent selection on quantitative characters provides a more natural setting for studying multilocus models of speciation, an approach that was pioneered by Alex Kondrashov (Kondrashov, 1983a, 1983b, 1986; Kondrashov & Kondrashov, 1999; Kondrashov & Mina, 1986). Kondrashov's models are not based on an explicit ecological embedding, and frequency-dependent disruptive selection is simply assumed to be operating on the quantitative trait at all times. Seger (1985) was perhaps the first to study the possibility of sympatric speciation based on an explicit ecological model for resource competition mediated by a quantitative character. Seger's models only involved a small number of loci, but a similar type of ecological setup was used in Doebeli (1996b) to investigate the case where the ecological trait is determined by many additive loci. In that paper, I used the so-called hypergeometric model (Barton, 1992; Kondrashov, 1984) to describe the dynamics of phenotype distributions determined by multiple E-loci. The model did not have any A-loci, and instead I first assumed fixed degrees of assortative mating with respect to the ecological trait, and then investigated the evolution of assortment by using a clonal invasion analysis, in which resident and mutant assortment types were compared successively. In essence, the ecological models I employed in that paper were equivalent to the models for frequency-dependent resource competition discussed in the previous chapter.

In Dieckmann & Doebeli (1999) we extended these models in various ways. We first established the connection to adaptive dynamics theory and evolutionary branching, as explained in the previous chapter. Rather than using the deterministic and phenotypic hypergeometric model, we then used individual-based simulations of explicit multi-locus models containing both E-loci and A-loci. In these stochastic models, each individual is characterized by its diploid genotype consisting of the alleles 0 or 1 at each locus. The E-loci define the ecological trait value additively, and two different cases were considered for the A-loci. In the first, there is one set of A-loci that additively determine the degree of assortment with respect to the ecological trait. For speciation to occur, assortment needs to be strong enough, which means that enough one-alleles need to be substituted in all individuals at the A-loci. Accordingly, this case represents a one-allele model. In the second case, there were two sets of

A-loci, one additively determining the degree of assortment as before, but this time with respect to a neutral marker trait, which was determined additively by the second set of A-loci. This case represents a two-allele model, because not only does assortment need to be strong enough at the first set of A-loci, but two different genetic clusters in the second set of A-loci, that is, in the marker trait, need to be established, so that assortative mating can latch on to the ecological trait via a linkage disequilibrium between the marker trait and the ecological trait.

In the past few years, a number of extensions and refinements of these multilocus models have been investigated (Bolnick, 2004, 2006; Bürger & Schneider, 2006; Bürger et al., 2006; Doebeli, 2005; Drossel & McKane, 2000; Gourbiere, 2004; Schneider, 2006, 2007; Schneider & Bürger, 2006). In terms of the genetic architecture of the ecological trait and the role of the strength of competition and assortment, the most detailed studies to date are those of Reinhard Bürger and colleagues, in particular Bürger et al. (2006). In terms of the genetic architecture of the assortative mating system, the most detailed study to date seems to be Doebeli (2005), in which I considered multilocus models with three different sets of A-loci: one for the strength of assortment, one for female preference, and one for a male marker trait.

Perhaps the most influential papers since Felsenstein (1981) were Kondrashov & Kondrashov (1999) and Dieckmann & Doebeli (1999). On the one hand, these papers showed that adaptive speciation is theoretically realistic even with two-allele mechanisms for assortative mating. On the other hand, we connected the speciation problem with adaptive dynamics and evolutionary branching, thereby showing that the emergence of the disruptive selection regimes necessary for sympatric speciation should be considered a generic feature of frequency-dependent ecological interactions (such as resource competition), rather than a feature seemingly requiring rather special assumptions, as in the original Maynard Smith model (Chapter 2). Thus, these models provide "an explanation for the nagging problem in other models of how the initial population comes to be in a state in which all phenotypes are intermediate and adaptation to the environment is suboptimal" (Tregenza & Butlin, 1999).

Much subsequent work has concentrated on the conditions of the genetic architecture and the assortment mechanisms that would allow adaptive speciation to occur under the assumption that a disruptive selection regime has been established. In fact, for multilocus models the situation is rather complicated, and whether speciation occurs can depend on many different model components (including the definition of speciation). For example, Bürger et al.

(2006) have emphasized that the interaction between the strength of frequency-dependent competition and the degree of assortment can be nonlinear, so that as the strength of frequency-dependence increases, the degree of assortment necessary for speciation first decreases, but then increases. This is because with very strong frequency-dependence, the given resource spectrum contains many niches separated by small phenotypic distances, so that the occupants of these niches can only be reproductively isolated by very strong assortment (see also Bürger & Schneider, 2006; Gourbiere, 2004; Schneider & Bürger, 2006). Also, the genetic architecture, for example, the number of loci, the ploidy level, and the distribution of allelic effects, can influence whether speciation occurs for a given ecological scenario. A potential hindrance to speciation is sexual selection due to assortative mating (Bolnick, 2004; Doebeli, 2005; Drossel & McKane, 2000; Gourbiere, 2004; Kirkpatrick & Nuismer, 2004), which could occur because more assortative individuals potentially have fewer suitable partners. However, both Doebeli (2005) and Bürger et al. (2006) have shown that as long as such costs are not very high, speciation is still a feasible outcome. The topic of costs of assortment will be discussed in more detail later in this chapter.

Bürger et al. (2006) gives an excellent summary of the various outcomes and complications that can arise in multilocus models of sympatric speciation due to assortment based on a quantitative trait determining frequency-dependent competition. Their basic conclusion is that speciation is a feasible evolutionary process in such models. This also seems to hold for multilocus models with more complicated two-allele mechanisms for assortative mating. For example, I have shown in Doebeli (2005) that adaptive speciation can occur even when assortative mating is mediated by female preference on a male marker trait, so that not only recombination between E-loci and A-loci, but also recombination between preference and marker trait hinders reproductive isolation. Moreover, in these models assortative mating can evolve even if it is costly. In general, Bürger et al. (2006) conclude that adaptive convergence to an evolutionary branching point generally leads to either ecological diversification, that is, speciation, or to genetic diversification, that is, within species polymorphism. They, as others (Rueffler et al., 2006b) also point out that in principle, other evolutionary responses to disruptive selection are possible, including the evolution of sexual dimorphism (Bolnick & Doebeli, 2003; Van Dooren et al., 2004), evolution of the genetic architecture (Kopp & Hermisson, 2006; Schneider, 2007; Van Doorn & Dieckmann, 2006) including the evolution of dominance (Durinx & Van Dooren, 2009; Peischl & Bürger, 2008), or the evolution of phenotypic plasticity (Leimar, 2005; Rueffler et al., 2006b;

Svanback et al., 2009; see Box 2.1 in Chapter 2 for a closer look at the latter possibility in an asexual model).

For speciation to be a possible consequence of convergence to a branching point, the question is of course not only what types of assortative mating are conducive to speciation, but also whether such assortment would evolve in the first place. Most work addressing this question has concentrated on the case where there is genetic variation for assortment with regard to the primary ecological trait, and has used numerical simulations. However, two recent papers derived analytical results for relatively simple genetic architectures, in which there is one diploid E-locus for the ecological trait and one diploid A-locus determining assortment, which is based on the E-locus. The two papers by Pennings et al. (2008) and Otto et al. (2008) reached remarkably similar conclusions. They proved that there is selection for stronger assortment as long as homozygotes at the E-locus are, on average, fitter than the heterozygote at the E-locus. They also showed that there are essentially three mechanisms that act against the evolution of assortment: sexual selection due to assortment, loss of polymorphism at the E-locus, and direct fertility costs of assortment (the latter was only considered in Otto et al. (2008)). However, both papers show that as long as direct costs are not very high, assortment can evolve despite sexual selection and lead to partial or complete reproductive isolation between the different ecological types. Another recent paper (De Cara et al., 2008) derived analytical results for models with many E-loci and assortment modifiers with weak effects and concluded that assortment is unlikely to evolve if it causes sexual selection. However, when incorporating sexual selection these authors only considered models in which the costs imposed by sexual selection were very severe (rather than a range of models in which the costs due to sexual selection vary gradually, as considered, e.g., in Otto et al. (2008), so their conclusions regarding the effects of sexual selection on the evolution of assortment should be viewed with caution). Assortment is ubiquitous in nature, and it is important to note that sexual selection may actually favor the evolution of assortment in certain circumstances, for example in the presence of Allee effects, as we will see later in this chapter. In particular, assortment may evolve even when the trait on which assortment is based is under stabilizing or directional selection, thus paving the way for speciation at later evolutionary stages when ecological selection becomes disruptive.

Concerning the likelihood of adaptive speciation due to competition, it should be kept in mind that all the explicit multilocus models that have been studied in this context probably still represent a vast simplification of the actual genetics of the traits involved in competition and assortment. For example, several papers (e.g., Bürger et al., 2006; Otto et al., 2008; Pennings et al., 2008) have pointed out that one obstacle to the evolution of assortment is that for weak to moderate assortment, genetic variation at the ecological trait can be depleted, which would impede speciation. But I find it rather unrealistic to imagine that sexual populations would almost completely lose genetic variation for ecologically important traits such as body size. While it is certainly informative to study the multilocus models referred to above, it is now common knowledge that in the real world, the determination of phenotype from genotype is typically much more complicated than envisaged in these models, and probably highly nonlinear. In fact, ten Tusscher & Hogeweg (2009) have recently argued that diversification and speciation will be substantially easier in models incorporating more realistic and complicated genetic architectures based on gene regulatory networks that involve transcription factors as promoters and repressors of structural genes determining ecological phenotypes. In my opinion, this seriously questions whether some of the more intricate and special results and conclusions drawn from simple multilocus models have any correspondence in the real world.

Consequently, I will pursue a different approach in the remainder of this chapter to show that evolutionary branching, that is, adaptive speciation, can occur in models for sexual populations. But rather than considering even more complicated genetic scenarios, I will simplify things by glossing over genetic details. Even without knowing the genetic details, it is often plausible that if two parents with values x and y of a quantitative trait, such as body size, produce an offspring, then the trait value of the offspring will be drawn from a certain distribution whose mean is somehow determined by the two parent values. This distribution is called the segregation kernel (Roughgarden, 1979). For example, in many situations it seems natural to assume that this distribution is centered at the mid-parent value $(x + y)/2$. In principle, the distribution could have any shape, and the shape could depend on the parent values x and y, as well as on the current phenotype distribution of the population. For example, one could expect that the variance of the segregation kernel is not constant and becomes smaller for matings between individuals with phenotypes at the margin of the trait interval considered, because such marginal phenotypes would have the same alleles at many loci. In the competition models to be considered below, this particular difficulty can be avoided by assuming that the trait interval is much larger than the width of the carrying capacity function. In that case the ecological conditions confine the phenotype distribution to an area in trait space that is much smaller than the interval of genetically possible traits. Thus, phenotypes that are marginal to the current phenotype distribution are never marginal to the genetic trait interval and hence do not necessarily share alleles at many loci. In principle, it should be possible to derive the segregation kernels for many different assumptions about the genetic architecture of the trait x.

Here, however, I will make the simple assumption that the segregation kernel is of Gaussian form, and that the variance of the segregation kernel does not depend on the parent phenotypes. These are the standard assumptions of the so-called "infinitesimal model" (Lynch & Walsh, 1998, Chap. 15). In explicit genetic terms, these assumptions can be shown to hold under certain selective conditions and in the limit where the quantitative trait is determined by effectively infinitely many unlinked loci of infinitesimal, additive effects (Lynch & Walsh, 1998, Chap. 15). However, for the present purposes I would like to view the assumptions made regarding the segregation kernel as purely heuristic, reflecting a simple scheme of "blending inheritance" at the phenotypic level.

As mentioned, the assumption of a constant variance of the segregation kernel is most likely to be applicable when the range of possible character values is much wider than the actual phenotype distribution of the population. This is the case in the models used below, in which the range of possible characters is essentially infinite. In fact, it is exactly this situation of unconstrained character ranges for which Polechova & Barton (2005) claimed that speciation due to competition does not occur. These authors argued that in multilocus models, speciation occurs because the genetic architecture is not flexible enough for generating the "right" phenotype distributions. Thus, under the constraints imposed by multilocus genetics, populations would be prevented from escaping disruptive selection through an adequately molded unimodal phenotype distribution with a large enough variance. In contrast, these authors argued that the flexibility of infinitesimal models with infinite character ranges would always allow for appropriate unimodal equilibrium distributions, representing polymorphic populations, but not speciation. However, Bürger et al. (2006) have shown that the arguments of Polechova & Barton (2005) regarding the multilocus models are flawed, and Doebeli et al. (2007) have shown that the results of Polechova & Barton (2005) regarding deterministic infinitesimal models are incorrect. This is further explained in Chapter 9, and corroborated by the individual-based models that I will investigate in the remainder of this chapter. In these models, individuals have ecological traits (E-traits) and assortative mating traits (A-traits) that are inherited according to the "blending inheritance" model. The ecological setup for the E-traits is the same as for the individual-based models in the previous chapter, but instead of reproducing clonally, individuals now mate according to various assortative mating mechanisms that are determined by the A-traits.

4.1 EVOLUTIONARY BRANCHING IN SEXUAL POPULATIONS WHEN ASSORTATIVE MATING IS BASED ON THE ECOLOGICAL TRAIT (ONE-ALLELE MODELS)

The individual-based models for sexual populations considered in this chapter are an extension of the asexual individual-based models used in the previous chapter. Individuals again have an ecological trait value x, which can influence birth and death rates. Specifically, let us again assume that the per capita birth rate is independent of the ecological phenotype and equal to a constant b for all x, and that the per capita death rate is $bN_{\mathrm{eff}}/K(x)$, where $K(x)$ is the carrying capacity of a population that is monomorphic for phenotype x, and N_{eff} is the effective density experienced by phenotype x. To calculate the effective density N_{eff}, we need to specify the competition kernel $\alpha(x, y)$, describing the competitive impact of phenotype y on phenotype x (see eq. (3.32) in Chapter 3). As in the asexual case, the per capita birth and death rates determine the probabilities per unit time with which individuals give birth or die. Because all individuals have the same basic birth rate b, the total birth rate at any given time t is $B = \sum_{i=1}^{N} b_i = Nb$, where N is the population size at time t. Similarly, the total death rate is $D = \sum_{i=1}^{N} d_i$, where d_i are the individual death rates, which are determined by the carrying capacity function and by the current phenotype distribution (see eq. (3.33) in Chapter 3). As in the previous chapter, the individual-based model is then implemented by choosing, at any given time t, the type of event that occurs next, birth or death, with probabilities proportional to the total rates B and D. If a death event occurs, individual i is chosen for the death event, and hence removed from the population, with probability d_i/D. If a birth event occurs, individual i is chosen to give birth with probability b_i/B. However, in sexual populations the chosen individual does not simply produce a copy of itself. Instead, it first searches for a mating partner, and the resulting mating pair then produces an offspring whose phenotype is drawn randomly from the distribution given by the segregation kernel. The mechanisms by which the mate search is determined are thus a crucial ingredient of the sexual model.

The simplest scenario occurs when mate search is determined by a fixed degree of assortment with respect to the ecological trait x. To model this, we introduce a preference function $A(x, y)$ that determines the relative preference an individual of a given phenotype x has for mates with phenotype y. For example, the preference function could be of normal form

$$A(x, y) = \frac{1}{\sqrt{2\pi}\sigma_A} \exp\left[-\frac{(x-y)^2}{2\sigma_A^2}\right]. \tag{4.1}$$

Here the parameter σ_A measures the degree of assortment: the smaller σ_A, the stronger the preference for phenotypically similar mating partners. For simplicity, I consider a hermaphroditic population, where all individuals are potential mating partners of all other individuals. Alternatively, this corresponds to a situation in which only female individuals are modeled under the assumption that males occur in the same phenotype distribution as females and the sex ratio is $1/2$. I further assume that matings are initiated unilaterally, so that only the preference of the females is taken into account. The probability of mating between individuals of phenotypes x and y is then proportional to $A(x, y)$. More precisely, once an individual with phenotype x has been chosen to give birth, the probability $P(x, y)$ that the offspring is generated by mating with a y-individual is given by

$$P(x, y) = \frac{A(x, y)}{\sum_z A(x, z)},\tag{4.2}$$

where the sum runs over all individuals in the population. Note that $\sum_y P(x, y) = 1$, hence the individual chosen to give birth is guaranteed to find a mate. This corresponds to the baseline case in which assortative mating incurs no costs to individuals exhibiting mate preferences. (Note, however, that when mating is assortative, rare phenotypes are less likely to be chosen as mates than common phenotypes!) Models incorporating explicit costs of assortment will be considered later. Also note that as long as the degree of assortment, σ_A, is a fixed parameter for the whole population, an assumption of bilateral mating preference, where the probability of mating between x and y would be $P(x, y) = \frac{A(x,y)A(y,x)}{\sum_z A(x,z)A(z,x)}$, would generate, up to a rescaling of the assortment parameter, exactly the same mating probabilities, and hence the same results, as the assumption of unilateral mating preferences used here. Once the focal individual x has chosen a mating partner y, one individual is added to the population whose phenotype is chosen from the normal distribution given by the segregation kernel for the ecological trait. This segregation kernel has mean $(x + y)/2$ and variance σ_{s_x}, which is a further parameter in the model.

Performing one birth or death event in the manner described completes one computational step in the individual-based model, which advances the system from time t to time $t + \Delta t$ in real time. To make the translation from discrete computational steps to continuous real time, Δt is drawn from an exponential probability distribution with mean $1/E$, where E is the total event rate $E = B + D$. Thus, if the total event rate E is high, the time lapse Δt between one event and the next is small, and vice versa if the total event rate is low. Starting from some initial population containing N_0 individuals with phenotypes $x_1^0, \ldots, x_{N_0}^0$ at time 0, iteration of the computational steps described above generates the stochastic evolutionary dynamics of a finite population in continuous time. It is straightforward to implement these individual-based models in a computer program. Note that extinction of the population is the only absorbing state of these individual-based models, that is, the only true equilibrium state, which will be reached eventually because the population is finite. However, for reasonable parameter choices, that is, for large enough population sizes (typically a few hundred individuals), this equilibrium will not be attained even after a very long (but finite) time. Rather, after a comparably short period of time, the system typically settles on a "stochastic (quasi-) equilibrium," representing a population distribution in phenotype space that subsequently does not change qualitatively over long periods of time.

The basic model has four key ingredients: the shapes, and in particular the relative widths, of the ecological functions, that is, the carrying capacity K and competition kernel α, as well as the width of the "genetic" functions, that is, the width σ_A of the preference function, and the width σ_{s_x} of the segregation kernel. We are interested in studying divergence in sexual populations in which the ecological conditions create the potential for disruptive selection. In general, this means that the competition kernel should be "narrower" than the carrying capacity. For example, if both these functions are Gaussian, I will assume that $\sigma_\alpha < \sigma_K$ (see Chapter 3). Also, it is intuitively clear that wide preference functions, that is, large σ_A implying low degrees of assortment, as well as wide segregation kernels, that is, large σ_{s_x}, tend to prevent divergence.

We first look at the case where the carrying capacity and the competition kernel are Gaussian functions. For fixed parameters $\sigma_\alpha < \sigma_K$ and σ_{s_x}, Figure 4.1 shows the effect of changing the degree of assortment σ_A. Panel (a) illustrates the outcome for a very large σ_A, corresponding to indiscriminate mate choice and hence random mating. As expected, at the stochastic equilibrium the population shows a unimodal phenotype distribution centered at the maximum of the carrying capacity, and no diversification occurs. The evolutionary dynamics change with increasing strength of assortment, as shown in Figure 4.1(b). For intermediate assortment strength, the population distributions become multimodal. Because mating is assortative, individuals belonging to different phenotypic modes are at least partially reproductively isolated, and gene flow between the different models is reduced. Therefore, the emergence of multimodality can be viewed as incipient speciation. With stronger assortment (panels (c) and (d)), the equilibrium phenotype distribution tends to be unimodal again, but this is likely a consequence of the assumption of Gaussian ecological functions. This will become more clear in Chapter 9, where we will investigate deterministic models for the dynamics of phenotype distributions in sexual populations that correspond to the individual-based

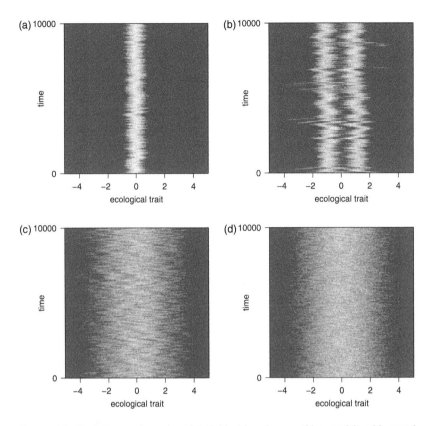

FIGURE 4.1. Evolutionary dynamics of individual-based competition models with sexual reproduction and Gaussian competition kernels and carrying capacities. Plots show the phenotype frequency distributions over time. (a) No diversification in the form of multimodal phenotype distributions is observed when mating is effectively random. (b) Diversification in the form of bimodal phenotype distributions can be observed for intermediate degrees of assortative mating. (c), (d) For Gaussian competition kernels and carrying capacity functions, strong degrees of assortment again tend to generate unimodal phenotype distributions, albeit with much larger variance than with random mating. Parameter values were $r = 1$, $K_0 = 250$, $\sigma_K = 2$, $\sigma_\alpha = 1$, $\sigma_{s_x} = 0.2$, and $\sigma_a = 0.65$ for (a), $\sigma_A = 0.45$ for (b), $\sigma_A = 0.25$ for (c), and $\sigma_A = 0.05$ for (d). Populations were initialized with a Gaussian phenotype distribution with variance 1 and centered at 0 (the maximum of the carrying capacity function).

models studied in the present chapter. These deterministic models are formulated by means of partial differential equations, and with Gaussian carrying capacities and competition kernels, these models always have unimodal equilibrium distributions, corresponding to polymorphic populations, but not to diversification into distinct phenotypic clusters. However, unimodal equilibrium distributions appear to be a special property of Gaussian deterministic models. Moreover, these equilibrium distributions need not be dynamically stable, and for intermediate degrees of assortment, instability of such unimodal equilibrium distributions can give rise to multimodal equilibrium distributions, similar to those shown in Figure 4.1(b). For now, it suffices to note that models with non-Gaussian ecological functions are in general not expected to have unimodal equilibrium distributions even for the case strong assortment (Doebeli et al., 2007).

This is illustrated by the example shown in Figure 4.2, where instead of a Gaussian carrying capacity, I used the "quartic" function

$$K(x) = K_0 \exp\left[-\frac{x^4}{2\sigma_K^4}\right].$$
(4.3)

This function is more "box-like" than the Gaussian, that is, platycurtic. In this case, it is still true that divergence does not occur when mating is random (Figure 4.2(a)), but when assortment is strong enough, diversification and incipient speciation in the form of multimodal phenotype distributions is readily seen (Figures 4.2(b)–(d)). Note that the model now exhibits multimodal equilibrium distributions even with very strong assortment (Figure 4.2(d)).

Interestingly, the sexual competition model can exhibit complicated, nonequilibrium dynamics of the phenotype distribution. This is illustrated in Figure 4.3, which shows an example in which the phenotype distribution continually fluctuates between different multimodal states. It is clear from the figure that these fluctuations are not merely due to stochastic drift in phenotype space. On the other hand, they also do not seem to be solely due to the ecological interactions, because the underlying logistic competition model is not expected to generate nonequilibrium dynamics. (Indeed, the corresponding asexual models would exhibit much more stable dynamics, showing only noise fluctuations.) Therefore, the fluctuations appear to be due to an interaction between the mixing of phenotypes through assortative mating and the ecological dynamics. We will encounter other examples of complicated dynamics of phenotype distributions in Chapter 9.

Challenge: Investigate the effects of varying the parameters σ_A and σ_{s_x} on the phenotype dynamics of sexual competition models with general ecological functions. For example, consider cases where the competition kernel and the carrying capacity are of the general forms

$$\alpha(x, y) = \exp\left[-\frac{(x - y)^{2+\epsilon_\alpha}}{2\sigma_\alpha^{2+\epsilon_\alpha}}\right]$$
(4.4)

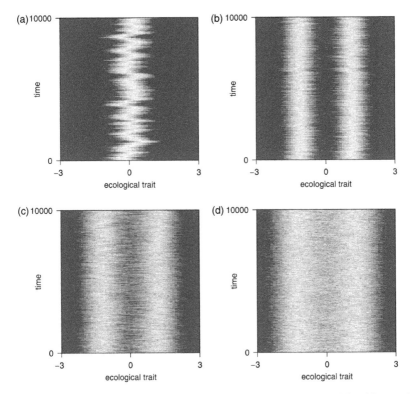

FIGURE 4.2. Evolutionary dynamics of individual-based competition models with sexual reproduction and non-Gaussian carrying capacities. Plots show the phenotype frequency distributions over time. (a) No diversification is observed when the strength of assortative mating is low. (b), (c), (d) Diversification in the form of bimodal phenotype distributions is observed for intermediate and strong degrees of assortative mating. The examples shown are for quartic carrying capacities, while the competition kernel was assumed to be Gaussian. Parameter values were $r = 1$, $K_0 = 250$, $\sigma_K = 2$, $\sigma_\alpha = 1$, $\sigma_{s_x} = 0.2$, and $\sigma_a = 0.65$ for (a), $\sigma_A = 0.45$ for (b), $\sigma_A = 0.25$ for (c), and $\sigma_A = 0.05$ for (d). Populations were initialized with a Gaussian phenotype distribution with variance 1 and centered at 0 (the maximum of the carrying capacity function).

and

$$K(x, y) = K_0 \exp\left[-\frac{x^{2+\epsilon_K}}{2\sigma_K^{2+\epsilon_K}} \right] \qquad (4.5)$$

(see also Doebeli et al., 2007 and Chap. 9). Also, investigate cases with asymmetric competition kernels such as those used in Chapter 3.

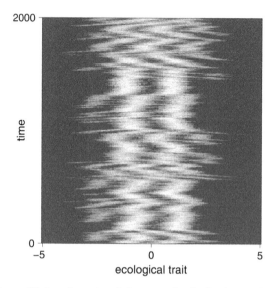

FIGURE 4.3. Nonequilibrium dynamics of phenotype distribution in sexual populations. The distribution alternates essentially between bimodal and trimodal states while undergoing significant fluctuations. In this example, both the carrying capacity and the competition kernel were Gaussian, and parameter values were $r = 1$, $K_0 = 600$, $\sigma_K = 2$, $\sigma_\alpha = 1$, $\sigma_{s_x} = 0.2$, and $\sigma_a = 0.4$. Populations were initialized with a Gaussian phenotype distribution with variance 1 and centered at 0 (the maximum of the carrying capacity function).

It may sometimes be reasonable to assume that assortment incurs costs. Allee effects due to difficulties of finding a mate are one particular source of costs that is relevant even in populations in which all individuals have the same degree of assortment. Such Allee effects can occur when choosiness makes it hard to find a mate for individuals with rare phenotypes. Here I follow the approach taken by Noest (1997) to model Allee effects for rare phenotypes, according to which an individual that is chosen to give birth based on the ecological functions (as described above) subsequently foregoes reproduction with a probability that is determined by the availability of suitable mates. For an individual of phenotype x, the quantity $T(x) = \sum_z A(x, z)$, with the sum running over all individuals in the population, can be viewed as a measure for the total amount of available mates (note that this quantity is the denominator in eq. (4.2)). We then assume that the probability that the focal individual actually finds a mate, which we denote by $f(x)$, is a saturating function of the amount of available mates:

$$f(x) = a + (1 - a) \left(\frac{T(x)}{\eta + T(x)} \right). \tag{4.6}$$

Here $\eta > 0$ is a parameter measuring the magnitude of the Allee effect, and $1 - a$ is the maximal reduction in the probability of finding a mate ($0 \leq a \leq 1$). Even if all individuals in a population exhibit the same degree of assortment, there is now a cost to having a rare phenotype, because the total amount of available mates, and hence the probability f, is generally smaller for rare phenotypes than for common phenotypes. The shape of the function $T(x)$, for example, its maximum value and how fast it declines as phenotypes become rare, depends on a population's current phenotype distribution, and it is therefore not straightforward to gauge the effects of the parameter η. To make the costs at least independent of the total population size, we can divide the amount of available mates by the (current) total population size N, so that the probability of mating becomes

$$f(x) = a + (1 - a) \left(\frac{\tilde{T}(x)}{\eta + \tilde{T}(x)} \right), \qquad (4.7)$$

where $\tilde{T}(x) = T(x)/N$. For any given phenotype frequency distribution and any given x (and hence for any given $\tilde{T}(x)$), the probability $f(x)$ is larger for smaller η. Thus, smaller values of the parameter η correspond to lower costs. Similarly, larger values of the parameter a correspond to lower costs. It is easy to incorporate the "fertility" $f(x)$ into the individual-based models considered so far by including a step in which an individual selected to give birth according to the various event rates actually only does so with probability $f(x)$.

Figure 4.4 shows examples of the phenotype dynamics when assortment is costly due to Allee effects. In these examples, the degree of assortment (measured by σ_A) was fixed, that is, the same for all individuals. It is important to note that the phenotype dynamics can now critically depend on the initial phenotype distribution. For example, even with very strong Allee effects the distribution can become multimodal if the initial frequency of phenotypes corresponding to different modes is high enough. This is illustrated in Figure 4.4(a) and reflects the fact that Allee effects tend to be severe only for rare phenotypes. Thus, even when initial phenotype distribution are unimodal and centered at the phenotype with maximal carrying capacity, speciation in the form of multimodal equilibrium distributions is often possible in the presence of significant Allee effects. But if the initial distribution is too narrow, the marginal phenotypes have too much of a disadvantage due to the Allee effects, and hence multimodality does not emerge (Figure 4.4(b)). In particular, the phenotype dynamics may depend on the initial conditions. I note

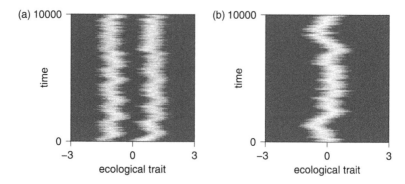

FIGURE 4.4. Diversification in individual-based models for sexual populations in the presence of Allee effects. In the examples shown, both the competition kernel and the carrying capacity function had quartic form (eqs. (4.4) and (4.5) with $\epsilon_K = \epsilon_\alpha = 2$). In (a), speciation occurs due to assortative mating despite Allee effects because the initial phenotype distribution is sufficiently wide. With the same parameter values, speciation does not occur in (b) even after long times, because the initial phenotype distribution is too narrow. Parameter values were $r = 1$, $K_0 = 600$, $\sigma_K = 2$, $\sigma_\alpha = 1$, $\sigma_A = 0.45$, $\sigma_{s_x} = 0.2$, $\sigma_a = 0.2$, $\eta = 1$, and $a = 0$. Populations were initialized with a Gaussian phenotype distribution centered at 0 with variance 1 for (a) and variance 0.5 for (b).

that this dependence on initial conditions is not just because with narrow initial conditions, rare phenotypes may simply not be generated by chance for a long time. Instead, it has a deterministic component due to Allee effects preventing the growth of rare phenotypes even if those phenotypes are present at small frequencies. As a consequence, the dependence on initial conditions due to Allee effects can also be seen in the deterministic models discussed in Chapter 9.

4.2   EVOLUTION OF ASSORTATIVE MATING

Given that under the right ecological conditions, adaptive speciation in the form of the evolution of multimodal phenotype distributions is possible when mating is assortative, but not possible when mating is random, it is natural to ask whether assortative mating would evolve in initially randomly mating populations. Many studies have addressed this question using a variety of different approaches (see Doebeli (2005), Otto et al. (2008), and Pennings et al. (2008), for some recent examples). A nice example of an analytical treatment of this question is given by Pennings et al. (2008). Even though a number of complications can arise when studying this problem, the general picture that emerged is that in one-allele models, in which assortment is based on the trait under disruptive selection, evolution of assortment that is strong enough to cause speciation is a theoretically plausible scenario, essentially because assortment allows individuals to alleviate the effects of frequency-dependent competition. This can, for example, be seen by extending the individual-based models used here to include the degree of assortment as an evolving trait.

To do this, we simply assume that the parameter σ_A is now a phenotypic trait, so that individuals are not only described by their ecological trait x, but also by their assortment trait σ_A. Per capita birth and death rates are the same as before, and if an individual is chosen to give birth, it choses a mate based on its assortment trait (but not that of its potential mating partners, since we assume unilateral mate choice; however, it would be easy to include bilateral mate choice in these models). Once the mate is chosen, the two trait values of the offspring are each chosen independently from a Gaussian distribution with mean the mid-parent value of the respective trait. The width of the segregation kernel in the assortment trait is denoted by σ_{s_A}. For the assortment trait, one has to avoid negative values, for example, by redrawing the offspring trait value from the segregation kernel in case the previous draw resulted in a negative number. Also note that it is not a priori clear how the genotype-phenotype map for the assortment trait (or any other trait, for that matter) should be constructed. In general, if y is the (continuous) genotype determining the assortment trait, then for any y the trait value $\sigma_A(y)$ will be some function of y. The nature of this function is important for determining the effect of segregation on σ_A. For example, if we assume a normal segregation kernel in y, nonlinearities in the function σ_A may generate substantially different segregation kernels for the actual phenotype σ_A. This may in turn affect the evolutionary dynamics of σ_A (see e.g., Bolnick (2004) for a study of related issues). Here I assume for simplicity that $\sigma_A(y) = y$, and hence that the segregation kernel in σ_A has the usual normal form.

The individual-based models resulting from the procedure described above yield the evolutionary dynamics of two-dimensional phenotype distributions. A typical example of such dynamics is shown in Figure 4.5(a), which illustrates a scenario that generally unfolds in cases in which the ecological conditions can induce speciation when mating is assortative enough. In such cases, assortment typically evolves from random mating to a degree that allows diversification. A comprehensive investigation of the ecological parameters for which assortment evolves to a degree that induces speciation is given in Figure 5.3 of Dieckmann & Doebeli (2004) for a related individual-based model with explicit multilocus genetics, and my simulations indicate that similar results hold for the models used here: if there are no explicit costs to assortment, and if the ecological interactions generate strong frequency-dependent

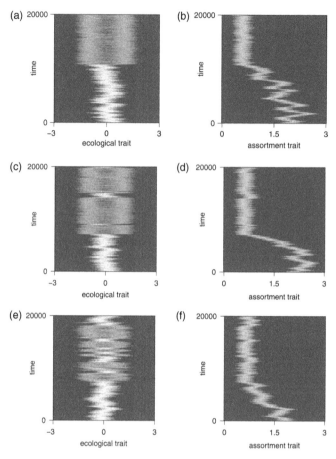

Figure 4.5. Evolution of assortative mating and diversification in individual-based models for sexual populations. In all panels, the competition kernel was Gaussian, the carrying capacity was quartic (eq. (4.3)), and the ecological conditions allow for diversification in asexual population. Sexual populations were initiated with a low degree of assortment corresponding to random mating, so that diversification in the form of bimodal phenotype distributions for the ecological trait is only possible once assortment evolves to a high enough degree. In (a) and (b) there is no cost to assortment. In (c) and (d) a cost of assortment $c = 0.1$ has been introduced according to eq. (4.8). In (e) and (f), there is a higher cost to assortment ($c = 0.3$), but Allee effects nevertheless cause the evolution of assortment to levels that are high enough to generate diversification (at least intermittently). Parameter values were $r = 1$, $K_0 = 250$, $\sigma_K = 2$, $\sigma_\alpha = 0.75$, $\sigma_{s_x} = 0.2$, $\sigma_{s_A} = 0.1$, and $n = 1$ (eq. (4.8)) for panels (a)–(d), in (a) and (b), $c = 0$, and in (c) and (d), $c = 0.1$. For (e) and (f), parameter values were the same, except $K_0 = 400$, $\eta = 0.25$, and $a = 0.5$ in eq. (4.7), and $c = 0.3$ in eq. (4.8). All evolving populations were initialized with a Gaussian distribution centered at 0 with variance 1 for the ecological phenotype, and such that all individuals had an assortment trait value of 1.9, corresponding to essentially random mating. To avoid unrealistically low values of assortment, the assortment trait was constrained to be ≥ 0.2 in the simulations.

selection, assortment often evolves to values allowing for multimodal equilibrium phenotype distributions.

In one-allele models, disruptive selection puts a direct premium on the evolution of assortment. It is therefore plausible that assortment can evolve even if it incurs direct costs, for example, in terms of individual birth rates. This can be modeled by assuming that individual birth rates are not constant, but instead depend on the degree of assortment, for example, through

$$\tilde{b} = b(1 - c/(1 + \sigma_A)^n). \tag{4.8}$$

Here $\tilde{b}$ is the per capita birth rate of an individual with assortment determined by σ_A, and the parameter c is a measure of the cost of assortment, with higher c corresponding to higher costs. The parameter n controls the shape of the cost function. Such costs imply that random mating, that is, the least amount of assortment, yields the highest per capita growth rate, and high degrees of assortment are selectively disadvantageous in terms of individual growth rates. However, disruptive selection affecting individual death rates and generated by frequency-dependent competition can still lead to the evolution of high degrees of assortment and subsequent speciation, as is illustrated in Figure 4.5(c), (d). Of course, if costs are high enough, assortment will not evolve, and hence speciation will not occur. But it is clear from many simulations that speciation can occur in the presence of moderate explicit costs to assortment. In addition, the negative consequences of direct costs to assortment can be alleviated by Allee effects, as we will now see.

Since Allee effects can affect the phenotype dynamics when the degree of assortment is fixed, it is natural to ask how Allee effects impinge on the evolution of assortment. In fact, Allee effects can promote the evolution of assortment, for the following intuitive reason. Let us assume that the current phenotype distribution is unimodal, which means that most individuals belong to phenotype classes that are relatively common. Such individuals can increase the total amount of available mates, $T(x)$, by being more assortative, because for such individuals being more assortative means an increased preference for other common phenotypes. Because higher $T(x)$ implies higher probabilities of giving birth (i.e., higher $f(x)$ in eqs. (4.6) and (4.7)), it is therefore advantageous for common phenotypes to be more assortative. By the same token, for rare phenotypes it pays to be less assortative, but because the advantage of assortment accrues to common phenotypes and hence to most individuals, it outweighs the disadvantage on average, and hence there is selection for increased assortment. This argument is explained in more detail in Box 4.1.

It is important to note that this argument relies on the fact that to calculate the total amount of available mates $T(x) = \sum_z A(x, z)$, we have used *normalized* mating preference functions $A(x, z)$. This means that in a hypothetical population in which all phenotypes have equal frequencies (i.e., the phenotype distribution is uniform), the total amount of available mates is independent of the degree of assortment. In contrast, with nonnormalized mating functions, for example, of the form $A'(x, y) = \exp\left[-\frac{(x-y)^2}{2\sigma_A^2}\right]$, the total amount of available mates would depend on the degree of assortment even in uniform distributions, with more assortative individuals having fewer available mates. As a consequence, without normalization assortment would incur additional implicit costs (Box 4.1). The magnitude of these costs would depend in a nonobvious way on the shape of the mating function A and on the current phenotype distribution. It seems preferable to not include such implicit costs a priori, and hence to work with normalized mating functions. Additional costs to assortment can then be introduced explicitly, for example, by using eq. (4.8). In the presence of Allee effects with normalized mating functions, such direct costs imply that there is an intermediate level of assortment that yields the highest effective growth rate, that is, the maximal average value of the product $\tilde{b} \cdot f$ (Box 4.1). Because Allee effects promote the evolution of assortment, assortative mating can evolve in the presence of Allee effects even if there are substantial direct costs to assortment, as is illustrated in Figure 4.5(e), (f).

As mentioned, with normalized mating functions Allee effects imply that common phenotypes have an advantage if they mate more assortatively. In fact, this may often be biologically realistic, and may be one of the reasons why assortative mating is common in nature, even in the absence of disruptive selection regimes. One scenario in which the positive effect of Allee effects on the evolution of assortment may be important occurs when assortment is based not on the ecological trait under disruptive selection, but on neutral marker traits, that is, in two-allele models. In such models, Allee effects can trigger speciation by promoting the evolution of assortment with respect to the marker traits, as we will see in the next section.

In this section we have studied the evolution of assortative mating in classical models for frequency-dependent competition, but of course similar questions arise in other models for adaptive diversification. For example, Van Doorn et al. (2009) studied one-allele mechanisms in sexual versions of the two-patch Levene models discussed in Chapter 2. These authors investigated the evolution of female preference for a male signal that indicates habitat viability, which is an interesting approach because assortment is not based directly on the ecological trait, but on the habitat viability conferred by this trait. Once the female preference and the male viability signal have evolved, mating is assortative for the ecological trait determining habitat viability, because only females that have high viability survive to mate in a given habitat patch, and when doing so prefer males that also have high viability in that habitat (as indicated by the signal). Van Doorn et al. (2009) showed that female preference and male signal can evolve even if these traits infer substantial costs for their carriers, and that the assortment generated by this mechanism can lead to adaptive speciation. The model describes a one-allele mechanism because in both females and males, substitution of one type of allele (preference alleles in females and signal allele in males) is enough for speciation to occur.

Challenge: It may be unreasonable to assume that choosy individuals simply forgo mating if they do not find a suitable mate. Instead, one could assume that individuals that mate sample a finite number of mates and then simply mate with the one that is most preferred among those sampled. Use individual-based models to investigate such scenarios.

Challenge*: As we have already seen in Chapter 2, evolutionary branching is not the only possible consequence of disruptive selection due to frequency-dependent selection. In sexual populations, the evolution of assortative mating and subsequent speciation is of course also not the only possible outcome of disruptive selection caused by competitive interactions. For example, Slatkin (1984) has already argued that competition can drive sexual dimorphism, and Bolnick & Doebeli (2003) have investigated the evolution of sexual dimorphism as an alternative to the evolution of assortative mating and speciation in an explicit multilocus competition model. In their model, one set of diploid E-loci determined the ecological trait in females, and another set of E-loci determined the ecological trait in males. The overlap between these two sets could vary from no overlap (completely separate loci determine the ecological traits in the two sexes) to complete overlap (exactly the same loci determine the ecological trait in the two sexes). In addition, a set of A-loci determined the degree of assortative mating, which was based on the ecological traits in males and females. Bolnick & Doebeli (2003) showed that with decreasing overlap in the E-loci, sexual dimorphism became a more likely outcome than speciation. In addition, they showed that when disassortative mating is possible (so that "large" females prefer "small" males and "small" females prefer "large" males), sexual dimorphism and speciation can occur concurrently, resulting in two species in each of which one sex is large and the other is small.

This interplay between adaptive speciation and sexual dimorphism can be investigated using the "blending inheritance" model based on

BOX 4.1

EVOLUTION OF ASSORTATIVE MATING DUE TO ALLELE EFFECTS

This box shows that Allee effects alone (i.e., in the absence of any ecological fitness differences between phenotypes) can select for increased assortment. The
box also illustrates that using nonnormalized mating functions in the presence
of Allee effects can entail unwanted implicit costs to assortative mating whose
magnitude is difficult to gauge. To streamline the arguments, we assume infinite
population sizes, which means that we will consider continuous phenotype distributions rather than a finite set of individual phenotypes (and integrals instead
of sums). We assume a fixed phenotype distribution $\phi(x)$, representing the current state of the population. The exact form of the distribution is not important
for the general gist of the argument, and it is most convenient to assume that the
distribution ϕ is unimodal and of normal form:

$$\phi(x) = \frac{1}{\sqrt{2\pi}} \exp\left(-\frac{x^2}{2}\right). \tag{B4.1}$$

Given this phenotype distribution, as well as a mating function $A(x, y)$, eq.
(4.1), we are interested in the quantity

$$T(x, \sigma_A) = \int_y A(x, y)\phi(y)\, dy \tag{B4.2}$$

which is a measure for the total amount of mates available in the population
described by the distribution $\phi(x)$. Here I have emphasized the dependence of
the function T on the quantity σ_A, which determines the strength of assortment
in the mating function A. (Note that there is no need to normalize the quantity T,
because the total population density is $\int \phi(x)dx = 1$ by assumption.)

First, we are interested in how the amount of available mates changes, for
any given phenotype x, with the degree of assortment σ_A. Thus, we are interested in the partial derivative $\partial T/\partial \sigma_A$. One can use a computer algebra program
to determine an explicit formula for this partial derivative, but it is more informative to plot this derivative graphically, as shown in Figure 4.B1. For normalized
mating functions

$$A(x, y) = \frac{1}{\sqrt{2\pi}\sigma_A} \exp\left[-\frac{(x-y)^2}{2\sigma_A^2}\right], \tag{B4.3}$$

it is clear from Figure 4.B1 that for common phenotypes x, that is, for x
close to the maximum at 0 of the population distribution $\phi(x)$, the partial derivative $\partial T(x, \sigma_A)/\partial \sigma_A$ is negative, which means that the amount of available mates
increases with decreasing σ_A, that is, with increasing strength of assortment.
Only for rare phenotypes x, that is, for x-values at the margins of the distribution

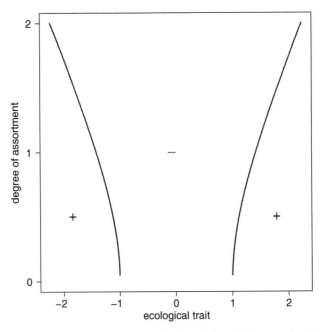

FIGURE 4.B1. The sign of the partial derivative $\partial T(x, \sigma_A)/\partial \sigma_A$ as a function of x and σ_A with normalized mating functions $A(x, y)$ and under the assumption that the current phenotype distribution is normal with variance 1. In common phenotypes, that is, for x close to 0, this partial derivative is always negative, indicating that the number of available mates increases with increasing degree of assortment, that is, with decreasing σ_A (y-axis).

$\phi(x)$ is the $\partial T(x, \sigma_A)/\partial \sigma_A$ positive, indicating that it is only the rare phenotypes that can increase the amount of available mates by being less assortative.

In contrast, with nonnormalized mating functions of the form

$$A(x, y) = \exp\left[-\frac{(x - y)^2}{2\sigma_A^2}\right],\tag{B4.4}$$

the partial derivative $\partial T(x, \sigma_A)/\partial \sigma_A$ is positive everywhere. This is because with nonnormalized mating functions, the maximum value of the mating function is independent of the degree of assortment σ_A, and hence the amount of available mates is always increased by widening the mating function, that is, by becoming less assortative. This is true independently of the phenotype x.

With Allee effects, the amount available mates T translates into a probability of finding a mate given by

$$f(x, \sigma_A) = \frac{T(x, \sigma_A)}{\eta + T(x, \sigma_A)},\tag{B4.5}$$

BOX 4.1 (*continued*)

where η determines the strength of the Allee effect (see eq. (4.7)). As mentioned in the main text, even if all individuals in the population have the same degree of assortment, such an Allee effect implies that rare phenotypes have a lower probability of finding a mate than common phenotypes, because rare x have a lower $T(x, \sigma_A)$. This is true for both normalized and nonnormalized mating functions.

But if we now assume that the parameter σ_A is an evolving trait and ask about the direction of selection on this trait, the answer is very different for normalized and nonnormalized mating functions. For nonnormalized functions, it is clear that $\partial f(x, \sigma_A)/\partial \sigma_A > 0$ for all x, because $\partial T(x, \sigma_A)/\partial \sigma_A > 0$ for all x, and because f is a monotonously increasing function of T. Thus, independent of the phenotype x, in terms of the probability of finding a mate f it is always better to have a larger σ_A, that is, to be less assortative.

In contrast, with the normalized mating functions given by eq. (B4.3), the dependence of f on σ_A is more complicated and reflects the dependence of T on σ_A. In fact, because f is a monotonic function of T, the sign of the partial derivative $\partial f(x, \sigma_A)/\partial \sigma_A$ is exactly the same as the sign of $\partial T(x, \sigma_A)/\partial \sigma_A$, and hence given as in Figure 4.B1, which shows that common phenotypes x can increase their mating probability by becoming more assortative ($\partial f(x, \sigma_A)/\partial \sigma_A < 0$), while the opposite is true for rare phenotypes. To get a sense of the total selection pressure on σ_A due to its effect on mating probabilities, one can average the partial derivative $\partial f(x, \sigma_A)/\partial \sigma_A$ over the population distribution $\phi(x)$, that is, one can consider the average selection pressure

$$S(\sigma_A) = \int_x \frac{\partial f(x, \sigma_A)}{\partial \sigma_A} \phi(x)\, dx. \qquad (B4.6)$$

The selection pressure $S(\sigma_A)$ is plotted as a function of σ_A in Figure 4.B2(a). In this example, $S(\sigma_A)$ is always negative, meaning that smaller σ_A, and hence stronger assortment are favored at any current level of assortment. It should be kept in mind that these calculations are always based on the same phenotype distribution $\phi(x)$ (which we assumed to be Gaussian with variance 1). In reality, the phenotype distribution would of course change as the level of assortment changes. Nevertheless, Figure 4.B2 illustrates that as long as phenotype distributions are unimodal and mating functions are normalized, Allee effects can be expected to select for higher levels of assortment. This is because in the presence of Allee effects, higher levels of assortment increase the probability of mating for common phenotypes. It is clear that direct costs to assortment as introduced by eq. (4.8) will counteract selection for positive assortment. But

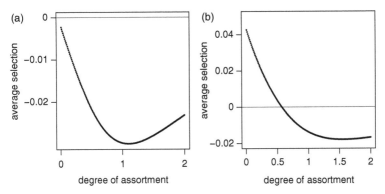

FIGURE 4.B2. The average selection $S(\sigma_A)$ on the degree of assortment in the presence of Allee effects in a population with a normal phenotype distribution with variance 1. Negative values of $S(\sigma_A)$ indicate selection for lower σ_A, and hence for stronger assortment. In panel (a) there is no direct cost of assortment, and in panel (b) there is a direct cost of assortment given by setting $c = 0.4$ and $n = 1$ in eq. (4.8).

even with relatively strong direct costs of assortment, assortment is still often selectively favored over random mating due to the mechanism described above, and only very high degrees of assortment will become disadvantageous. This is illustrated in Figure 4.B2.

■

segregation kernels used here by incorporating different ecological traits in males and females. The ecological traits of offspring are then drawn from a two-dimensional distribution that reflects the genetic covariance between male and female traits, which plays the role of the genetic overlap in the model of Bolnick & Doebeli (2003). If there is no genetic covariance between the ecological traits in males and females, offspring traits are drawn from a bivariate normal distribution, and with increasing covariance, offspring traits are drawn from a bivariate distribution that is increasingly centered along the "diagonal" of equal male and female trait values. The challenge is to formulate and analyze the corresponding models, and to investigate how the relative likelihoods of sexual dimorphism and speciation depend on the genetic covariance between male and female ecological traits. Can sexual dimorphism and speciation occur simultaneously with disassortative mating?

4.3 EVOLUTIONARY BRANCHING IN SEXUAL POPULATIONS WHEN ASSORTATIVE MATING IS NOT BASED ON THE ECOLOGICAL TRAIT (TWO-ALLELE MODELS)

We now turn to models of adaptive speciation in which assortment is based on marker traits that have no direct ecological significance. In this case, speciation requires diversification in the mating traits, that is, in the A-traits. In fact, it is of considerable interest to understand whether sexual selection alone, that is differential reproductive success imposed solely by the mating system, can lead to speciation. In models for speciation by sexual selection, mating partners become the limiting "resource," and competition for mating partners is the mechanism generating disruptive selection. These issues are briefly discussed in the last section of this chapter. In this section we follow the thread of the current chapter by concentrating on models in which competition for food resources, rather than competition for mates, is the primary driver of disruptive selection. In this case, speciation requires an association, or linkage, between the ecological trait and the mating traits, that is, between E-traits and A-traits. Without such linkage, there would be no selection for diversification in the mating traits, and there could be no diversification in the ecological trait even if the mating traits diversified. Thus, linkage between E-traits and A-traits is required both to translate disruptive selection from the E-traits to the A-traits, and to translate diversification in the A-traits back to the E-traits. In general, recombination between E-traits and A-traits acts against the establishment of linkage between these groups of traits, which is the reason why speciation is, a priori, a more complicated and impeded process in two-allele models than in one-allele models.

Among the first to study the process of speciation in two-allele models were Udovic (1980), Felsenstein (1981), and Seger (1985). In particular, Felsenstein's seminal paper drew attention to the fact that with the rather simple and coarse genetic architecture he considered, speciation in two-allele models seemed unlikely because of the detrimental effects of recombination between E-loci and A-loci. It took almost two decades until this statement was fundamentally challenged in two papers published in the same issue of *Nature* in 1999. Both papers used multilocus genetic models to argue that speciation can occur relatively easily in two-allele models. Kondrashov & Kondrashov (1999) used a model with one E-trait and two A-traits. They assumed that marginal E-traits are favored at all times (i.e., selection is always disruptive by assumption), and that assortment is determined by a female preference trait and a male marker trait. In their model the degree of assortment is a fixed parameter describing how strong a female with a certain preference favors males with the preferred trait. If disruptive selection and assortment are strong enough, the population can easily split into two distinct clusters containing different E-traits, different female preferences and different male marker traits, and thus corresponding to two different (incipient) species. The model by Dieckmann & Doebeli (1999) also had two A-traits, but one of these was the degree of assortment, while the other one was sex-specific: in females it determined preference, and in males it determined the marker trait on which female preference was based. The E-trait was modeled along the lines described earlier in this chapter, except that the underlying genetic architecture consisted of a finite number of additive loci. In this model, speciation requires the evolution of strong assortment in the first A-trait before linkage between the second A-trait and the E-trait can induce speciation, but our simulations showed that even under these hindering circumstances, speciation can occur relatively easily, given the right ecological setting.

In Doebeli (2005) I extended these multilocus models to cases with three A-traits: one for the female preference, one for the male marker trait, and one for the degree of assortment (including, as in Dieckmann & Doebeli (1999), the possibility of disassortative mating, that is, of females preferring males that are dissimilar to the females' "preference" trait). In these models, diversification can take new forms. For example, a split in the male marker trait may not be accompanied by a corresponding split in the female preference trait, but rather by a split in the degree of assortment, with one cluster mating strongly disassortatively and the other cluster mating strongly assortatively. Both these clusters had the same female "preference" trait, but preferred different male marker clusters because of their different assortment strategies. Even when there are costs of assortment, the results in Doebeli (2005) showed that speciation is a feasible scenario in two-allele models with three A-traits. I will illustrate this by extending the blending inheritance models used in this chapter to two-allele scenarios.

Incorporating additional A-traits into the individual-based models is again straightforward. In addition to the ecological trait, x, and the degree of assortment, σ_A, individuals now also have a trait p determining female preference, and a marker trait m acting as the target for female preference in males. Thus, p is only expressed in females, and m is only expressed in males. For simplicity, we again assume that individuals are hermaphroditic, or equivalently, that the model only follows the evolutionary dynamics in females, with males assumed to have the same phenotypic distribution as females at all times. Birth and death rates are determined in exactly the same way as before. When an individual with preference trait p is chosen to give birth, it choses a mating partner with marker trait m with a probability that is proportional to the mating function

$$A(p, m) = \frac{1}{\sqrt{2\pi}\sigma_A} \exp\left[-\frac{(p - m)^2}{2\sigma_A^2}\right]. \qquad (4.9)$$

The degree of assortment σ_A can be either a fixed parameter or an evolving trait. In the latter case, σ_A is given by the assortment trait of the individual chosen to give birth (and having preference trait p). More precisely, once an individual with phenotype p has been chosen to give birth, the probability $P(p, m)$ that the offspring is generated by mating with an m-individual is given by

$$P(p, m) = \frac{A(p, m)}{\sum_{m'} A(p, m')}, \qquad (4.10)$$

where the sum runs over all individuals in the population. Without further assumptions we again have the baseline case with $\sum_m P(p, m) = 1$, hence the individual chosen to give birth is guaranteed to find a mate. Once the focal individual has chosen a mating partner, one individual is added to the population whose phenotypes are chosen from the segregation kernels corresponding to the various traits. These segregation kernels have means $(i + j)/2$, where i and j are the parental values of the various evolving traits. The variances of the segregation kernels, denoted by σ_{s_x}, σ_{s_A}, σ_{s_p}, and σ_{s_m}, are additional system parameters.

We first look at the case where the degree of assortment is a fixed parameter (i.e., there are three evolving traits: the ecological trait, the preference trait, and the marker trait). Simulations confirm that in this case, adaptive speciation can occur if the ecological conditions are conducive to strong disruptive selection, that is, if σ_α is sufficiently smaller than σ_K, and if assortment is strong enough, that is, if σ_A is small enough. Two examples of evolutionary dynamics leading to speciation are shown in Figure 4.6. In these scenarios, speciation is hindered by recombination between the preference trait and the marker trait, and by recombination between the A-traits and the E-trait. If assortment were based on the ecological trait, speciation would be straightforward, as shown in the previous section. But with assortment based on the marker trait, linkage between the A-traits and the E-trait is necessary for assortment to be able to latch on to the ecological trait and induce diversification. Initially, small linkages arise due to chance events, and once assortment can latch onto the ecological trait enough to initiate divergence, the amount of linkage increases, which in turn increases the effects of assortment. Thus, speciation is mediated by a positive feedback between linkage and divergence. In general, larger widths of the segregation

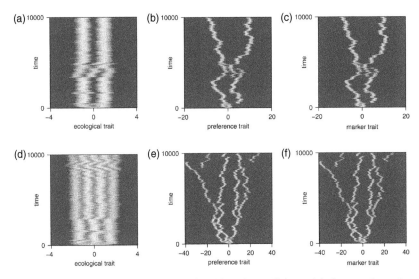

FIGURE 4.6. Adaptive speciation in individual-based two-allele models for sexual populations with a fixed degree of assortative mating. In the first example shown (panels (a)–(c)), the competition kernel was Gaussian and the carrying capacity function had quartic form. In the second example shown, both the competition kernel and the carrying capacity were quartic. In each case, the three panels show the evolution of the ecological trait (panels (a) and (d)), the preference trait, (panels (b) and (e)) and the marker trait (panels (c) and (e)). Multimodal pattern formation, and hence speciation, is evident in both examples. Note the very concerted evolution of the preference and the marker traits. Parameter values were $r = 1$, $K_0 = 200$, $\sigma_K = 2$, $\sigma_\alpha = 0.5$, $\sigma_A = 0.45$, and $\sigma_{s_x} = 0.2$ in both examples, and $\sigma_{s_p} = 0.35$, $\sigma_{s_m} = 0.35$ in the first case, and $\sigma_{s_p} = 0.4$, $\sigma_{s_m} = 0.4$ in the second case. Populations were initialized with Gaussian phenotype distributions that were centered at 0 in the ecological direction and at 2 in the preference and marker traits.

kernels for the preference and the marker trait promote the initiation of this feedback mechanism due to stochastic events.

We now turn to the evolution of assortment in two-allele models. In some sense, this is a catch-22 problem: if there is no assortment, linkage between A-loci and E-loci will not develop, and if there is no such linkage, there is no selection for increased assortment even if the E-loci are under disruptive selection. There are at least two possible escapes from this catch-22. One occurs when there are independent mechanisms that favors the evolution of assortment, such as Allee effects. Another possible group of mechanisms can be categorized under "large mutations": If, for one reason or another, at least some strongly assortative types are present in the population, then this would allow to kickstart the positive feedback between linkage disequilibrium and

divergence. The emerging linkage between marker traits and diverging ecological modes would in turn select for more assortment, thus reinforcing the
process of adaptive diversification. For example, in the multilocus models of
Dieckmann & Doebeli (1999) and Doebeli (2005), which incorporate the evolution of the degree of assortment in two-allele models, assortment still easily
evolves, essentially because the multi-locus genetic architecture used in these
models puts constraint on the possible assortment types and allows for at least
some strongly assortative genotypes to be generated fairly easily even in populations with a low average degree of assortment. However, in the present
setup, in which new variants are produced from the segregation kernel for the
assortment phenotype, the probability of production of very assortative types
in randomly mating populations is very low unless the width of this segregation
kernel, σ_{a_A}, is very large.

Here I will illustrate two alternative mechanisms for the evolution of assortative mating, one from each of the two categories mentioned above. First,
we can introduce Allee effects in two-allele models in the same way as in the
one-allele models of the previous section, that is, by using the total amount of
available mates, $T(p)$, for any given individual with preference phenotype p:

$$T(p) = \sum_m A(p, m), \qquad (4.11)$$

where the sum runs over all individuals (or all males) in the population. We
again assume that once an individual with preference phenotype p is chosen
to give birth, the probability that it actually finds a mate, $f(p)$, is a saturating
function of the (normalized) amount of available mates:

$$f(p) = a + (1 - a) \left(\frac{\tilde{T}(p)}{\eta + \tilde{T}(p)} \right), \qquad (4.12)$$

where η is the parameter measuring the magnitude of the Allee effect, $1 -$
a is the maximal reduction in probability of mating, and $\tilde{T}(p) = T(p)/N$
with N the current total population size. Even if all individuals in a population exhibit the same degree of assortment, there is now a cost to having a
preference for rare marker phenotypes, because the total amount of available
mates, and hence the probability f, is generally smaller for females with a
preference for rare phenotypes than for females with a preference for common
phenotypes. Therefore, the emergence of multimodal phenotype distribution
from initially unimodal populations becomes harder. Without going into further details, I note again that because Allee effects induce costs only for rare

(preference) phenotypes, the evolutionary dynamics can be strongly influenced by the initial conditions. In fact, Allee effects may actually promote speciation if the initial phenotype distributions have a large variance or are already multimodal, whereas Allee effects can prevent diversification when the initial distributions are unimodal and strongly peaked.

When the mating function $A(p, m)$ is normalized (as is assumed here), Allee effects can promote speciation when assortment, that is, the quantity σ_A, is an evolving trait. As explained in Box 4.1, Allee effects can promote the evolution of assortment even in unimodal populations that do not experience disruptive selection, essentially because common individuals that mate more assortatively can increase the total amount of mates available to them, and hence decrease the costs due to Allee effects. In two-allele models, this implies that Allee effects can induce the evolution of assortment even if there is no linkage between the A-traits and the E-trait. Once assortment is strong enough, speciation can occur through the establishment of such a linkage (and despite Allee effects disfavoring rare phenotypes!). An example of this evolutionary dynamics is shown in Figure 4.7.

To illustrate the large mutation mechanism, let us consider a model in which the degree of assortment is controlled by a single haploid locus with two alleles, one coding for a very low degree of assortment (i.e., for essentially random mating), and the other coding for a high degree of assortment. During sexual reproduction, offspring inherit the allele from either parent with probability $1/2$, and large mutations from one allele to the other occur with a certain probability in each offspring. In this situation, the positive feedback between linkage and selection on assortment can get started due to the continued presence (even if initially only at low frequency) of strongly assortative types. An example of this evolutionary dynamics is shown in Figure 4.8.

To end this chapter, I will briefly address the evolutionary relationship between one-allele and two-allele models. Starting with a two-allele model in which the ecological conditions are conducive to adaptive speciation, one could ask the question whether the mating system would evolve in such way as to transform the two-allele model into a one-allele model. If mating is initially based on an ecologically neutral marker trait, and if there is genetic variability for shifting the preference onto the ecological trait, one might expect that, under suitable circumstances, the mating preference evolves to be based on the ecological trait, thus allowing diversification based on one-allele mechanisms. To investigate this scenario, I used two-allele models with a fixed degree of assortment (i.e., σ_A is not evolving), and with only one A-trait, which serves as both the preference trait in females and as the marker trait in males. Thus, in expression (4.15) for the mate choice function, $A(p, m)$ is replaced by $A(p, q)$,

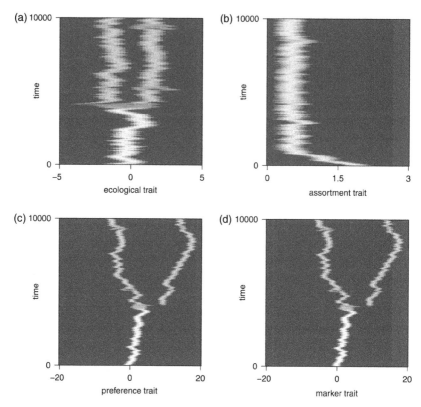

FIGURE 4.7. Evolution of assortment due to Allee effects in individual-based two-allele models for sexual populations. In the example shown, the competition kernel was Gaussian and the carrying capacity function had quartic form. The four panels show (a) the evolution of the ecological trait, (b) the assortment trait, (c) the preference trait, and (d) the marker trait. After the assortment trait has evolved to low enough values (so that the degree of assortment is high enough), bimodal pattern formation in the other traits, and hence speciation, occurs. Note again the very concerted evolution of the preference and the marker traits. Parameter values were $r = 1$, $K_0 = 250$, $\sigma_K = 4$, $\sigma_\alpha = 0.5$, $\sigma_{s_x} = 0.2$, $\sigma_{s_A} = 0.1$, $\sigma_{s_p} = 0.4$, and $\sigma_{s_m} = 0.4$. Populations were initialized with Gaussian phenotype distributions that were centered at 0 in the ecological, preference, and marker traits. All individuals initially had an assortment trait value of 2.

where q is the preference trait of potential mating partners. This corresponds to the two-allele preference scenarios used in Dieckmann & Doebeli (1999). Note that if there were two A-traits, a female preference trait and male marker trait, then shifting the female preference from the marker onto the ecological trait would still result in a two-allele model, because with female preference

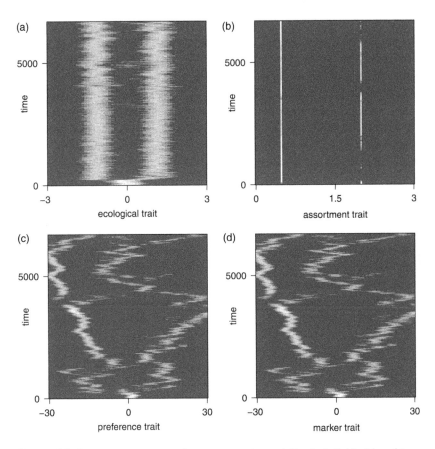

FIGURE 4.8. Large assortment mutation can generate speciation in individual-based two-allele models for sexual populations. In the example shown, the competition kernel was Gaussian and the carrying capacity function had quartic form. The four panels show (a) the evolution of the ecological trait, (b) the assortment trait, (c) the preference trait, and (d) the marker trait. After the assortment trait has evolved to low enough values (so that the degree of assortment is high enough), bimodal pattern formation in the other traits, and hence speciation, occurs. Note again the very concerted evolution of the preference and the marker traits. Parameter values were $r = 1, K_0 = 250, \sigma_K = 2, \sigma_\alpha = 1, \sigma_{s_x} = 0.2, \sigma_{s_p} = 0.4$, and $\sigma_{s_m} = 0.4$. The two possible assortment trait values were $\sigma_A = 2$ and $\sigma_A = 0.5$, with the lower one inducing strong enough assortment for speciation to occur. The probability of mutation between the two assortment trait values was 0.001. Populations were initialized with Gaussian phenotype distributions that were centered at 0 in the ecological trait and centered at 2 in the preference and marker traits. All individuals initially had an assortment trait value of 2.

based on the ecological trait, the female preference still needs to diversify for the population to speciate (i.e., recombination between the ecological and the preference trait would still hinder speciation). Thus, to study the evolutionary shift from two-allele to one-allele models, a two-allele model with a single A-trait (and a fixed degree of assortment) is an appropriate starting point.

To enable a potential evolutionary shift of assortment from being based on the preference to being based on the ecological trait, I assume that there is an additional haploid locus with two "choice" alleles: allele 1 codes for assortment being based on the preference trait, and allele 0 codes for assortment being based on the ecological trait. Then, if an individual with ecological trait x and preference (and marker) trait p has choice allele 1, it chooses mates based on the mating function $A(p, q)$, where q is the preference trait of potential mating partners. However, if an individual with ecological trait x and preference (and marker) trait p has choice allele 0, it chooses mates based on the mating function $A(x, y)$, where y is the ecological trait of potential mating partners. Thus, if an individual with ecological trait x, preference (and marker) trait p and choice allele z is chosen to give birth, it choses individuals with ecological trait y, preference (and marker) trait q and "choice" trait w with probability

$$P(x, p, w; y, q, v) = \frac{(1 - z)A(x, y) + zA(p, q)}{\sum_{y', q'} (1 - z)A(x, y') + zA(p, q')}. \tag{4.13}$$

Here the degree of assortment σ_A is a fixed parameter in the mating function $A(\cdot, \cdot)$. Note that by normalizing the mating probabilities (i.e., by dividing by the sum in the denominator of eq. (4.19), which runs over the whole population), I again assume that every individual chosen to give birth is assured to produce an offspring. (Note also that the mating probability does not depend on the value v of the choice trait of the mating partner.) As usual, when an offspring is produced its two continuous phenotypes are chosen from the segregation kernels corresponding to the evolving traits, having the midparent values as means and variances given by model parameters. Offspring inherit the choice allele from either parent with probability $1/2$, and mutations from one allele to the other occur with a certain probability in each offspring.

Starting with a population in which all individuals have choice allele 1 and hence base their mate choice exclusively on the marker trait, one can now ask whether the choice allele 0 can invade and take over, that is, whether the mate choice shifts onto the ecological trait, thus transforming the initial two-allele model into a one-allele model. In particular, in situations in which the ecological conditions favor speciation, one can ask whether the shift in mate choice onto the ecological trait evolves before speciation occurs in the initial two-allele model. Once the shift has occurred, one would then expect to see

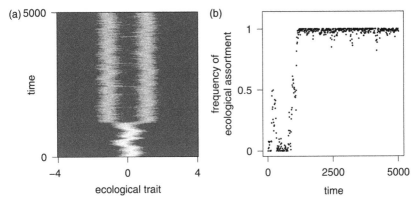

FIGURE 4.9. Evolutionary change in the basis of assortment in individual-based two-allele models for sexual populations. In the example shown, the competition kernel was Gaussian and the carrying capacity function had quartic form. Panel (a) shows the evolution of the ecological trait, and panel (b) shows the frequency of the choice allele coding for assortment being based on the ecological trait as a function of time. (Thus, if one imagines mating divided into one-allele and two-allele mechanisms, panel (b) shows the fraction of the one-allele mechanism.) Over time, the frequency of the choice allele for assortment being based on the ecological trait increases, enabling the population to undergo adaptive speciation. Parameter values were $r = 1$, $K_0 = 250$, $\sigma_K = 2$, $\sigma_\alpha = 1$, $\sigma_A = 0.45$, $\sigma_{s_x} = 0.2$, and $\sigma_{s_p} = 0.2$. The population was initialized with Gaussian phenotype distributions that were centered at 0 in the ecological trait and in the preference trait (the dynamics of the distribution of the preference trait is not shown). All individuals initially had the choice allele for assortment being based on the preference trait. Mutations between the two choice alleles occurred with a probability 0.001.

speciation as in one-allele models. Figure 4.9 illustrates just such a scenario. In fact, this scenario is akin to the "large mutation" scenarios mentioned earlier, because the choice allele 0 has a large effect on the linkage between A-traits and E-traits. The model indicates that if ecological traits are under disruptive selection, there is selection for shifting mate choice from neutral marker traits to these ecological traits. Extrapolating from this, one can expect that given suitable ecological circumstances, there will be an evolutionary tendency toward increasing the linkage between traits determining mate choice and traits that are important ecologically (the spread of some chromosome inversions could be examples of this (Kirkpatrick & Barton (2006)). This would allow adaptive diversification to run its course, resulting in species in which mating traits and ecological traits are tightly linked.

Challenge: Extend the two-allele models described here to scenarios that allow for the possibility of disassortative mating and show that speciation can occur in the form of a split in the trait determining the degree of assortment, with one cluster mating assortatively and the other cluster mating disassortatively (see Fig. 2(c) in Doebeli, 2005).

Challenge[*]: Investigate the models presented in this chapter for segregation kernels whose variances depend on parent values. For example, one could assume that variances decrease as the midparent values become more extreme (i.e., more distant from the trait value optimizing the carrying capacity). Also, investigate scenarios with non-Gaussian segregation kernels.

I have spent quite some time in this chapter to discuss how evolutionary branching can occur in sexual populations through various mechanisms of assortative mating. However, as explained in Chapter 1, addressing the various genetic mechanisms of assortment enabling adaptive speciation is not the main purpose of this book, and hence the discussion in this chapter must suffice (other models for diversification in sexual population will be mentioned, albeit briefly, in later chapters, including a model for adaptive speciation due to postzygotic isolation in Box 9.1). Before I return to the main theme of the book, that is, to the primary problem of identifying ecological scenarios that generate disruptive selection and hence are conducive to adaptive diversification, I close this chapter with a brief digression about the scope of the ecology of mate choice for adaptive speciation.

4.4   A FOOTNOTE ON ADAPTIVE SPECIATION DUE
TO SEXUAL SELECTION

In principle, the various mechanisms of assortative mating discussed in this chapter can give rise to sexual selection in the form of competition for mating partners. In theory, this can for example lead to Fisherian runaway processes (Fisher, 1930), in which sexual selection leads to ever increasing preference and marker traits on which preference is based. Sexual selection has long been implicated as a possible mechanism for allopatric speciation, according to which different runaway processes occurring in different geographical locations may produce populations that are reproductively isolated due to their differently evolved preference and marker traits (Lande, 1981). However, sexual selection could potentially also lead to adaptive speciation due to frequency-dependent interactions. Without going into any technical details, I very briefly summarize here the current state of theoretical thinking on this topic.

It is intuitively obvious that sexual selection may often be frequency-dependent, because success in choosing or being chosen as mate may depend on the phenotypes of the other individuals that choose or are being chosen. Thus, competition for mates may often be frequency-dependent and lead to disruptive selection, much like frequency-dependent competition for food resources. Indeed, sexual selection has received considerable attention from theoreticians as a potential mechanism for adaptive speciation (e.g., Almeida & de Abreu, 2003; Arnegard & Kondrashov, 2004; Higashi et al., 1999; Takimoto, 2002; Takimoto et al., 2000; Turner & Burrows, 1995; Van Doorn et al., 2004; Van Doorn et al., 2001; see Gavrilets & Waxman (2002) for a model of adaptive speciation due to sexual conflict, which is explained in Dieckmann & Doebeli (2004) in the context of adaptive dynamics). However, with regard to adaptive speciation there is a fundamental difference between competition for food resources and competition for mates: for adaptive speciation to occur due to sexual selection, competition for mates should lead to disruptive selection in both males and females. Thus, rather than requiring one instance of disruptive as under competition for resources, adaptive speciation due to sexual selection requires two instances of disruptive selection to occur.

To date, Van Doorn et al. (2004) is the most comprehensive theoretical analysis of adaptive speciation due to sexual selection. These authors have argued that mate choice alone is unlikely to generate disruptive selection in both males and females simultaneously. Their arguments are mainly based on one particular way of modeling frequency-dependent mate choice (Fig. 4 in Van Doorn et al. (2004)), and it remains to be seen whether other ways of modeling mate choice are more conducive to produce disruptive selection in both sexes (see e.g., Almeida & de Abreu, 2003). In addition, due to differentiation between the sexes mate choice may not be the only mechanism generating frequency-dependent sexual selection, and competition for mates can be caused by different mechanisms in males and females. For example, females may compete for access to unmated males, while males may compete (in a frequency-dependent manner) for territories that allow them to mate (Van Doorn et al., 2004). Van Doorn et al. (2004) have argued that scenarios involving independent mechanisms for competition for mates in females and males are more conducive to adaptive speciation by sexual selection than scenarios involving only mate choice. Nevertheless, at present the conclusion seems to be that in theory, competition for mates is not as conducive to adaptive speciation as ecological competition for resources, essentially because competition for mates needs to generate diversity in both males and females.

This conclusion is supported by other theoretical studies (e.g., Arnegard & Kondrashov, 2004; Kondrashov & Shpak, 1998), as well as by empirical data (Ritchie, 2007), which together generate the perspective that sexual selection is most likely to promote adaptive speciation when it operates in conjunction with frequency-dependent ecological interactions (Ritchie, 2007; Van Doorn et al., 2004).

Adaptive Diversification Due to Predator-Prey Interactions

Up to now we have mainly considered adaptive diversification in consumer species undergoing evolutionary branching due to competition for resources. (I say "mainly" because in the classical Levene models discussed in Chapter 2, the mechanisms leading to frequency-dependent adaptation to different habitats could in principle involve various types of ecological interactions.) It has of course long been recognized that consumption, that is, predation, can not only exert strong selection pressure on the consumer, but also on the consumed species. However, predation has traditionally received much less attention than competition as a cause for the origin and maintenance of diversity (Vamosi, 2002). Abrams et al. (1993) were perhaps the first to explicitly point out, using mathematical theory, that coevolution of predators and their prey could pave the way for adaptive diversification in the prey. In fact, their examples of evolutionary fitness minimization in a prey species could be viewed as among the very first examples of evolutionary branching points. Abrams et al. (1993) did not use this terminology, but they showed, using quantitative genetic models, how gradual evolution in a continuous character determining susceptibility to predation in a prey species can converge to a point in phenotype space at which fitness is minimized due to frequency-dependent interactions, that is, to an evolutionary branching point.

In this chapter, I will first use adaptive dynamics theory as well as individual-based models to illustrate that adaptive diversification in prey species due to frequency-dependent predator-prey interactions is a theoretically plausible scenario. I will describe conditions for diversification due to predator-prey interactions in classical Lotka-Volterra models, which requires analysis of coevolutionary dynamics between two interacting species, and hence of adaptive dynamics in two-dimensional phenotype spaces. As we have seen in Box 2.1, increasing the dimension can considerably increase the complexity of adaptive dynamics, and in the context of predator-prey coevolution it adds some theoretical twists to the problem of evolutionary branching. The chapter concludes with an example of evolution in another important class of predator-prey models.

5.1   ADAPTIVE DIVERSIFICATION IN CLASSICAL PREDATOR-PREY MODELS

Dieckmann et al. (1995) and Marrow et al. (1996) were the first to use adaptive dynamics models to investigate coevolutionary dynamics of prey and predator populations. Their analysis concentrated on the existence of cyclic adaptive dynamics representing oscillatory coevolutionary arms races between predator and prey. Doebeli & Dieckmann (2000) used essentially the same types of models to give examples of evolutionary branching in predator-prey systems, and Dercole et al. (2003) provided the most complete analysis of the adaptive dynamics of such predator-prey models to date. Here I first follow Dieckmann et al. (1995) and Doebeli & Dieckmann (2000) in setting up the basic model.

For monomorphic populations, let $N(x)$ be the density of a prey population in which all individuals have trait value x, where x is some quantitative trait (e.g., body size) that determines ecological interactions in a way that will be specified shortly. Similarly, let $P(y)$ be the density of a predator population in which all individuals have trait value y, where y is some quantitative trait in the predator. For the ecological dynamics, I assume that in the absence of the predator, $N(x)$ grows logistically to a carrying capacity $K(x)$ that depends on the trait value x (see Chap. 3 for a discussion of how the function $K(x)$ can be interpreted as a property of individuals). Thus, when $P(y) = 0$,

$$\frac{dN(x)}{dt} = rN(x)\left(1 - \frac{N(x)}{K(x)}\right),\tag{5.1}$$

where the carrying capacity function is, as so often, assumed to be unimodal and of Gaussian type:

$$K(x) = K_0 \exp\left[\frac{-x^2}{2\sigma_K^2}\right].\tag{5.2}$$

However, the precise form of the carrying capacity is not important for the main results reported below, as long as it is a unimodal differentiable function. The parameter $r > 0$ in eq. (5.1) is the intrinsic per capita growth rate of the prey. If the predator is present, its interaction with the prey is described by a linear functional response, so that the full ecological model for $N(x)$ and $P(y)$ is

$$\frac{dN(x)}{dt} = rN(x)\left(1 - \frac{N(x)}{K(x)}\right) - \beta(x, y)N(x)P(y) \tag{5.3}$$

$$\frac{dP(y)}{dt} = c\beta(x, y)N(x)P(y) - dP(y). \tag{5.4}$$

Here $\beta(x, y)$ describes the efficiency with which a predator of type y attacks and kills prey of type x, the parameter c is the conversion efficiency of the predator (i.e., the rate at which captured prey is converted into predator offspring), and the parameter d is the per capita death rate in the predator population. We will first assume that the function β has the following form:

$$\beta(x, y) = b \exp\left[\frac{-(x - y)^2}{2\sigma_\beta^2}\right]. \tag{5.5}$$

Biologically, this implies that the prey and predator traits can be appropriately scaled in such a way that for any given prey trait x, the predator with trait $y = x$ has the maximal attack rate. For example, if the two traits are body size in the prey and the predator, respectively (measured, e.g., in body weight), then it may often be reasonable to assume that for any given prey body size, there is an optimal predator body size in terms of attack rate: if the predator is too small, it cannot efficiently attack the prey, and if it is too large, then the prey might be able to more effectively evade the predator. A priori one would not expect that the predator body size that optimizes attack rate on a given prey body size is exactly equal to that prey body size in the original units of measurements (e.g., kg), but what is assumed by using the function β above is that predator body size can be rescaled so that β holds with the rescaled predator trait. (Technically speaking, if z is the original predator trait, then there should be an invertible function $x = h(z)$ such that the attack rates are given by β.)

It is well known (Kot, 2001) that the ecological model given by (5.3) and (5.4) has a stable equilibrium $(\hat{N}(x, y), \hat{P}(x, y))$ at which both species persist if the predator death rate is not too large. More precisely, if $d < c\beta(x, y)K(x)$, then

$$(\hat{N}(x, y), \hat{P}(x, y)) = \left(\frac{d}{c\beta(x, y)}, \frac{r}{\beta(x, y)}\left(1 - \frac{d}{c\beta(x, y)K(x)}\right)\right), \tag{5.6}$$

and all ecological trajectories starting out with positive prey and predator densities converge to $(\hat{N}(x, y), \hat{P}(x, y))$. If we define $\gamma = c\beta(x, y)K(x)/r$, then

$(\hat{N}(x, y), \hat{P}(x, y))$ is a stable node if $1 > d/c\beta(x, y)K(x) > 4\gamma/(1 + 4\gamma)$, and a stable focus if $1 > 4\gamma/(1 + 4\gamma) > d/c\beta(x, y)K(x)$ (Kot, 2001). It follows that the ecological equilibrium is a stable focus if the predator death rate d is small enough. Note that both the prey and the predator equilibrium densities are functions of the two trait values x and y for which the prey and predator are assumed to be monomorphic.

To derive the adaptive dynamics of the traits x and y, we assume that these traits are confined to regions in trait space in which the conditions for the existence of the stable equilibrium $(\hat{N}(x, y), \hat{P}(x, y))$ are satisfied (i.e., we assume that $d < c\beta(x, y)K(x)$ always holds). There are interesting evolutionary scenarios in which the trait values evolve to the boundary of the coexistence region, thus leading to the extinction of the predator (e.g., Dieckmann et al., 1995). However, here I concentrate on diversification in the prey, for which coexistence is a prerequisite.

We first determine the invasion fitness functions for mutant traits in both prey and predator. Assuming that the resident is at the equilibrium $(\hat{N}(x, y), \hat{P}(x, y))$, the per capita growth rate of a rare prey mutant type x' is

$$f_{prey}(x, y, x') = r\left(1 - \frac{\hat{N}(x, y)}{K(x')}\right) - \beta(x', y)\hat{P}(x, y). \qquad (5.7)$$

The first term of the right-hand side of this equation reflects intrinsic growth of the mutant as well as intraspecific competition that the mutant experiences from the resident prey type (but not from itself, because the mutant is assumed to be rare). Note that in contrast to Chapters 3 and 4, the resident prey density is not discounted according to phenotypic distance between the resident and the mutant (i.e., between x and x'), and hence competition is not explicitly frequency-dependent. (Note too, however, that there is implicit frequency-dependence, because the equilibrium $\hat{N}(x, y)$ depends on the resident type x.) To determine the effects of competition, the resident density is scaled by the mutant's carrying capacity, $K(x')$. The second term in eq. (5.7) reflects per capita predation pressure of the resident predator population density $\hat{P}(x, y)$ on the mutant, on which the resident has an attack rate $\beta(x', y)$.

Assuming the same resident (x, y), the per capita growth rate of a rare predator mutant type y' is

$$f_{pred}(x, y, y') = c\beta(x, y')\hat{N}(x, y) - d. \qquad (5.8)$$

The first term on the right-hand side is the per capita production of mutant offspring as a consequence of predation on the resident prey, while the second term is simply the phenotype-independent per capita rate of death.

Given the invasion fitness functions, the selection gradients in the prey and the predator are determined as

$$D_{prey}(x, y) = \left. \frac{\partial f_{prey}(x, y, x')}{\partial x'} \right|_{x'=x}$$

$$= \frac{r\hat{N}(x, y)K'(x)}{K(x)^2} - \frac{\partial \beta}{\partial x}(x, y)\hat{P}(x, y) \qquad (5.9)$$

$$D_{pred}(x, y) = \left. \frac{\partial f_{pred}(x, y, y')}{\partial y'} \right|_{y'=y} = c\frac{\partial \beta}{\partial y}(x, y)\hat{N}(x, y). \qquad (5.10)$$

The adaptive dynamics of the traits x and y are then given by

$$\frac{dx}{dt} = m_{prey}\hat{N}(x, y)D_{prey}(x, y) \qquad (5.11)$$

$$\frac{dy}{dt} = m_{pred}\hat{P}(x, y)D_{pred}(x, y). \qquad (5.12)$$

Here the parameters m_{prey} and m_{pred} reflect the rate and distributions of individual mutations, so that $m_{prey}\hat{N}(x, y)$ and $m_{pred}\hat{P}(x, y)$ describe the total mutation production in the two species (Dieckmann & Law, 1996). In the single species competition models of Chapter 3, the mutation production was given by a positive parameter that scaled time, but otherwise did not affect any of the main conclusions. Here things are different, because we are dealing with a two-dimensional adaptive dynamical system, in which the mutational parameters can greatly affect the evolutionary dynamics. In particular, the relative magnitude of the mutational parameters can determine whether the adaptive dynamics have a convergence stable equilibrium or a stable limit cycle (Doebeli & Dieckmann, 2000).

Equilibria of the adaptive dynamics (5.11) and (5.12), that is, singular points, are obtained by setting the right-hand side of these equations equal to 0. Since we assume coexistence of prey and predator at all times, this is equivalent to requiring that the fitness gradients in both species vanish. Thus, equilibria (x^*, y^*) of the adaptive dynamics are given as solutions to the equations

$$\frac{r\hat{N}(x^*, y^*)K'(x^*)}{K(x^*)^2} = \frac{\partial \beta}{\partial x}(x^*, y^*)\hat{P}(x^*, y^*) \qquad (5.13)$$

$$c\frac{\partial \beta}{\partial y}(x^*, y^*)\hat{N}(x^*, y^*) = 0. \qquad (5.14)$$

Given the functional form of $\beta(x, y)$, eq. (5.5), it is immediately clear from the second of these equations that an equilibrium (x^*, y^*) must satisfy $y^* = x^*$. But then the right-hand side of the first equation is zero, and hence the prey equilibrium must satisfy $K'(x^*) = 0$ as well. Given the unimodal function $K(x)$, eq. (5.2), it follows that $x^* = 0$, that is, the prey equilibrium trait must be the trait value maximizing the prey carrying capacity. Thus, the adaptive dynamics has a unique singular point at $(x^*, y^*) = (0, 0)$.

Basic theory of differential equations (e.g., Edelstein-Keshet, 1988) tells us that to determine convergence stability of this singular point, we need to calculate the Jacobian matrix at the singular point. That is, we need to calculate the matrix of derivatives

$$
J(x, y) = \begin{pmatrix} \dfrac{\partial}{\partial x}\left[m_{prey}\hat{N}(x, y)D_{prey}(x, y)\right] & \dfrac{\partial}{\partial y}\left[m_{prey}\hat{N}(x, y)D_{prey}(x, y)\right] \\ \dfrac{\partial}{\partial x}\left[m_{pred}\hat{P}(x, y)D_{pred}(x, y)\right] & \dfrac{\partial}{\partial y}\left[m_{pred}\hat{P}(x, y)D_{pred}(x, y)\right] \end{pmatrix}
$$

(5.15)

and evaluate it at $(x, y) = (0, 0)$. Taking into account that the selection gradients vanish at the singular point, we find from (5.9) and (5.10) that

$$
J(0, 0) = \begin{pmatrix} m_{prey}\hat{N}(0, 0)\dfrac{\partial D_{prey}}{\partial x}(0, 0) & m_{prey}\hat{N}(0, 0)\dfrac{\partial D_{prey}}{\partial y}(0, 0) \\ m_{pred}\hat{P}(0, 0)\dfrac{\partial D_{pred}}{\partial x}(0, 0) & m_{pred}\hat{P}(0, 0)\dfrac{\partial D_{pred}}{\partial y}(0, 0) \end{pmatrix}. \quad (5.16)
$$

The singular point is convergence stable if and only if both eigenvalues of J have negative real part, which is the case if and only if the determinant of J is positive and the trace of J is negative. Using the functions $K(x)$ and $\beta(x, y)$ given by eqs. (5.2) and (5.5), as well as expression (5.6) for the equilibrium densities $\hat{N}(0, 0)$ and $\hat{P}(0, 0)$ and expressions (5.9) and (5.10) for the selection gradients, one can calculate that the determinant of J is a positive multiple of $cbK_0 - d$, where $b = \beta(0, 0)$ and $K_0 = K(0)$. Therefore, the determinant of J is always positive by assumption. Also, the trace of J is a positive multiple of the expression $(cm_{pred} - m_{prey})\sigma_K^2(d - cbK_0) - dm_{prey}\sigma_\beta^2$. Thus (and because we always assume $d - cbK_0 < 0$), $cm_{pred} > m_{prey}$ guarantees that the trace is negative, and hence that $(0, 0)$ is convergence stable. Since c is the conversion efficiency of captured prey to predator offspring, this last condition can be interpreted as requiring that production of mutant predators per (captured) prey individual is faster than production of mutant prey. In particular, scenarios in which the prey evolves much faster than the predator tend to destabilize the singular point, which makes it clear that the parameters describing the mutational process in the two species can greatly affect the coevolutionary dynamics. Note that these statements about convergence stability of $(0, 0)$ do not depend on the particular form of the functions $K(x)$ and $\beta(x, y)$, as long as these functions are unimodal, and as long as β is symmetric in x and y.

Challenge: Prove the previous assertions.

Dieckmann et al. (1995) and Doebeli & Dieckmann (2000) have shown that if the singular point in the previously mentioned system is unstable, the adaptive dynamics can have stable limit cycles describing cyclic arms races between predator and prey. Here I concentrate on the case where the singular point is convergent stable to investigate the conditions leading to adaptive diversification. It is easy to calculate the second derivatives of the invasion fitness functions at the singular point. In fact, these second derivatives are simply the two diagonal elements of the Jacobian matrix J at the singular point:

$$\left.\frac{\partial^2 f_{prey}(x, y, x')}{\partial x'^2}\right|_{x=y=x'=0} = -\frac{r\hat{N}(0, 0)}{\sigma_K^2 K_0} + \frac{\hat{P}(0, 0)}{\sigma_\beta^2} \qquad (5.17)$$

$$\left.\frac{\partial^2 f_{pred}(x, y, y')}{\partial y'^2}\right|_{x=y=y'=0} = -\frac{c\hat{N}(0, 0)}{\sigma_\beta^2}. \qquad (5.18)$$

Clearly, at the singular point the predator always sits on a fitness maximum, since $\left.\frac{\partial^2 f_{pred}(x,y,y')}{\partial y'^2}\right|_{x=y=y'=0} < 0$. This makes sense of course, because if the prey trait is 0, then attack rates are by assumption maximized for the same predator trait value. However, it is also clear from the previous expressions that if σ_β is small enough, then the invasion fitness of the prey has a minimum at the singular point, that is, $\left.\frac{\partial^2 f_{prey}(x,y,x')}{\partial x^2}\right|_{x=y=x'=0} > 0$. (Note that by definition of the singular point, the first derivatives of the invasion fitness functions are of course zero.) This also makes sense intuitively: in the prey, deviations from the singular point have the disadvantage of incurring a smaller carrying capacity $K(x)$, but the advantage of leading to smaller attack rates; this advantage becomes larger the smaller the value of σ_β, and hence outweighs the disadvantage if σ_β is small enough.

In Doebeli & Dieckmann (2000) we concluded that therefore the adaptive dynamical system (5.11) and (5.12) has an evolutionary branching point for the prey under suitable conditions. Later, Eva Kisdi kindly pointed out to us that this is not so. If the point $(0, 0)$ is a convergent stable singular point that is also a fitness minimum in the direction of the prey trait, why is it not a branching point for that trait? The problem is that convergence stability refers to the whole two-dimensional coevolutionary system, whereas evolutionary branching only refers to the one-dimensional subsystem consisting of the prey alone, while the predator is held fixed at its singular trait value zero. Viewed thus, evolutionary branching in the prey indeed requires that the prey's invasion fitness function has a minimum at zero, but this is not enough. What is also needed is that the singular point zero in the prey is convergent stable in the subsystem consisting of the prey alone, and that's the snag: two-dimensional convergence stability does not imply convergence stability along the prey trait axis (Geritz et al., 1998). Indeed, with the predator trait fixed at zero, the adaptive dynamics along the prey axis is given by the prey's selection gradient, that is, by

$$
D_{prey}(x, 0) = \left. \frac{\partial f_{prey}(x, 0, x')}{\partial x'} \right|_{x'=x} = \frac{r\hat{N}(x, 0)K'(x)}{K(x)^2} - \frac{\partial \beta}{\partial x}(x, 0)\hat{P}(x, 0).
$$

(5.19)

Of course, $D_{prey}(0, 0) = 0$, and convergence stability along the prey axis is (up to a positive constant) determined by

$$
\left. \frac{\partial D_{prey}(x, 0)}{\partial x} \right|_{x=0}.
$$

(5.20)

But this is (again up to a positive constant) just the upper left entry of the Jacobian matrix $J(0, 0)$ of the full system at the singular point, which, as we have seen, is also the second derivative of the prey's invasion fitness function at the singular point. Therefore, if the singular point is evolutionarily unstable in the prey direction, then it is also convergence unstable in that direction, and vice versa. Another way of seeing this is by noting that

$$
\left. \frac{\partial^2 f_{prey}(x, 0, x')}{\partial x \partial x'} \right|_{x'=x=0} = 0.
$$

(5.21)

But according to general adaptive dynamics theory (Geritz et al., 1998), we always have

$$
\left. \frac{\partial D_{prey}(x, 0)}{\partial x} \right|_{x=0} = \left. \frac{\partial^2 f_{prey}(x, 0, x')}{\partial x^2} \right|_{x'=x=0} + \left. \frac{\partial^2 f_{prey}(x, 0, x')}{\partial x \partial x'} \right|_{x'=x=0}.
$$

(5.22)

Therefore, (5.21) implies that

$$\left.\frac{\partial D_{prey}(x, 0)}{\partial x}\right|_{x=0} = \left.\frac{\partial^2 f_{prey}(x, 0, x')}{\partial x'^2}\right|_{x'=x=0}. \qquad (5.23)$$

It follows that along the one-dimensional prey trait axis, evolutionary stability and convergence stability of the singular prey trait are the same if the predator traits is fixed at its singular value. In particular, evolutionary instability of the singular prey value implies that the singular prey trait is not convergence stable. As is explained in Section A.3 in the Appendix, this in turn implies that if the singular prey trait is evolutionarily unstable, then the condition of mutual invasibility of prey traits on either side of the singular value is not satisfied. Thus, if the singular point $(0, 0)$ is convergent stable in the full predator-prey system and evolutionarily unstable along the prey trait, then the singular prey trait itself can be invaded by all nearby trait values, but nearby prey traits on either side of the singular point cannot invade each other, and hence cannot coexist, so that evolutionary branching should not occur.

But why do the numerical simulations in Figure 3 of Doebeli & Dieckmann (2000) show evolutionary branching? Using pairwise invasibility plots (PIPs), Eva Kisdi also provided an answer to that question. Figure 5.1(a) shows an example of a PIP for the prey in which the predator trait is at its singular value zero, and that the singular prey trait is evolutionarily unstable. As argued before, the singular prey trait is then not convergence stable, which is reflected in the PIP by the configuration of $+$ and $-$ regions near the singular point. However, the PIP also shows that if the resident is away from the singular point, then mutants that are substantially different can invade and coexist with the resident. In other words, the PIP reveals regions of coexistence, as shown in Figure 5.1(b). Moreover, it can be shown that in such regions of coexistence, adaptive dynamics would converge to either of the dots indicated in Figure 5.1(b). These points lie on the boundary of the region of coexistence, and for the corresponding trait values the two prey types coexist neutrally, that is, their frequencies drift arbitrarily. Thus, diversification is possible if mutations are not assumed to be small in the prey.

Challenge: Prove the previous assertions about evolution of coexisting prey types when the predator is fixed at its singular value.

Nevertheless, even with large mutations such diversification would only result in neutral coexistence and would therefore unlikely be permanent. Eva Kisdi also provided a more robust mechanism for prey diversification, which is based on the assumption that the predator population itself exhibits some variability due to mutation-selection balance, so that the predator population

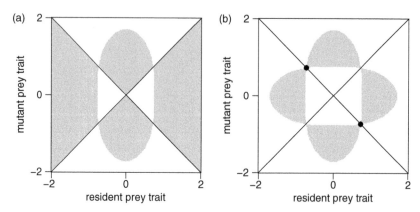

FIGURE 5.1. Pairwise invasibility plot (PIP) for the prey when the predator is at its singular value. (a) is the PIP, in which gray indicates a positive invasion fitness of the mutant in the corresponding resident, and white indicates a negative invasion fitness. (b) is the PIP superimposed with its mirror image along the diagonal; gray indicates regions of resident-mutant pairs in which both types have a positive invasion fitness when invading the other type. Thus, (a) reveals the one-dimensional evolutionary dynamics of the prey trait (when the predator trait is fixed at the singular value), while (b) reveals regions of coexistence between two prey traits. (a) shows that the singular point is not an attractor, because for resident values slightly smaller than the singular value, the region immediately above the diagonal is white, indicating that the corresponding mutants cannot invade, and vice versa for resident values slightly larger than the singular value. Thus, there is no convergence to the singular point. (b) shows that there are indeed regions where both types can invade each other. The black dots indicate the points in two-dimensional phenotype space to which the two-dimensional adaptive dynamics would converge when started from a pair of coexisting types. Parameter values were $r = 1$, $K_0 = 1$, $\sigma_K = 2$, $a = 1$, $d = 1$, $\sigma_\beta = 1$, and $c = 10$.

consists of two types $-\gamma$ and γ that lie close to the singular point on either side. In this case, if there are two prey types, ϵ and $-\epsilon$, and if, say, ϵ is common, then the corresponding (matching) predator type γ will be more common than the predator type $-\gamma$, which in turn confers an advantage to the other prey type $-\epsilon$, enabling invasion of that type. Thus, small amounts of variation in the predator can provide the frequency dependence necessary for mutual invasibility and coexistence of prey types on either side of the singular point. Moreover, the adaptive dynamics in two coexisting prey types would then converge to a singular coalition at which ecological coexistence is stable rather than neutral. Note, however, that the singular prey trait for the one-dimensional adaptive dynamics would still not be convergent stable, and hence even with variation in the predator, this singular point for the prey would not be an evolutionary branching point in the technical sense.

Challenge: Prove the previous assertions for the adaptive dynamics of the prey species under the assumption that the predator population exhibits variation and consists of two types $-\gamma$ and γ.

The preceding analysis shows that for adaptive dynamics in trait spaces of more than one dimension, convergence stability of a singular point together with evolutionary instability of the singular point along a given trait dimension is not enough to ensure evolutionary branching. Technically speaking, in higher dimensions convergence stability and evolutionary stability do not generally imply mutual invasibility, as they do in one-dimensional trait spaces (see Appendix). In the present context, an obvious question is whether the predator-prey system can be altered so that it exhibits true evolutionary branching points.

For example, one could try using a nonlinear functional response in the underlying ecological model, for example, the commonly used type II functional response, according to which the amount of prey consumed per predator is a saturating function of the prey density. This changes the ecological dynamics (5.3) and (5.4) to

$$\frac{dN(x)}{dt} = rN(x)\left(1 - \frac{N(x)}{K(x)}\right) - \frac{\beta(x, y)N(x)}{1 + h\beta(x, y)N(x)}P(y) \tag{5.24}$$

$$\frac{dP(y)}{dt} = c\frac{\beta(x, y)N(x)}{1 + h\beta(x, y)N(x)}P(y) - dP(y) \tag{5.25}$$

This model, which is known as the Rosenzweig-MacArthur model, can be derived by assuming that predators need a time h to handle and process a captured prey individual, which generates saturating capture rates (see, e.g., Britton, 2003). Nonlinear functional responses can generate new dynamic regimes in the ecological predator-prey model. In particular, the population dynamics of the two species can now undergo permanent cycles (Kot, 2001). Nonlinear functional responses also considerably widen the scope for the evolutionary dynamics. Even when the traits are constrained to regions in which the underling ecological dynamics have a stable equilibrium, many mathematically different dynamic regimes can be observed (Dercole et al., 2003). Interesting evolutionary dynamics can also be observed when traits are not constrained to regions of stable ecological coexistence. In this case, the traits can, for example, oscillate between regions of stable ecological equilibrium coexistence and regions of stable ecological limit cycles, leading to recurrent drastic changes in the ecological dynamics as a consequence of the evolutionary dynamics (Dercole et al., 2006). Also, the evolutionary dynamics can converge to predator-prey systems that are poised on the boundary between ecologically stable and unstable regions, in which small environmental changes could lead to large changes in the ecological dynamics (Dercole et al., 2006; Dercole & Rinaldi, 2008). Despite all these potential complications, the adaptive dynamics derived from the ecological model (5.24) and (5.25) do again not yield any evolutionary branching points.

To see this, one goes through the same analytical steps as with linear functional responses and checks that $(0, 0)$ is still the unique singular point, and that if the predator trait is held fixed at the singular value 0, the invasion fitness function of the prey still satisfies the "degeneracy condition"

$$\left. \frac{\partial^2 f_{prey}(x, 0, x')}{\partial x \partial x'} \right|_{x'=x=0} = 0. \tag{5.26}$$

Consequently, evolutionary and convergence stability in the direction of the prey trait are the same, so that if the singular prey trait is evolutionarily unstable, it is also convergence unstable, even though the singular point $(0, 0)$ may be convergent stable for the full predator-prey adaptive dynamics. This suggests that the functional response is not the crucial factor determining the existence of evolutionary branching points. Instead, what matters is the details of how the evolving traits are assumed to impinge on the ecological parameters.

Later I describe three ways in which the degeneracy of the invasion fitness function described by (5.21) can be removed. First, one could assume that competition in the prey species is frequency-dependent, and that the prey trait not only determines interactions with the predator, but also the strength of competition between prey individuals. This can be done along exactly the same lines as in the competition models studied in Chapter 3, that is, by assuming that the strength of competition between prey individuals with phenotypes x and x' is given by a competition kernel

$$\alpha(x, x') = \exp\left(\frac{-(x - x')^2}{2\sigma_\alpha^2}\right). \tag{5.27}$$

One can easily see that this changes the invasion fitness function of the prey from eq. (5.7) to

$$f_{prey}(x, y, x') = r\left(1 - \frac{\alpha(x, x')\hat{N}(x, y)}{K(x')}\right) - \beta(x', y)\hat{P}(x, y). \tag{5.28}$$

The invasion fitness function of the predator is unchanged and given by (5.8). It is easy to see that $(0, 0)$ is still the unique singular point of the adaptive dynamics resulting from these invasion fitness functions, and that the determinant of the Jacobian matrix at the singular point is always positive. As before, the diagonal elements of the Jacobian matrix are given by the derivatives of the selection gradients $\partial D_{prey}/\partial x(0, 0)$ and $\partial D_{pred}/\partial y(0, 0)$, where the latter expression is the same as before. For the former (5.22) is again valid, but now it is easy to see that

$$\left.\frac{\partial^2 f_{prey}(x, 0, x')}{\partial x \partial x'}\right|_{x'=x=0} = \frac{rd}{cbK_0\sigma_\alpha} < 0. \qquad (5.29)$$

Therefore, it is possible that $\left.\frac{\partial^2 f_{prey}(x,0,x')}{\partial x'^2}\right|_{x'=x=0} > 0$ and $\left.\frac{\partial D_{prey}(x,0)}{\partial x}\right|_{x=0} < 0$, that is, that with the predator trait fixed at its singular value, the singular value of the prey is evolutionarily unstable, but convergence stable. Moreover, if this is the case the trace of the Jacobian matrix is negative (because both diagonal elements are negative), and hence the singular point $(0, 0)$ is convergent stable for the full adaptive dynamics. Thus, with frequency-dependent competition in the prey, it is now possible to have convergent stable singular point for the predator-prey evolution that is an evolutionary branching point in the direction of the prey trait.

Of course, in some sense this does not come as a big surprise. After all, we already know from Chapter 3 that frequency-dependent competition alone can generate evolutionary branching in the prey, which happens if the ecological parameters in the prey satisfy $\sigma_\alpha < \sigma_K$. However, it is important to note that due to the presence of the predator, evolutionary branching in the prey can occur even if $\sigma_\alpha > \sigma_K$, that is, when competition alone is not enough to cause diversification in the prey. Thus, even weak frequency-dependent competition in the prey removes the degeneracy in the predator-prey adaptive dynamics and can lead to the existence of a convergent stable singular point of the full system that is a true evolutionary branching point for the prey trait, and hence the starting point for adaptive diversification in the prey.

Challenge: Prove all claims made before for the predator-prey system with frequency-dependent competition in the prey.

A second way to remove the degeneracy of the singular point consists of assuming that just as in the prey, the evolving trait in the predator not only affects predator-prey interactions, but also some other ecological parameter. For example, following Dercole et al. (2006) one could assume that the death rate is a function $d(y)$ of the predator trait y with a global minimum at some value $y^* \neq 0$. In that case, even if the prey trait were fixed at zero (i.e., at the maximum of the carrying capacity function), the predator trait would not evolve to zero, that is, to matching the prey trait, because there is now a trade-off between matching the prey and minimizing the death rate. Thus, at a singular point the predator trait is not expected to match the prey trait, which in turn implies that the prey trait is not expected to be at the maximum of the carrying capacity, because there is now a trade-off in the prey between maximizing the carrying capacity and increasing the phenotypic distance to the predator. As a result, singular points of the adaptive dynamics will not be at $(0, 0)$ any longer, which removes their degeneracy. Unfortunately, this also makes the model analytically intractable in general. Nevertheless, it can be shown by way of examples that under the given assumptions, it is possible that the adaptive dynamics of the predator-prey system can have convergent stable singular points that are evolutionary branching points in the direction of the prey trait (i.e., when the predator trait is fixed at the singular value, then the singular prey trait both convergent stable and evolutionarily unstable).

Challenge: Prove this assertion by showing that if $d(y)$ is a parabola with a global minimum $y^* \neq 0$, there exist scenarios in which the two-dimensional adaptive dynamics has a convergent stable singular point that is an evolutionary branching point for the prey trait.

So far, we have assumed that for any given prey trait x, the maximal attack rate, obtained when the predator trait is equal to the prey trait, is independent of trait values and equal to the parameter b appearing in the function β given by eq. (5.5). This suggests that a third way of removing the degeneracy of the singular point $(0, 0)$ consists of assuming that the maximal attack rate is not a constant, but instead a function $b(x)$ of the prey trait. For example, one could assume

$$b(x) = b \exp\left(\frac{-(x - b_0)^2}{2\sigma_b^2}\right), \tag{5.30}$$

so that the attack rate of predator type y on prey type x is given

$$\beta(x, y) = b(x) \exp\left(\frac{-(x - y)^2}{2\sigma_\beta^2}\right). \tag{5.31}$$

This means that the maximal attack rate has itself a maximum at the prey trait $x = b_0$, which is a global maximum for the function $\beta(x, y)$, and there is no a priori reason to assume that $b_0 = 0$, that is, that the trait value maximizing the maximal attack rate is the same as the trait value maximizing the carrying capacity. The parameter σ_b describes how fast the maximal attack rate decreases with increasing distance between the prey trait and b_0. This setup is essentially the same as that in Dercole et al. (2003), who used predator-prey models with saturating (type II) functional responses and showed the existence of convergence stable singular points for the predator-prey adaptive dynamics that are evolutionary branching points for the prey. I have argued above that nonlinear functional responses are not essential for the existence of such points. Indeed, one can remove the degeneracy of the singular point in the adaptive dynamics based on linear functional responses by changing only the attack rate function $\beta(x, y)$ to the new expression (5.31). Again, this change in the setup makes the model analytically intractable in general, but it is easy to find numerical examples of convergent stable singular points that are evolutionary branching points for the prey. For a similar model with type II functional response, Dercole et al. (2003) have shown that evolutionary branching occurs in large regions of parameter space.

Challenge: Prove, using numerical examples, that the adaptive dynamics based on linear functional responses and an attack rate function $\beta(x, y)$ given by expression (5.31) can have convergent stable singular points that are evolutionary branching points for the prey.

To complement the analytical theory, one can use stochastic individual-based simulations to demonstrate adaptive diversification due to predator-prey coevolution. These simulations are set up as in previous chapters, that is, using the Gillespie algorithm (Erban et al., 2007; Gillespie, 1976, 1977; Pineda-Krch, 2010) based on event rates. There are four basic types of events: birth and death in the prey, and birth and death in the predator. The per capita birth rate in the prey is $B_{prey}(x) = r$, independent of the prey phenotype x. The per capita death rate of a prey individual with phenotype x is given by

$$D_{prey}(x) = \frac{rN_{\text{eff}}}{K(x)} - \sum_y \beta(x, y), \qquad (5.32)$$

where the sum runs over all predator individuals. Here N_{eff} is the effective prey population size experienced by an individual with phenotype x. Thus, if competition is not frequency-dependent, then $N_{\text{eff}} = N$, the current number of prey individuals. When competition is frequency-dependent, then just as in the competition models of Chapter 3 (eq. 3.32), $N_{\text{eff}} = \sum_z \alpha(x, z)$, where the sum runs over all prey individuals.

The per capita birth rate of a predator individual with phenotype y is given by

$$B_{pred}(y) = c \sum_x \beta(x, y), \qquad (5.33)$$

where the sum runs over all prey individuals. The per capita death rate in the predator is $D_{pred}(y) = d$, independent of the predator phenotype y.

At any given point in time, the total birth rate in the prey population is $B_{prey,tot} = \sum_x B_{prey}(x) = rN$, the total death rate in the prey is $D_{prey,tot} = \sum_x D_{prey}(x)$, the total birth rate in the predator is $B_{pred,tot} = \sum_y B_{pred}(y)$, and the total death rate in the predator is $D_{pred,tot} = \sum_y D_{pred}(y) = dP$. The total event rate is $E_{tot} = B_{prey,tot} + D_{prey,tot} + B_{pred,tot} + D_{pred,tot}$. The type of event occurring at the next computational step is chosen with probabilities $B_{prey,tot}/E_{tot}$, $D_{prey,tot}/E_{tot}$, $B_{pred,tot}/E_{tot}$ and $D_{pred,tot}/E_{tot}$, respectively. Once an event type is chosen, the individual undergoing this event is chosen according to probabilities $B_{prey}(x)/B_{prey,tot}$, $D_{prey}(x)/D_{prey,tot}$, $B_{pred}(y)/B_{pred,tot}$ and $D_{pred}(y)/D_{pred,tot}$, respectively. If a death occurs, the chosen individual is removed, and if a birth occurs, the population is augmented by an individual whose phenotype is drawn from a Gaussian distribution with mean equal to the phenotype of the chosen parental individual, and with a mutational variance of $\sigma_{m,prey}$ and $\sigma_{m,pred}$, respectively. The lapse in real time between one computational step and the next is drawn from a negative exponential distribution with mean E_{tot}.

When simulating finite predator-prey populations, additional complications concerning the ecological dynamics arise. McKane & Newman (2005) and Pineda-Krch et al. (2007) have shown that even when a deterministic ecological predator-prey model, such as the Rosenzweig-MacArthur model (eqs. (5.24) and (5.25)), has a stable equilibrium, the corresponding individual-based model can exhibit sustained oscillations in the population size of both prey and predator (in fact, an example of this has already been noted by Renshaw (1991)). This generally happens when the stable equilibrium is a stable focus, and hence when the deterministic ecological model exhibits damped oscillations toward the equilibrium. It is important to note that in this case, the dynamics of the individual-based model is not just a noisy version of the damped deterministic oscillations, but instead exhibits sustained oscillations of substantial amplitude that can for example be detected using power spectrum analysis (Pineda-Krch et al., 2007).

This phenomenon can be seen in the individual-based model defined above by assuming that the prey and predator populations are monomorphic for trait values x and y, respectively, and that there are no mutations (which corresponds to assuming $\sigma_{m,prey} = \sigma_{m,pred} = 0$), so that all offspring are identical to their parent. If the trait values and the system parameters were chosen such that the resulting ecological model falls into the region of stable foci in the Rosenzweig-MacArthur model given by eqs. (5.24) and (5.25) then the individual-based model would generally exhibit sustained oscillations. However, the adaptive dynamics analyzed above was developed under the assumption that the resident populations are at a stable equilibrium, and it

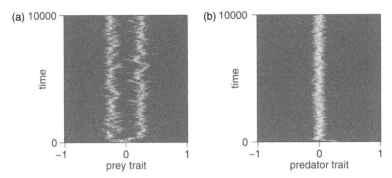

FIGURE 5.2. Evolutionary dynamics in asexual individual-based predator models with frequency-dependent competition in the prey. In the example shown, the singular point is an evolutionary branching point in the prey (and an evolutionarily stable point in the predator). Consequently, adaptive diversification occurs in the prey, after which the prey converges to an evolutionarily stable coalition of two prey phenotypes (while the predator stays at the original singular value). Parameter values were $r = 1$, $K_0 = 2000$, $\sigma_K = 0.25$, $d = 1$, $\sigma_\beta = 0.45$, $\sigma_\alpha = 0.3$, and $e = 0.001$, $\sigma_{m,prey} = \sigma_{m,pred} = 0.1$. The populations were initialized with Gaussian distributions with a small variance and means 0.4 and 0.42 in the prey and the predator, respectively.

is in general not clear how oscillatory ecological dynamics will affect the evolutionary dynamics, and in particular adaptive diversification. Below I will give an example of evolutionary branching in individual-based models in which the parameter and trait values are in regions where the deterministic ecological dynamics exhibits stable foci. In view of what was just said, this shows that evolutionary branching is also possible in finite populations exhibiting sustained ecological fluctuations. However, a systematic analysis of the effect of fluctuating population dynamics on adaptive diversification in predator-prey models is not available at this point.

Figure 5.2 shows a stochastic simulation of the adaptive predator-prey dynamics under the assumption that there is frequency-dependent selection in the prey. In the scenario shown evolutionary branching in the prey occurs despite the fact that frequency-dependent competition by itself is not sufficient to cause diversification. After branching has occurred in the prey, the system settles at a new evolutionary equilibrium at which the prey consists of two different phenotypic clusters equidistant from the branching point zero, while the phenotype distribution in the predator remains unimodal with mean ~ 0. Thus, the predator phenotypes are intermediate between the two prey clusters, and the predator can be viewed as a generalist. After branching in the prey, the two prey clusters effectively become different "habitat patches" for the predator, and it is in principle possible that this niche separation generates diversification in the predator as well. More precisely, much like in the classical niche models discussed in Chapter 2, secondary evolutionary branching into two predator specialists can occur if the prey equilibrium clusters are far enough apart (see Doebeli & Dieckmann, 2000).

Challenge: Find scenarios in which branching in the prey generates secondary branching in the predator, resulting in two specialist predator lineages. Find examples of evolutionary branching in finite populations in the other scenarios discussed before, that is, when the predator death rate d is a function of the predator trait y, and when the attack rate function $\beta(x, y)$ has a global maximum.

As in the previous chapter, the individual-based models can be extended to sexual populations by using segregation kernels to describe the distribution of offspring phenotypes produced by a given mating pair. To obtain evolutionary branching in a sexual prey population there must be some form of assortative mating, and I consider the simplest case where mate search is determined by a fixed degree of assortment with respect to the ecological trait x. We again use the preference function

$$A(x, x') = \frac{1}{\sqrt{2\pi}\sigma_A} \exp\left[-\frac{(x - x')^2}{2\sigma_A^2}\right], \tag{5.34}$$

where the parameter $\sigma_{A,prey}$ measures the degree of assortment (the smaller $\sigma_{A,prey}$, the stronger the preference for phenotypically similar mating partners). If matings are initiated bilaterally, the probability of mating between prey individuals of phenotypes x and x' is proportional to the product $A(x, x')$ $A(x', x)$, and assuming no explicit costs to assortment, this probability is given by

$$P(x, x') = \frac{A(x, x')}{\sum_z A(x, z)}, \tag{5.35}$$

where the sum runs over all individuals in the prey population. Since $\sum_y P(x, y) = 1$, the individual chosen to give birth is guaranteed to find a mate. Once the focal individual x has chosen a mating partner x', one individual is added to the population whose phenotype is chosen from the normal distribution given by the segregation kernel for the ecological trait, which is a Gaussian function with mean equal to $(x + x')/2$ and variance equal to $\sigma_{s,prey}$. Sexual reproduction in the predator is described in the same way by two parameters $\sigma_{A,pred}$ and $\sigma_{s,pred}$.

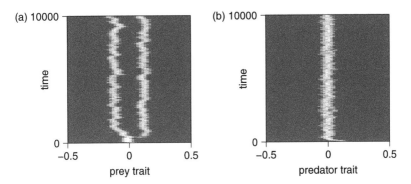

FIGURE 5.3. Same as Figure 5.2, but with sexual reproduction and assortative mating in the prey. Additional parameter values were $\sigma_s = 0.02$ and $\sigma_A = 0.03$.

Figure 5.3 shows the evolutionary dynamics of sexual prey populations in the ecological setting that was used in Figure 5.2 to illustrate evolutionary branching due to predation in the asexual model. In sexual prey populations, speciation does not occur if mating is random. However, strong assortative mating can enable evolutionary branching in the prey and generate two phenotypically distinct subpopulations with little mating between them, as shown in Figure 5.3(a). In the example shown, the predators are assumed to be asexual, and their distribution remains unimodal and centered in between the two emerging prey species. This outcome would also occur for sexually reproducing predators with either random or assortative mating, as it corresponds to the fact that there is no secondary branching in the predator in the corresponding asexual predator-prey model (Fig. 5.2).

In principle, one could now proceed to study secondary evolutionary branching in sexual predator populations, as well as the interplay of assortative mating and evolutionary branching in predator-prey systems under more complicated mating assumptions, such as costs to assortment and assortment based on marker traits. As I have argued in Chapter 4, it seems clear that a sizable subset of biologically feasible assumptions would lead to speciation in these more complicated models. As long as the ecological interactions generate evolutionary branching points and hence persistent disruptive selection in the corresponding asexual models, the adaptive escape will at least sometimes result in speciation. I refer to Chapter 9 for examples of diversification in deterministic sexual predator-prey models and leave more detailed explorations of the conditions for evolutionary branching and adaptive speciation in individual-based models to the reader. To conclude this chapter, the next section describes an interesting model for evolutionary branching in a host-pathogen model.

Challenge*: Construct and investigate adaptive dynamics models of prey and predator traits determining discrete-time predator-prey population dynamics based on the Nicholson-Bailey equations (Edelstein-Keshet, 1988).

5.2   AN EXAMPLE OF EVOLUTIONARY BRANCHING IN HOST-PATHOGEN MODELS

Host-pathogen interactions are a very important and widely studied class of predator-prey interactions. In particular, the epidemiological dynamics of infectious diseases have received a lot of attention in recent years (see Anderson & May (1992) and Ewald (1996) for seminal contributions), and many mathematical models have been developed in this area (see, e.g., Edelstein-Keshet (1988) and Otto & Day (2007) for introductory models, and Brauer & Castillo-Chavez (2000) for a more advanced treatment). One of the central and most enduring evolutionary questions in this context concerns the evolution of virulence (Ewald, 1996), which is usually thought to be constrained by a fundamental trade-off: more virulent pathogens, that is, pathogens with a higher growth rate inside their host, have a higher chance of being transmitted between hosts, but suffer the cost of increasing their hosts' death rate, thus shortening the period of time during which the pathogen is transmitted. Pathogen diversity is a central problem for the control of infectious diseases (e.g., HIV (Regoes & Bonhoeffer, 2005) and influenza (Regoes & Bonhoeffer, 2006)). It is therefore important to understand the evolutionary origin and maintenance of pathogen diversity, and it is interesting to investigate the role of the trade-off between transmission rate β and pathogen-induced host death rate α in this context. Boldin & Diekmann (2008) have recently produced an interesting model for evolutionary branching in the within-host pathogen growth rate (for an earlier model of evolutionary branching in virulence see Koella & Doebeli (1999)). Their model is based on earlier work by Gilchrist & Coombs (2006) and combines within-host dynamics of pathogen growth with epidemiological dynamics of infections in a population of hosts.

In this model, the evolving trait is the within-host reproduction rate p, and to describe the ecological dynamics of a pathogen of type p inside a single host, it is assumed that the state of a single host is described by three variables: the density of uninfected target cells in a host individual, T, the density of infected hosts cells, T_I, and the density of free pathogens, V. The within-host dynamics of these variables is given by the following system of differential equations:

$$\frac{dT}{dt} = \lambda - kVT - dT$$

$$\frac{dT_I}{dt} = kVT - (\mu(p) + d)T_I \qquad (5.36)$$

$$\frac{dV}{dt} = pT_I - kVT - cV.$$

In the absence of the pathogen, target cells are produced a rate λ and die at a per capita rate d. The mass action term kVT describes the rate at which free pathogens encounter and enter uninfected cells. Infected host cells produce free pathogen at a per capita rate p (the phenotype of the pathogen), and they die at a per capita rate that is increased by some function $\mu(p)$ compared to uninfected cells. Here $\mu(p)$ is a nonnegative increasing function of the production rate p. Finally, the per capita death rate of free pathogens is c.

Boldin & Diekmann (2008) define the burst size $B_0(p)$ as the expected number of pathogens produced by one infected host cell and show that

$$B_0(p) = \frac{p}{\mu(p) + d}. \qquad (5.37)$$

The burst size, multiplied by the probability with which a pathogen infects a host cell in a previously uninfected host, $k\lambda/(k\lambda + cd)$, is the expected number of new pathogens produced by a single pathogen introduced into a uninfected host:

$$R_0^w(p) = \frac{k\lambda}{k\lambda + cd} B_0(p). \qquad (5.38)$$

Here the superscript indicates that the basic reproduction ratio $R_0^w(p)$ refers to within-host dynamics. If $R_0^w(p) > 1$, the ecological dynamics (5.36) of a single pathogen strain p has a globally attracting, nontrivial equilibrium $\hat{T}(p), \hat{T}_I(p), \hat{V}(p)$ (see eq. (3.4) in Boldin & Diekmann (2008)).

Based on this, one can easily show that the evolutionary dynamics of the trait p is based on a pessimization principle (Mylius & Diekmann, 1995): in the long run, the evolutionary dynamics will converge to that trait value p^*, which minimizes $\hat{T}$, that is, to the trait value that depresses the resource consisting of uninfected hosts cells to its lowest density. The value of p^* depends on the trade-off function $\mu(p)$, but based on this simple within-host model, the adaptive dynamics of the trait p does not yield an evolutionary branching point.

The situation changes when the within-host model is combined with a simple model for the dynamics of infection in a host population. In this between-host model, the transmission rate $\beta(p)$ and the pathogen-induced host death rate $\alpha(p)$ are functions of the within-host production rate p, that is, of the pathogen phenotype. However, to take the within-host dynamics explicitly into account, it is assumed that this dependence of β and α on p is mediated mechanistically in terms of the within-host steady state $\hat{T}(p)$, $\hat{T}_I(p)$, $\hat{V}(p)$. Thus,

$$\alpha(p) = A(\hat{T}(p), \hat{T}_I(p), \hat{V}(p)) \tag{5.39}$$

$$\beta(p) = B(\hat{T}(p), \hat{T}_I(p), \hat{V}(p)) \tag{5.40}$$

for some functions A and B.

Competition between two different pathogen strains p and q on the epidemiological (i.e., host population) scale is not only influenced by the quantities $\beta(p)$, $\beta(q)$, $\alpha(p)$, and $\alpha(q)$, but also by the possibility of superinfection. Superinfection occurs when a pathogen infects a host that is already infected by another pathogen. For simplicity, it is assumed that when a pathogen attacks an already infected host, that pathogen either takes over and becomes the sole occupant of the attacked host, or it cannot establish itself at all in the attacked host. This allows one to model superinfection by a function $\Phi(p, q)$ that gives the probability with which a pathogen with phenotype q, upon transmission to host individual that is infected with pathogens of type p, eliminates p and establishes itself in the host individual. Since the ability of a q type to invade an individual host infected by p is solely determined by the quantities $\hat{T}(p)$ and $\hat{T}(q)$ (see before), it is reasonable to assume that the function $\Phi(p, q)$ can be written in the form

$$\Phi(p, q) = \Psi(\hat{T}(p), \hat{T}(q)). \tag{5.41}$$

It turns out that the form of the function Ψ is crucial for the existence of evolutionary branching points for the pathogen phenotype p.

To determine the evolutionary dynamics of p, Boldin & Diekmann (2008) proceed to specify the superinfection model of two competing pathogen strains p and q at the level of the host population as follows:

$$\frac{dS}{dt} = b - \beta(p)SI_p - \beta(q)SI_q - \delta S$$

$$\frac{dI_p}{dt} = \beta(p)SI_p - (\alpha(p) + \delta)I_p + (\beta(p)\Phi(q, p) - \beta(q)\Phi(p, q))I_pI_q \tag{5.42}$$

$$\frac{dI_q}{dt} = \beta(q)SI_q - (\alpha(q) + \delta)I_q + (\beta(q)\Phi(p, q) - \beta(p)\Phi(q, p))I_pI_q.$$

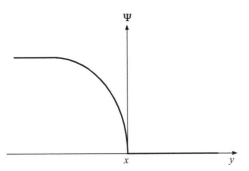

FIGURE 5.4. Schematic representation of the function $\Psi(x, y)$ as a function of y for a given x-value.

Here S, I_p, and I_q are the densities of uninfected hosts, hosts infected by pathogen p, and hosts infected by pathogen q, respectively. b is the host population birth rate, and δ is the per capita host death rate in the absence of the disease. All interactions are assumed to obey mass action, $\beta(p)$ and $\beta(q)$ are the transmission rates, and $\alpha(p)$ and $\alpha(q)$ are the disease-induced per capita death rates in the host. Finally, when two host individuals infected by p and q meet, the individual infected by p loses p and becomes infected by q at rate $\beta(q)\Phi(p, q)$, while the individual infected by q loses q and becomes infected by p at rate $\beta(p)\Phi(q, p)$.

Based on this between-host ecological dynamics for competition between two pathogen strains, it is straightforward to derive the invasion fitness function $f(p, q)$ of a rare mutant pathogen q into a resident pathogen strain p (eq. (3.10) in Boldin & Diekmann (2008)). This invasion fitness can then be used to derive the adaptive dynamics of the trait p. Boldin & Diekmann (2008) show that for a class of functions $\Psi(\hat{T}(p), \hat{T}(q))$, evolutionary branching and the maintenance of diversity in the trait p is indeed a possible outcome. This class is exemplified by the following function:

$$\Psi(x, y) = \begin{cases} 1 - \left(1 - \dfrac{kx}{c + kx} + \dfrac{ky}{c + ky}\right)^n & \text{for } y < x \\ 0 & \text{otherwise} \end{cases} \tag{5.43}$$

(Here k and c are the parameters used in eq. (5.36) above.) This function has the following properties, which are illustrated in Figure 5.4. First, $\Psi(x, y) = 0$ for $y > x$, which translates into $\Phi(p, q) = \Psi(\hat{T}(p), \hat{T}(q)) = 0$ if $\hat{T}(q) > \hat{T}(p)$, which in turn corresponds to the pessimization principle for the within-host dynamics. Second, $\Psi(x, y) > 0$ for $y < x$, which translates into a positive probability of a q-pathogen taking over a p-host if $\hat{T}(q) < \hat{T}(p)$. Third, as a

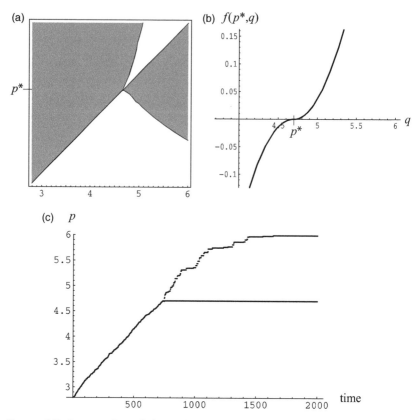

FIGURE 5.5. Asymmetric evolutionary branching in pathogen virulence. Panel (a) shows the pairwise invasibility plot, indicating convergence to the singular point p^*, which can only be invaded by mutants $q > p^*$, as shown by the invasion fitness function $f(p^*, q)$ in panel (b). As a result, the adaptive dynamics of the pathogen virulence p exhibits asymmetric evolutionary branching, as shown in panel (c). With kind permission from Springer Science+Business Media: Journal of Mathematical Biology, *Superinfections can induce evolutionarily stable coexistence of pathogens*, Volume 56, 2008, Pages 635–672, by Barbara Boldin and Odo Diekmann, Figures 8 and 9 (see source for parameter values).

function of y, $\Psi(x, y)$ is continuous at $y = x$, but not differentiable. Boldin & Diekmann (2008) call functions Ψ with these properties "mechanistic" and argue that such functions represent the biologically most realistic scenarios for superinfection. In particular, functions of the form eq. (5.43) above can be derived from describing the initial stages of superinfection as a stochastic birth-death process (Boldin & Diekmann, 2008).

Figure 5.5 illustrates a typical case of evolutionary branching in the production rate p for a mechanistic superinfection function Ψ as before. In fact, a very interesting type of asymmetric evolutionary branching occurs in this model, because the invasion fitness function need not have a local minimum at the singular point (Fig. 5.5(b)). Instead, if p^* is the evolutionary branching point, then as a function of the mutant trait value q, the invasion fitness $f(p^*, q)$ may not be twice differentiable at $q = p^*$ (essentially because $\Psi(x, y)$ is not differentiable at $y = x$). Moreover, the second derivatives of $f(p^*, q)$ at $q = p^*$ may have different signs depending on whether the limit is taken when q approaches p^* from below or from above. This implies that the singular point p^* can only be invaded by mutants on one side of p^*. In the example shown in Figure 5.5, the singular point p^* can only be invaded by mutants $q > p^*$. However, if q is close enough to p^*, such an invasion leads to coexistence, and as a consequence, the fitness landscape changes in such a way that selection favors divergence in the two coexisting strains. More precisely, after invasion of $q > p^*$ leads to coexistence between q and p^*, selection in the p^*-strain only favors mutants from a small interval $(p^* - \epsilon, p^*)$ (ϵ small and positive), wheres in the q-strain, selection favors any mutant that is larger than q (see Boldin & Diekmann (2008) for details). This asymmetry in the invasion fitness function generates the asymmetric adaptive dynamics shown in Figure 5.5, which results in the permanent coexistence of two pathogen strains with very different within-host production rates, and hence very different virulences.

In sum, Boldin & Diekmann (2008) showed that frequency-dependent host-pathogen interactions can lead to the emergence and persistence of diversity in pathogen virulence. It is important to note that superinfection is an essential ingredient for evolutionary branching. In fact, without superinfection at the epidemiological level of the entire host population, that is, when the function $\Phi(p, q)$ is 0 everywhere in system (5.42), the trade-off between transmission rate and pathogen-induced host death rate implicitly given by the functions $\beta(p)$ and $\alpha(p)$ produces only convergent stable strategies $\hat{p}$ that are also evolutionarily stable. In particular, this trade-off itself is not enough to generate diversity. On the other hand, in the example shown in Figure 5.5 selection at the within-host level alone produces a single winning strategy $\bar{p}$ with $\bar{p} > \hat{p}$ (Boldin & Diekmann, 2008). It is superinfection that provides the essential connection between the within-host and the between-host level and generates a trade-off between what is good for the pathogen at the within-host level and what is good for the pathogen at the epidemiological level. Moreover, the pros and cons of a given pathogen strategy at the two different levels depend on the other strategies present in the pathogen population, so that selection is frequency-dependent. Together with the trade-off mediated by superinfection, this frequency dependence can generate diversity through evolutionary branching. In many real host-pathogen systems, selective forces on pathogens are likely to be generated both within single hosts and through epidemiological dynamics in the entire host population, and the work of Boldin & Diekmann (2008) has opened very interesting perspectives on understanding the origin and maintenance of pathogen diversity through the interplay between these selective forces. Recently, Boldin et al. (2009) have extended this analysis to find necessary and sufficient conditions for evolutionary branching in host-pathogen models with super infections for any transmission-virulence trade-off. In contrast to the class of functions shown schematically in Figure 5.4, Boldin et al. (2009) assume differentiable superinfection functions. This allows them to use the method of critical function analysis introduced by de Mazancourt & Dieckmann (2004) to provide a comprehensive classification of evolutionary outcomes, including adaptive diversification through evolutionary branching, for the adaptive dynamics of pathogen virulence in a class of simple and general host-pathogen models (see Geritz et al. (2007) for another interesting application of critical function analysis to investigate evolutionary branching in a predator population due to predator-prey interactions).

Adaptive Diversification Due to Cooperative Interactions

If predation has received less attention than competition as a cause for the origin and maintenance of diversity, mutualistic interactions have fared even worse. There is quite a substantial theoretical literature on the ecology of mutualistic interactions (e.g., Boucher, 1985; Bronstein et al., 2003; Jones et al., 2009; Vandermeer & Boucher, 1978; Wolin & Lawlor, 1984), but only a few studies have investigated mutualism as a potential driver of diversification (e.g., Doebeli & Dieckmann, 2000; Ferrière et al., 2002). There is of course a rather huge literature on the evolution of intraspecific cooperation, and many of these models implicitly address the problem of coexistence between cheaters and cooperators, and hence the maintenance of diversity. However, the origin of diversity in cooperative contributions has only recently been investigated (Doebeli et al., 2004; Koella, 2000). Most models of cooperation assume that cooperators make a costly contribution to a public good, which is then distributed among certain members of the population. In this chapter, I will first describe how such simple scenarios can give rise to evolutionary branching in a single evolving population. I will then extend these models to cooperation between different species and show how interspecific mutualism can give rise to adaptive diversification.

6.1 DIVERSIFICATION IN MODELS FOR INTRASPECIFIC COOPERATION

The Snowdrift Game (Sugden, 1986) is arguably the simplest model for the maintenance of polymorphism in cooperative interactions. The name of this game derives from its anthropocentric interpretation of a situation in which two drivers are caught on either side of a Snowdrift and have the options to either stay in the car and wait, or get out and start shoveling away the snow. If the other driver is shoveling, it is better to stay put to save energy, but if the other driver stays put, it is better to start working, because even though work is costly, at least one gets home eventually. In this situation, whether cooperation (shoveling) is advantageous depends on the actions of the other player. In the framework of evolutionary game theory, in which the Snowdrift game is known as the Hawk-Dove game (Maynard Smith, 1982), this generates coexistence between the cooperative and noncooperative types, because each type has an advantage when rare. This is in contrast to the famous Prisoner's Dilemma game, in which each of two prisoners can either act as informants to the police, or not. Here the cooperative type is the one that is not an informant (and hence does not rat on the other prisoner). Defection, that is, being an informant, results in a reduced sentence independent of the other prisoner's action, and hence defection is always advantageous. The social dilemma is that if both prisoners cooperated (i.e., did not inform), then they would both get free (due to a lack of evidence). In terms of evolutionary game theory, this implies that cooperation cannot be maintained in the Prisoner's Dilemma game, even though individual payoffs would be highest in a population of cooperators.

To investigate the gradual evolutionary emergence of diversity in cooperative investments, Doebeli et al. (2004) have extended the Snowdrift game to models with quantitative cooperative investments. In the resulting Continuous Snowdrift game, individuals are given by a quantitative phenotype $x \geq 0$, which determines their contribution to a common good. Here I will assume that individuals interact pairwise, so that if two individuals with phenotypes x and y interact, the public good they generate is $x + y$. However, the Continuous Snowdrift game can easily be extended to interaction groups of arbitrary size, as explained in the supplement of Doebeli et al. (2004) (see also Box 6.1). With pairwise interactions, each of the interacting partners derives a benefit $B(x + y)$ from this public good, where B is the benefit function, which is typically nonlinear and saturating. For biological intuition, one can for example assume that x and y describe the amount of a chemical compound produced by microbes or plants and secreted into the environment, where it serves to process environmental components for consumption. Typically, the benefit that accrues due to presence of such compounds saturates eventually, because consumption is limited by other processes as well. Note, however, that the benefit could also be accelerating over a certain range of compound abundance. For the continuous Snowdrift game, it is also assumed that the production of the public good is costly, so that in an interaction between x and y, the individuals pay costs $C(x)$ and $C(y)$, respectively, where C is the cost function. Again, this function is generally nonlinear, and both accelerating and decelerating costs functions are plausible under certain conditions. In particular, decelerating cost functions are plausible if costs are initially high because individuals need to undergo regulatory and metabolic changes to turn on production of the public good, whereas further increases in the output of the public good would be relatively less costly.

Adding benefits and costs, the payoffs from a pairwise interaction of individuals with phenotypes x and y are $P_x = B(x + y) - C(x)$ and $P_y = B(x + y) - C(y)$, respectively (note that strictly speaking, the benefits should be multiplied by a factor of 1/2 in these expressions, because benefits are shared by the two interaction partners, but this factor can simply be absorbed in the benefit function B). For the benefit and cost functions we first of all assume that $B'(0) > C'(0)$. This assumption corresponds to the basic notion of the Snowdrift game that (small) cooperative investments should be advantageous in a population of defectors, because it ensures that if most individuals in a population do not contribute anything to the public good, then a rare mutant phenotype with a small positive investment x has an advantage. This is because such a mutant would essentially only play against noninvesting residents and hence receive an approximate (average) payoff $B(x) - C(x) = B(0) + xB'(0) - C(0) - xC'(0) = B(0) - C(0) + x(B'(0) - C'(0)) > B(0) - C(0)$, whereas the noninvesting resident would essentially only play against itself and hence receive an (average) payoff $B(0) - C(0)$. Thus, cooperative investments would evolutionarily increase at least initially.

It is worth noting that the crucial ingredient for the Continuous Snowdrift game is the assumption that the benefit received by each individual is a function not only of the investment made by the interaction partners, but also of the investment made by the individual itself. In contrast, in the Continuous Prisoner's Dilemma one would assume that the benefit received from an interaction is solely a function of the partner's investment, so that the payoffs from a single interaction between x and y would be $P_x = B(y) - C(x)$ and $P_y = B(x) - C(y)$, respectively. In this case the phenotype only affects an individual's costs in any given interaction, and it easily follows that in well-mixed populations, cooperation always evolves to 0 (which corresponds to the fact that defection always dominates in the classical Prisoner's Dilemma game). However, it is worth noting that nonzero levels of cooperative investments can evolve when the Continuous Prisoner's Dilemma is played in spatially structured populations (Doebeli & Knowlton, 1998; Killingback et al., 1999; Koella, 2000).

In the sequel, the adaptive dynamics of the Continuous Snowdrift game is derived under the assumption that the population size is constant, and that the outcome of the evolutionary game between different cooperation strategies is determined by the replicator equation, as in standard evolutionary game theory (Hofbauer & Sigmund, 1998). In particular, if a population contains individuals of only two types x and y, then the dynamics of the frequency q of type y is given by

$$\frac{dq}{dt} = q \left(P_y - \bar{P} \right), \tag{6.1}$$

where P_y is the average payoff for y-individuals, and $\bar{P}$ is the average payoff in the population. In general, both these averages are of course functions of the frequency q. However, to calculate the invasion fitness function, we are interested in the replicator dynamics under the additional assumption that x is a common resident and y is a rare mutant, that is, that $q \approx 0$. Then y essentially only interacts with resident types, and hence the average payoff for the mutant y is $P_y = B(x+y) - C(y)$. Similarly, residents only interact with other residents if the mutant is rare, so that their payoff, which is also the average payoff in the population (again because the mutant is rare) is $\bar{P} = B(2x) - C(x)$. Thus, the per capita growth rate of the mutant y in the resident x, that is, the invasion fitness $f(x, y)$ is given by:

$$f(x, y) = \frac{1}{q}\frac{dq}{dt} = (B(x+y) - C(y) - B(2x) + C(x)). \tag{6.2}$$

From this, we derive the selection gradient as

$$D(x) = \left.\frac{\partial f}{\partial y}\right|_{y=x} = B'(2x) - C'(y), \tag{6.3}$$

and the adaptive dynamics as

$$\frac{dx}{dt} = mD(x), \tag{6.4}$$

where m is a parameter describing the effects of mutation and thus influencing the speed of evolution. Equilibria of the adaptive dynamics are given by solutions of $D(x^*) = 0$, that is, of $B'(2x^*) = C'(x^*)$, and an equilibrium is convergent stable if (and only if)

$$\left.\frac{dD}{dx}\right|_{x=x^*} = 2B''(2x^*) - C''(x^*) < 0. \tag{6.5}$$

On the other hand, an equilibrium is evolutionarily stable if (and only if)

$$\left.\frac{\partial^2 f}{\partial y^2}\right|_{y=x=x^*} = B''(2x^*) - C''(x^*) < 0 \tag{6.6}$$

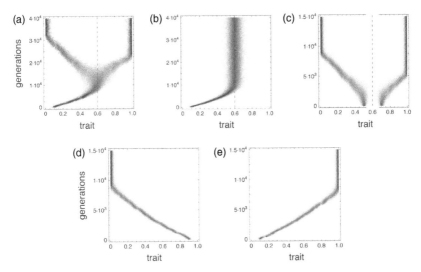

FIGURE 6.1. Different regimes of the evolutionary dynamics of cooperative investments for quadratic cost and benefit functions $B(z) = b_2z^2 + b_1z$ and $C(z) = c_2z^2 + c_1z$. Darker shades indicate higher frequencies of a trait value. The singular strategies are indicated by dashed vertical lines where appropriate. (a) Evolutionary branching. (b) Evolutionarily stable singular strategy. (c) Evolutionary repellor: depending on the initial conditions, the population either evolves to full defection or to full cooperation (two distinct simulations are shown). (d) and (e) Unidirectional evolutionary dynamics in the absence of singular strategies; in (d), cooperative investments decrease to zero, in (e), full cooperation evolves. Results were obtained from numerical simulations of the Continuous Snowdrift game in finite populations. From Doebeli et al. (2004). Reprinted with permission from AAAS, to which I refer for a description of the simulation procedure and for parameter values.

(see Appendix). The difference between the two stability conditions stems from the factor 2 in front of the second derivative of the benefit function in the derivative of the selection gradient. Clearly, if $B''(2x^*) < 0$ and $C''(x^*) > 0$, it is in principle possible that an equilibrium x^* is convergent stable $(2B''(x^*) - C''(x^*) < 0)$ but evolutionarily unstable $(B''(2x^*) - C''(x^*) > 0)$, and hence that the equilibrium is an evolutionary branching point. Indeed, Doebeli et al. (2004) have shown that this can already happen for quadratic cost and benefit functions. This is illustrated in Figure 6.1, which also shows the other scenarios possible with quadratic cost and benefit functions: directional selection toward ever increasing or ever decreasing cooperative investments, and evolutionary repellors (i.e., convergent unstable equilibria). More complicated scenarios are possible with more complicated benefit and cost functions, as shown in Figure 6.2.

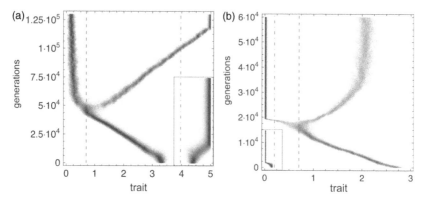

FIGURE 6.2. Examples of simultaneous occurrences of a branching point (dashed line) and a repellor (dash-dotted line) in the Continuous Snowdrift game. In panel (a), if the population starts to the right of the repellor, cooperative investments continue to increase until the upper limit of the trait interval is reached (see inset). However, if started below the repellor, the population evolves to the evolutionary branching point, where it splits into coexisting high-investing cooperators and low-investing defectors. In the defector branch, investments do not evolve to zero, and due to the existence of the defector branch, the cooperator branch no longer feels the repellor. In panel (b), the defective state is locally convergent stable (see inset). Only populations that start out above the repellor evolve toward the branching point and undergo diversification. From Doebeli et al. (2004), to which I refer for a description of the simulation procedure and for parameter values. Reprinted with permission from AAAS.

The conditions given above imply that evolutionary branching requires accelerating benefits and decelerating costs at the singular point. Note that this does not mean that benefits have to be accelerating over the whole phenotypic domain, nor that costs have to decelerate everywhere. Instead, there has to be a range of phenotypes where this is the case, and that contains a singular point. Given the typically nonlinear nature of benefits and costs, this does not seem to be unrealistic a priori. Nevertheless, it should be pointed out that there are scenarios of evolutionary branching in the Continuous Snowdrift game that do not require accelerating costs and decelerating benefits. This can occur when the benefits are not shared equally among interaction partners, that is, when there is asymmetric competition for the benefits generated by an interaction. Following Ferrière et al. (2002), one could, for example, assume that individuals making higher cooperative investments have an advantage in the form of getting a disproportionately higher share of the benefits produced by an interaction. Thus, if $\sigma(z)$ is a monotonically increasing function of its argument (e.g., a sigmoid function), one could assume that the payoffs to phenotypes x

and y in a pairwise interaction are

$$P_x = \frac{\sigma(x)}{\sigma(x) + \sigma(y)} B(x+y) - C(x) \tag{6.7}$$

$$P_y = \frac{\sigma(y)}{\sigma(x) + \sigma(y)} B(x+y) - C(y). \tag{6.8}$$

For this Asymmetric Continuous Snowdrift game, one can show that evolutionary branching is possible under more general conditions than with symmetric assumptions. In particular, evolutionary branching is now possible even with linear benefit functions.

Challenge*: Prove this claim by providing a general analysis of the adaptive dynamics of the Asymmetric Continuous Snowdrift game.

Returning to evolutionary branching in the symmetric case with quadratic benefit and cost functions, it is possible to determine analytically the adaptive dynamics ensuing after evolutionary branching. In general, the equilibrium frequencies of two coexisting strategies $x > x^* > y$ (where x^* is the branching point), can be calculated using replicator dynamics. According to the replicator equation (6.1), the equilibrium frequency of x is determined by the condition that the average payoffs of the two strategies are equal to the average population payoff, that is, by $P_x = \bar{P} = P_y$. If p^* is the frequency of x, this condition becomes $P_x = p^* [B(2x) - C(x)] + (1 - p^*) [B(x+y) - C(x)] = p^* [B(x+y) - C(y)] + (1 - p^*) [B(2y) - C(y)] = P_y$. Given p^*, the invasion fitness of a rare mutant v in each of the two resident branches x and y is given by

$$f(x, y, v) = p^* [B(x+v) - C(v)]$$
$$+ (1 - p^*) [B(y+v) - C(v)] - \bar{P}(x, y), \tag{6.9}$$

where $\bar{P} = P_x = P_y = p^* [B(2x) - C(x)] + (1 - p^*) [B(x+y) - C(x)]$ is the population mean payoff. The adaptive dynamics in the two branches is then determined by the two selection gradients

$$D_x(x, y) = \left. \frac{\partial f(x, y, v)}{\partial v} \right|_{v=x} \tag{6.10}$$

$$D_y(x, y) = \left. \frac{\partial f(x, y, v)}{\partial v} \right|_{v=y} \tag{6.11}$$

With quadratic benefit and cost functions $B(z) = b_1 z + b_2 z^2$ and $C(z) = c_1 z + c_2 z^2$, solving for p^* yields

$$p^* = \frac{c_1 - b_1 + x(c_2 - b_2) + y(c_2 - 3b_2)}{2b_2(x - y)}, \qquad (6.12)$$

which results in the selection gradients

$$D_x(x, y) = (b_2 - c_2)(x - y) \qquad (6.13)$$

$$D_y(x, y) = -(b_2 - c_2)(x - y). \qquad (6.14)$$

It follows that after evolutionary branching, selection is always directional in the two branches, with the upper branch (starting at some trait value $x > x^*$) increasing toward the upper boundary of feasible trait values, and the lower branch (starting at some trait value $y < x^*$), decreasing to 0. With more complicated benefit and cost functions, analytical solutions for the equilibrium p^*, and hence for the selection gradients determining the two-dimensional adaptive dynamics, can in general not be obtained. It is interesting to note, however, that in any case, the two coexisting phenotypes emerging from evolutionary branching always exhibit mutual invasibility, the defining characteristic of the standard Snowdrift game with two strategies. Viewed thus, evolutionary branching in the Continuous Snowdrift game into a type making low cooperative investments and a type making high cooperative investments yields an evolutionary explanation for the origin of the two strategies Cooperate and Defect in the standard Snowdrift game.

The Continuous Snowdrift game admits a dual formulation, describing not a situation in which individuals make contributions to a public good, but instead a situation in which phenotypes determine how much an individual takes from a common pool of resources. In this case, the benefit is an increasing function of a single individual's trait, whereas the cost now becomes a function of the traits of both individuals in a pairwise interaction, corresponding to the assumption that the damage to the common resource resulting from an interaction is a function of the sum of what is taken by the two interacting individuals. This dual game, which could be called a Continuous Tragedy of the Commons, exhibits very similar evolutionary dynamics as the continuous Snowdrift game. The Continuous Tragedy of the Commons is explained in Box 6.1 for interactions in groups of arbitrary size.

In sum, adaptive diversification into cooperators and defectors can easily occur in both the Continuous Snowdrift Game and the Continuous Tragedy of the Commons. Because public goods and common resource use are extremely common in nature, this could have significant implications for our

BOX 6.1

THE CONTINUOUS TRAGEDY OF THE COMMONS

The treatment here follows Killingback et al. (2010). To analyze the evolution of quantitative traits determining exploitation of common resources, consider a game that is played in interaction groups of N individuals that share a common resource, of which individual i consumes an amount $x_i \geq 0$ $(i = 1, \ldots, N)$. Each individual benefits from consuming the resource, but the costs incurred from consumption depend on the total consumption of all individuals. Therefore, the payoff to individual i from the given interaction group is $P(x_i; x_1, \ldots, x_N) = B(x_i) - C(x_1 + \cdots + x_N)$. Here $B(x)$ and $C(x)$ are benefit and cost functions determining the gain of individual i due to its own consumption, as well as its loss due to the accumulated consumption of all individuals. We assume that $B(x)$ and $C(x)$ are smooth, strictly increasing functions, with $B(0) = C(0) = 0$ such that no consumption returns no benefits and incurs no costs. As in the Continuous Snowdrift game, we also assume that $B'(0) > C'(0)$, so that $B(x) > C(x)$ at least for small x, reflecting an incentive to consume the common resource if nobody else does.

To determine the adaptive dynamics of the consumption level, we assume that a resident population is monomorphic for trait value x. The growth rate of a rare mutant trait y is then given by the invasion fitness $f_x(y) = P(y; y, x, \ldots, x) - P(x; x, \ldots, x)$. This is because as long as the mutant is rare, and as long as the interaction groups are formed randomly, a mutant individual always plays in groups containing one mutant and $N - 1$ residents, and resident individuals always play in groups containing N residents. The selection gradient is then $D(x) = \partial f_x / \partial y|_{y=x} = B'(x) - C'(Nx)$, with x obeying the equation $\dot{x} = mD(x)$, where m is a parameter describing how mutations influence the speed of evolution, for which we assume $m = 1$.

Singular points, that is, equilibrium points of the adaptive dynamics, are solutions x^* of $D(x^*) = 0$, that is, points in phenotype space at which the selection gradient vanishes (see Appendix). It is interesting to note that when the population is at a singular point, the classical Tragedy of the Commons is recovered—the population over-consumes the resource, to the detriment of individuals. This follows from observing that at x^* each individual receives a payoff $B(x^*) - C(Nx^*)$, but if all individuals reduced their consumption by a small amount $\epsilon > 0$ to $x^* - \epsilon$, then each individual's payoff changes to $B(x^* - \epsilon) - C(Nx^* - N\epsilon)$, which, to first order in ϵ, is equal to $B(x^*) - C(Nx^*) + \epsilon(N - 1)C'(Nx^*)$ (taking into account that $B'(x^*) - C'(Nx^*) = 0$ by definition of the singular point of variation and x^*). Because $C(x)$ is a strictly increasing function,

BOX 6.1 (*continued*)

a uniform reduction in the consumption level by ϵ thus results in an increased payoff for all individuals.

Convergence and evolutionary stability of singular points x^* are determined in the standard way (see Appendix). x^* is convergence stable, and hence an attractor for the adaptive dynamics, if $dD/dx|_{x=x^*} = B''(x^*) - NC''(Nx^*) < 0$. x^* is evolutionarily stable if $\partial^2 f_{\tilde{x}^*}/\partial y^2|_{y=x^*} = B''(x^*) - C''(Nx^*) < 0$. A singular point is an evolutionary branching point if it is convergence stable, but evolutionarily unstable, that is, if $NC''(Nx^*) > B''(x^*) > C''(Nx^*)$. It follows easily that for evolutionary branching to occur, the cost function must be concave up at Nx^*, and the benefit function must be concave up at x^*. Note that these conditions are "dual" to the conditions for evolutionary branching in the Continuous Snowdrift game, where the cost and benefit function must have negative curvature at the singular point for evolutionary branching to be possible.

Suitable cost and benefit functions permit an analysis of the adaptive dynamics. For example, the functions $B(x) = -ax^3 + 2bx^2 + ax$ and $C(x) = bx^2$ with $a, b > 0$ satisfy our assumptions in the trait interval $x \in \left[0, (2b + \sqrt{4b^2 + 3a^2})/(3a)\right]$. For pairwise interactions, $N = 2$, the adaptive dynamics has a unique and parameter independent singular point at $x^* = 1/\sqrt{3}$, which is always convergent stable. The singular trait is evolutionarily unstable, and hence an evolutionary branching point, if $\sqrt{3}\, a/b < 1$.

∎

understanding diversification in cooperative investments in many natural populations. Moreover, since these models are based on game theory and on the replicator equation, they cannot only be interpreted in the context of genetic evolution, but also in the context of cultural evolution, in which the replicator dynamics corresponds to learning from successful individuals and adopting strategies yielding high payoffs (Hofbauer & Sigmund, 1998). Thus, these cooperation games can also serve as paradigm for cultural diversification of cooperation in human societies. Cultural evolution, and in particular adaptive cultural diversification, will be discussed in more detail in Chapter 8.

6.2 DIVERSIFICATION IN COEVOLUTIONARY MODELS OF COOPERATION

We now extend the framework of continuous cooperative investments to mutualistic interactions between two separate species. In such models, costly cooperative investments in one species generate a benefit for individuals of the other species, and in contrast to the Continuous Snowdrift game, individuals do not benefit anymore from their own investments. Based on a fairly long tradition in ecological modeling of mutualism (e.g., Boucher, 1985; Vandermeer & Boucher, 1978), a few attempts have been made to model the evolutionary dynamics of quantitative traits affecting mutualistic interactions (e.g., Bever, 1999; Doebeli & Dieckmann, 2000; Doebeli & Knowlton, 1998; Ferrière et al., 2002; Kiester et al., 1984). For example, Ferrière et al. (2002) described evolutionary branching in such traits. In their models, ecological dynamics are driven by asymmetric competition for the mutualistic help provided by the other species, so that individuals that make a higher investment into helping the mutualistic partner also have a higher chance of getting access to the help provided by the partner. In this setting, diversification is therefore akin to the diversification observed in models for asymmetric resource competition (Chapter 3). Doebeli & Dieckmann (2000) described (co-)evolutionary branching in models in which interspecific frequency dependence was incorporated by assuming that the amount of help provided by an individual of one species to an individual of the other species depends on the phenotypic distance between the interacting individuals. Their models were based on explicit logistic dynamics for mutualism (Boucher, 1985), and they assumed that the effect of mutualistic help is to increase the per capita birth rate in the other species.

In contrast, here I consider simpler, more generic and hence in some sense more general models for the coevolution of mutualistic interactions. These models assume constant population sizes and hence are not based on explicit ecological dynamics. The models are set in discrete time, and I assume that the effect of mutualism is to increase the probability of survival in the other species. Specifically, I assume that individuals in the two mutualistic species are characterized by continuously varying traits x_1 and x_2, respectively. In the absence of the other species, in each generation the per capita probability of survival is a function of the mutualistic trait x_i given by

$$\frac{1}{1 + \dfrac{1}{K_i(x_i)}},$$

(6.15)

where K_i is unimodal function describing stabilizing selection for the trait value x_i^0 maximizing the function K_i ($i = 1, 2$). In other words, according to (6.15) the survival probability is highest for individuals with trait value x_i^0. The form (6.15) of the survival probability is chosen for mathematical convenience and mimics corresponding expressions for survival probabilities in difference equations describing density-dependent ecological dynamics of single species (see e.g., eq. (3.42) in Chapter 3 and eq. (7.4) in Chapter 7). Note that if K_i were increased to larger and larger values, the survival probability (6.15) would approach 1. I further assume that in each generation, the per capita expected number of offspring in species i is proportional to a constant λ_i ($i = 1, 2$) that is independent of the mutualistic phenotype. Thus, in the absence of mutualism the relative fitness of an x_i-individual in species i is

$$\frac{\lambda_i}{1 + \dfrac{1}{K_i(x_i)}} \quad (i = 1, 2). \tag{6.16}$$

To model mutualistic interactions, I will assume that mutualistic help increases the probability of survival. To quantify mutualism, I assume that every individual provides a constant amount of help to the other species. This per capita amount of help provided is denoted by a_1 for species 1 and a_2 for species 2. I further assume that the distribution of help provided by species 1 is given by a function $\alpha_{12}(x_1 - x_2)$, which describes the relative contribution that an x_1-individual in species 1 makes to an x_2-individual in species 2. Similarly, the distribution of help provided by species 2 is described by a function $\alpha_{21}(x_2 - x_1)$, which describes the relative contribution that an x_2-individual in species 2 makes to an x_1-individual in species 1.

The functions $\alpha_{12}(z)$ and $\alpha_{21}(z)$ are assumed to have a maximum at $z = 0$, corresponding to the assumption that mutualistic help increases with similarity between interacting individuals. (Note that it is implied here that the mutualistic phenotypes in each species can be rescaled so that they can be measured and compared along a single dimension.)

Assuming infinite population sizes, let $\phi_1(u_1)$ be the density distribution of trait u_1 in species 1, and let $\phi_2(u_2)$ be the density distribution of trait u_2 in species 2. Then $N_1 = \int \phi_1(x_1) dx_2$ is the total population size of species 1 (where the integral is taken over all possible phenotypes), and $N_2 = \int \phi_2(x_2) dx_2$, is the total population size of species 1. Note that both N_1 and N_2 are constant by assumption. Given the phenotype distribution ϕ_1, the proportion of help that an x_1-individual in species 1 receives from the total help a_2 provided by an x_2-individual in species 2 is then

$$a_2 \frac{\alpha_{21}(x_2 - x_1)}{\int \alpha_{21}(x_2 - u_1)\phi_1(u_1) du_1}, \tag{6.17}$$

where the integral in the denominator runs over all possible phenotypes in species 1. We note that when expression (6.17) is integrated over all individuals of species 1, we get

$$\int \phi_1(x_1) a_2 \frac{\alpha_{21}(x_2 - x_1)}{\int \alpha_{21}(x_2 - u_1)\phi_1(u_1)\, du_1} dx_1 = a_2, \tag{6.18}$$

that is, the total amount of help provided by an individual of species 2, as it should be. Expression (6.17) thus measures how the total help a_2 is distributed among species 1 individuals, depending on the phenotype x_2 of the helping individual, and on the phenotype distribution ϕ_1 in species 1. In particular, expression (6.17) is the source of within-species frequency dependence in this model, because it implies that the amount of help a species 1-individual with phenotype x_1-receives depends on the phenotype distribution ϕ_1 in species 1.

By integrating expression (6.17) over all species 2 individuals, we obtain the total amount of help received by an individual of species 1 with phenotype x_1 from all individuals in species 2, a quantity we denote by $T_1(x_1)$:

$$T_1(x_1) = \int \phi_2(x_2) \frac{a_2 \alpha_{21}(x_2 - x_1)}{\int \alpha_{21}(x_2 - u_1)\phi_1(u_1)\, du_1} dx_2. \tag{6.19}$$

Finally, to describe the effect of mutualistic help on survival probability, I simply assume that $T_1(x_1)$ is added to the quantity $K_1(x_1)$ in expressions (6.15) and (6.16). In other words, I assume that when the mutualistic help from species 2 with phenotype distribution $\phi_2(u_2)$ is taken into account, the survival probability of a species 1 individual with phenotype x_1 is

$$\frac{1}{1 + \dfrac{1}{T_1(x_1) + K_1(x_1)}}. \tag{6.20}$$

Completely analogous assumptions are made for the effect of mutualistic help on survival in species 2. Specifically, the proportion of help that an x_2-individual in species 2 receives from the total help a_1 provided by an x_1-individual in species 1 is

$$a_1 \frac{\alpha_{12}(x_1 - x_2)}{\int \alpha_{12}(x_1 - u_2)\phi_2(u_2)\, du_2}, \tag{6.21}$$

and the total amount of help received by an individual of species 2 with phenotype x_2 from all individuals in species 1 is:

$$T_2(x_2) = \int \phi_1(x_1) \frac{a_1 \alpha_{12}(x_1 - x_2)}{\int \alpha_{12}(x_1 - u_2)\phi_2(u_2)\, du_2} dx_1. \tag{6.22}$$

Finally, taking mutualistic help into account, the survival probability of a species 2 individual with phenotype x_2 is

$$\frac{1}{1 + \dfrac{1}{T_2(x_2) + K_2(x_2)}}. \tag{6.23}$$

Based on these assumptions, we can proceed with an adaptive dynamics analysis by considering rare mutants z in species 1 populations that are monomorphic for trait value x_1 and in species 2 populations that are monomorphic for trait value x_2. First we note that under the assumptions of monomorphic populations and rare mutants, the integrals needed for calculating the total help are easily evaluated, because in that case the distributions ϕ_1 and ϕ_2 are Dirac delta functions that are centered at x_1 and x_2, respectively, with total weights the population sizes N_1 and N_2. For example, we then have

$$\int \alpha_{21}(x_2 - u_1)\phi_1(u_1)\,du_1 = N_1\alpha_{21}(x_1 - x_2), \tag{6.24}$$

and the total help received by a mutant z-individual from a given x_2-individual is

$$\frac{a_2\alpha_{21}(x_2 - z)}{N_1\alpha_{21}(x_2 - x_1)}. \tag{6.25}$$

Therefore, the total help received by a mutant z-individual form all individual in species 2 is

$$T_1(z) = \frac{a_2 N_2 \alpha_{21}(x_2 - z)}{N_1\alpha_{21}(x_2 - x_1)}. \tag{6.26}$$

It follows that the relative fitness of a rare mutant z in species 1 is

$$\frac{\lambda_1}{1 + \dfrac{1}{\dfrac{a_2 N_2 \alpha_{21}(x_2 - z)}{N_1\alpha_{21}(x_2 - x_1)} + K_1(z)}}. \tag{6.27}$$

Since the model is set in discrete time, it further follows that the invasion fitness of a rare mutant z appearing in species 1 when species 1 is monomorphic for trait x_1 and species is monomorphic for trait x_2 is given by the quotient

$$\hat{f}_1(x_1, x_2, z) = \cfrac{\cfrac{\lambda_1}{1 + \cfrac{1}{\cfrac{a_2 N_2 \alpha_{21}(x_2 - z)}{N_1 \alpha_{21}(x_2 - x_1)} + K_1(z)}}}{1 + \cfrac{\lambda_1}{a_2 \cfrac{N_2}{N_1} + K_1(x_1)}}. \tag{6.28}$$

The denominator in this expression is simply the relative fitness of the resident phenotype x_1. In fact, it is easy to see that this denominator is irrelevant for the sign of the selection gradient, for the determination of singular points, as well as for the determination of convergence and evolutionary stability of singular points. Therefore, to determine the salient features of the adaptive dynamics of the trait x_1 in species 1 it is enough to consider the simplified invasion fitness

$$f_1(x_1, x_2, z) = \cfrac{\lambda_1}{1 + \cfrac{1}{\cfrac{a_2 N_2 \alpha_{21}(x_2 - z)}{N_1 \alpha_{21}(x_2 - x_1)} + K_1(z)}}. \tag{6.29}$$

Challenge: Prove this claim regarding the irrelevance of the denominator in expression (6.28) for $\hat{f}_1$.

In a completely analogous way, the adaptive dynamics of the trait x_2 in species 2 is determined by the invasion fitness

$$f_2(x_1, x_2, z) = \cfrac{\lambda_2}{1 + \cfrac{1}{\cfrac{a_1 N_1 \alpha_{12}(x_1 - z)}{N_2 \alpha_{12}(x_1 - x_2)} + K_2(z)}}, \tag{6.30}$$

where z is now the trait value of rare mutants appearing in species 2. The adaptive dynamics in the two species are then given by

$$\frac{dx_1}{dt} = m_1 \frac{\partial f_1(x_1, x_2, z)}{\partial z}\bigg|_{z=x_1} = m_1 D_1(x_1, x_2) \tag{6.31}$$

$$\frac{dx_2}{dt} = m_2 \frac{\partial f_2(x_1, x_2, z)}{\partial z}\bigg|_{z=x_2} = m_2 D_2(x_1, x_2), \tag{6.32}$$

where $D_1(x_1, x_2)$ and $D_2(x_1, x_2)$ are the selection gradients. The parameters m_1 and m_2 describe the mutational process in the two species and influence the speed of evolution, and we assume $m_1 = m_2 = 1$ in the following.

As usual, we are interested in trait pairs (x_1^*, x_2^*) for which $D_1(x_1^*, x_2^*) = D_2(x_1^*, x_2^*) = 0$, representing singular points of the adaptive dynamics. For convergence stability of a singular point (x_1^*, x_2^*), we have to determine the eigenvalues of the Jacobian matrix of the selection gradients at the singular point:

$$J = \begin{pmatrix} \dfrac{\partial}{\partial x_1} D_1(x_1, x_2) \Big|_{(x_1,x_2)=(x_1^*,x_2^*)} & \dfrac{\partial}{\partial x_2} D_1(x_1, x_2) \Big|_{(x_1,x_2)=(x_1^*,x_2^*)} \\ \dfrac{\partial}{\partial x_1} D_2(x_1, x_2) \Big|_{(x_1,x_2)=(x_1^*,x_2^*)} & \dfrac{\partial}{\partial x_2} D_2(x_1, x_2) \Big|_{(x_1,x_2)=(x_1^*,x_2^*)} \end{pmatrix}. \quad (6.33)$$

Finally, to determine evolutionary stability of singular points in each species, we need to determine the partial derivatives

$$\frac{\partial^2 f_1(x_1^*, x_2^*, z)}{\partial z^2} \Big|_{z=x_1^*} \quad (6.34)$$

$$\frac{\partial^2 f_2(x_1^*, x_2^*, z)}{\partial z^2} \Big|_{z=x_2^*}. \quad (6.35)$$

So far, we have not yet chosen specific forms of the functions K_1 and K_2 determining stabilizing selection, and of the functions α_{12} and α_{21} determining the distribution of mutualistic help. However, even without specifying these functions one can see relatively easily that it will be generally hard to analytically determine singular points, convergence and evolutionary stability. Nevertheless, it is possible to describe what one can expect to happen on intuitive grounds. First, it is clear that if the optima imposed by stabilizing selection are the same in the two species, then in both species the mutualistic trait will simply evolve to that optimum, which will at the same time maximize mutualistic help. In other words, if the optima of stabilizing selection are the same in both species, then there is no conflict, that is, no trade-offs, that could lead to interesting evolutionary dynamics. On the other hand, if the optima of stabilizing selection are different in the two species, then there is a trade-off between having trait values close to this optimum and having trait values close to the other species. Without loss of generality we can assume that the two optima of the unimodal functions K_1 and K_2 are $x_1^0 = \delta$ and $x_2^0 = -\delta$. Then it is intuitively clear that singular points (x_1^*, x_2^*), that is, equilibrium points of the two-dimensional adaptive dynamics (6.31) and (6.32), should lie in between the two optima, such that $-\delta < x_2^* < x_1^* < \delta$. In particular, in the symmetric case, in which the two species are identical except for the the position of the optima of stabilizing selection (so that the $\alpha_{12} = \alpha_{21}$, $a_1 = a_2$, $N_1 = N_2$, and K_1 and K_2 have the same functional form), a singular point should also be symmetric, that is, should satisfy $-\delta < x_2^* = -x_1^* < 0 < x_1^* < \delta$.

It seems reasonable that such singular points would generally be convergent stable, that is, that both eigenvalues of the Jacobian (6.33) have negative real parts. (In fact, one would not expect any cyclic components of the evolutionary dynamics in this simple system, and hence both eigenvalues of the Jacobian should be negative real numbers.) It is more difficult to intuitively understand evolutionary instability, but based on explicit expressions of (6.34) and (6.35), it can be shown in general that evolutionary instability is enhanced if the second derivatives of both the stabilizing selection functions K_i and the distribution functions α_{ij} are positive at the singular points. For example, let's assume that these functions have Gaussian form:

$$K_i(x_i) = K_{0,i} \exp\left(-\frac{(x - x_i^0)^2}{2\sigma_{K_i}^2}\right) \quad (i = 1, 2) \tag{6.36}$$

$$\alpha_{ij}(x_i - x_j) = \exp\left(-\frac{(x_i - x_j)^2}{2\sigma_{\alpha_{ij}}^2}\right) \quad (ij = 12 \text{ or } 21). \tag{6.37}$$

Here $K_{0,i}$, σ_{K_i} and $\sigma_{\alpha_{ij}}$ are (positive) parameters, with σ_{K_i} determining the strength of stabilizing selection and $\sigma_{\alpha_{ij}}$ determining the strength of frequency dependence. With these functions, and still in the symmetric case, if the singular values x_1^* and x_2^* are far enough apart, but also far enough from δ and $-\delta$, the second derivatives of the functions (6.36) and (6.37) will be negative at the singular point, which will increase the tendency toward evolutionary instability. It is intuitively clear that such situations should be obtained by increasing the distance between the optima of stabilizing selection in the two species, that is, by increasing δ, a finding that is analogous to the results reported in Doebeli & Dieckmann (2000). Roughly speaking, increasing the distance between the stabilizing optima increases the evolutionary tension between the stabilizing and the frequency-dependent components of selection, and if as a result the singular point comes to lie in regions where the functions determining stabilizing and frequency-dependent selection have positive curvature, evolutionary branching, and hence evolutionary diversification, becomes more likely.

It is easy to give numerical examples of this by solving the adaptive dynamics given by (6.31) and (6.32) using a computer algebra program. For example, for $\lambda_i = 1$, $N_1 = N_2 = 500$, $K_{0,i} = 1$, $\sigma_{K_i} = 1$, $a_i = 0.1$, $\sigma_{\alpha_{ij}} = 0.4$

$(i, j = 1, 2)$, the adaptive dynamics has a convergent and evolutionarily stable singular point $(x_1^*, x_2^*) \sim (0.18, -0.18)$ for $\delta = 0.4$, but for $\delta = 1$ the adaptive dynamics has a convergent stable singular point $(x_1^*, x_2^*) \sim (0.4, -0.4)$, which is evolutionarily unstable. More precisely, in this case x_1^* is an evolutionary branching point in species 1, and x_2^* is an evolutionary branching point in species 2. It is interesting to note that evolutionary branching is not only promoted by increasing distances between the optima of stabilizing selection in the two species, but also by decreasing, at least over a certain range, the total amount of mutualistic help provided by each individual, that is, by decreasing the parameters a_1 and a_2. For example, if in the above set of parameters one retains the distance $\delta = 1$ but increases the total help provided by individuals to $a_i = 0.4$, the singular values change from evolutionary branching points to evolutionarily stable strategies.

> *Challenge**: For the functions K_i and α_{ij} given by (6.36) and (6.37), prove the existence of evolutionary branching points analytically. Investigate the feasibility of analytical proofs for the existence, convergence stability and evolutionary instability of singular points for other functional forms for K_i and α_{ij}.

It is straightforward to implement the previous setup into individual-based models, which, with the evolutionary dynamics after evolutionary branching, can be investigated. To do this, one needs to replace the integrals appearing in the expressions for $T_1(x_1)$ and $T_2(x_2)$ by appropriate sums that run over the entire populations. For individuals of phenotype x_i in species i, this yields a relative fitness value of

$$\frac{\lambda_i}{1 + \dfrac{1}{T_i(x_i) + K_i(x_i)}}. \tag{6.38}$$

Given a species 1 population of N_1 individuals with phenotypes $x_{1,1}, \ldots, x_{1,N_1}$ and a species 2 population of N_2 individuals with phenotypes $x_{2,1}, \ldots, x_{2,N_2}$, one first calculates the relative fitness values (6.38) based on all phenotypes present, and one then simply creates the next generation in each species by drawing N_i new individuals whose mothers are chosen according to the relative fitness values in each species, and whose phenotype is drawn from a normal distribution with mean equal to the mother's phenotype and variance equal to an additional parameter σ_M. The ensuing stochastic dynamics is shown in Figure 6.3 for the functions (6.36) and (6.37) and the parameter values discussed above. In particular, the individual-based model exhibits evolutionary branching once the distance δ is large enough. Note

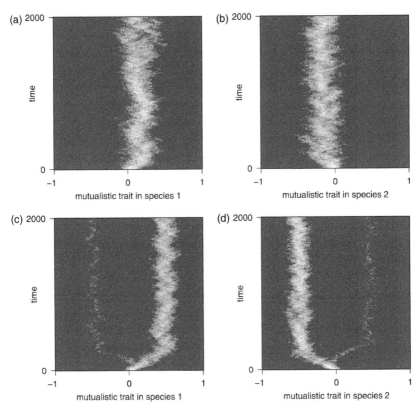

FIGURE 6.3. Coevolutionary dynamics in individual-based models of two symmetric mutualistic species. Panels (a) and (b) show the dynamics of the phenotype distribution when the distance between the optima of the carrying capacity functions for the two species is not large enough to induce evolutionary branching. Panels (c) and (d) show the dynamics of the phenotype distribution when the distance between the optima of the carrying capacity functions for the two species is large enough to induce evolutionary branching. In this case, evolutionary branching occurs in both species, in each of which it leads to the emergence and coexistence of two phenotypic clusters, a more abundant one lying close to the resource optimum, and a less abundant one lying close to the resource optimum of the other species, where it can take advantage of the help provided by the abundant branch of the other species lying close to that optimum. Thus, evolutionary branching results in two species pairs in which one species does not provide much help and only survives due to the help provided by the other species in the pair. Parameter values were $a_1 = a_2 = 0.1$, $K_{0,1} = K_{0,2} = 1$, $\sigma_{K_1} = \sigma_{K_2} = 1$, $\sigma_{\alpha_{12}} = \sigma_{\alpha_{21}} = 0.4$, $\lambda_1 = \lambda_2 = 1$, and $\sigma_M = 0.02$ for all panels, and $\delta = 0.5$ for panels (a) and (b), and $\delta = 1$ for panels (c) and (d). The populations were initialized with Gaussian distributions with a small variance and with mean -0.05 in species and 0.05 in species 2.

that diversification results in asymmetric branches in each species: one of the branches, containing most of the individual of a given species, stays close to the optimum of the stabilizing component of selection, whereas the other branch, which contains far fewer individuals, evolves away from the stabilizing optimum and thus increases the mutualistic help received from that branch in the other species that stays close to its respective optimum of stabilizing selection. Thus, evolutionary branching results in two species pairs, in each of which one species is close to the optimum of stabilizing selection and provides a substantial amount of mutualistic help to the other species, which in turn is far away from the optimum of stabilizing selection and survives essentially only due to the help received. These results are similar to the results obtained by Doebeli & Dieckmann (2000).

It is also interesting to consider asymmetric scenarios, in which parameter values are not the same for the two species. In this case, the outcome of the evolutionary dynamics is not symmetric anymore, as is illustrated in Figure 6.4. While it is still true that increasing the distance between the optima for stabilizing selection can promote diversification, it is now possible that a singular point (x_1^*, x_2^*) is an evolutionary branching point for only one of the species, so that frequency-dependent mutualism results in adaptive diversification in only one of the interacting species, as shown in Figure 6.4(c) and (d).

It is straightforward to extend the individual-based models to sexual populations. In this case, once a mother is chosen for a given new individual of the next generation (see above), the mother choses a mating partner to produce the offspring. For simplicity, I assume that mating partners occur in the same frequencies as given by the current population distribution (as would for example be the case with hermaphroditism), so that one only has to model females. As in previous chapters, mating partners are chosen according to Gaussian preference functions

$$A_i(x_i, y_i) = \frac{1}{\sqrt{2\pi}\sigma_{A_i}} \exp\left[-\frac{(x_i - y_i)^2}{2\sigma_{A_i}^2}\right] \qquad (i = 1, 2). \qquad (6.39)$$

Here x_i and y_i are the phenotypes of potential mating partners, and the parameters σ_{A_i} determine the strength of mating assortment in the two species, with large σ_{A_i} corresponding to random mating, and very small σ_{A_i} corresponding to strongly assortative mating. For a given x_i-female, the probability of mating with a y_i-male is then given by

$$\frac{A_i(x_i, y_i)}{\sum_z A_i(x_i, z)}, \qquad (6.40)$$

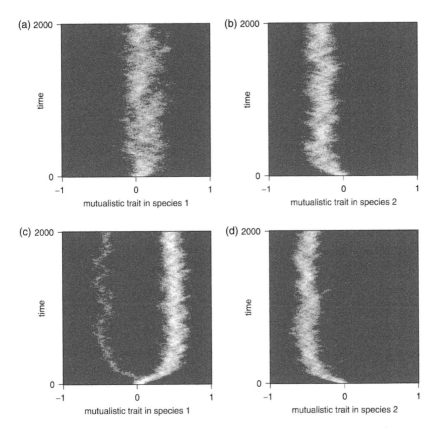

FIGURE 6.4. Coevolutionary dynamics in individual-based models of two asymmetric mutualistic species. Panels (a) and (b) show the dynamics of the phenotype distribution when the distance between the optima of the carrying capacity functions for the two species is not large enough to induce evolutionary branching. Panels (c) and (d) show the dynamics of the phenotype distribution when the distance between the optima of the carrying capacity functions is large enough to induce evolutionary branching. In this case, evolutionary branching only occurs in one of the interacting species due to asymmetry between the two species. Parameter values were the same as in Figure 6.3, except that an asymmetry was introduced by setting $a_2 = 0.02$.

where the normalization by the sum over all potential mating partners reflects the assumption that there is no cost to assortment due to Allee effects (see Chap. 4 for a detailed discussion of assortative mating and related issues). Also as in previous chapters, once a mating partner has been chosen the offspring phenotype is drawn from the segregation kernel, a normal distribution with mean equal to the average of the parental phenotypes and variance equal to a

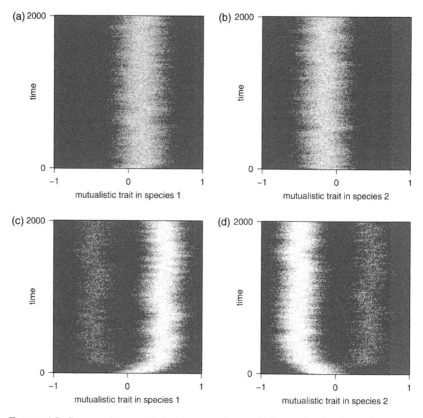

FIGURE 6.5. Same as Figure 6.3, but for sexual mutualistic species. Again, evolutionary branching, that is, adaptive speciation, occurs in both species when the distance between the optima of the carrying capacity functions for the two species is large enough (panels (c) and (d)). As in the asexual case, in each species diversification leads to the emergence and coexistence of two phenotypic clusters, resulting in two species pairs with asymmetric interactions. Parameter values were the same as for Figure 6.3, with the addition of $\sigma_{A_1} = \sigma_{A_2} = 0.2$ and $\sigma_{S_1} = \sigma_{S_2} = 0.15$.

parameter σ_{S_i}. Under these assumptions, and if assortment is strong enough, adaptive speciation in sexual populations can be readily observed. This is illustrated in Figure 6.5, which shows the evolutionary dynamics in sexual populations for the same parameter values as were used in Figure 6.3 for asexual populations, and in Figure 6.6, which shows the evolutionary dynamics in sexual populations for the same parameter values as were used in Figure 6.4 for asexual populations. These examples show that when mating is assortative, sexual populations can exhibit evolutionary dynamics that is

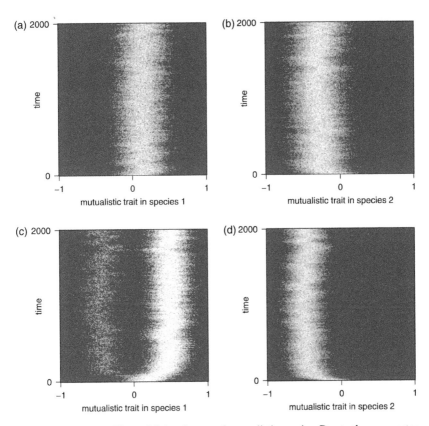

FIGURE 6.6. Same as Figure 6.4, but for sexual mutualistic species. Due to the asymmetry between the species, evolutionary branching, that is, speciation, occurs only in one species when the distance between the optima of the carrying capacity functions for the two species is large enough (panels (c) and (d)). Parameter values were the same as for Figure 6.4, with the addition of $\sigma_{A_1} = \sigma_{A_2} = 0.2$ and $\sigma_{S_1} = \sigma_{S_2} = 0.15$.

very similar to that in asexual populations. Overall, and in agreement with earlier results of Doebeli & Dieckmann (2000) and Ferrière et al. (2002), we conclude that in the minimalistic models presented here, adaptive diversification due to frequency-dependent mutualistic interactions is a plausible evolutionary outcome.

*Challenge**: Based on the ecological assumptions for the asexual models, and using the preference functions (6.39) in conjunction with segregation kernels to describe sexual reproduction, formulate analytical partial differential equation models for the evolutionary dynamics of mutualism between sexual species (see Chap. 9).

*Challenge**: Extend the analysis of adaptive speciation due to mutualism between sexual populations by investigating Allee effects and the evolution of assortative mating (cf. Chap. 4).

CHAPTER SEVEN

More Examples: Adaptive Diversification in Dispersal Rates, the Evolution of Anisogamy, and the Evolution of Trophic Preference

In the previous chapters, I have explained how the three fundamental types of ecological interactions—competition, predation, and mutualism—can drive adaptive diversification. To further illustrate the diversifying force of frequency-dependent interactions, I discuss three more examples that all arise in the context of fundamental ecological and evolutionary questions. The first example concerns the dynamics of spatially structured populations and serves as an excellent case study for illustrating the feedback between ecological and evolutionary dynamics. The second example concerns the evolution of asymmetry in gamete size between the sexes, which sets the stage for the "paradox of sex" (Maynard Smith, 1978). The third example concerns the fundamental question of the evolution of trophic levels in food webs, that is, the evolution of complexity in ecosystems.

7.1   DIVERSIFICATION IN DISPERSAL RATES

While most classical ecological and evolutionary theory assumes well-mixed populations, it has long been realized that spatial structure plays a crucial role in population biology. For example, there is a large literature on the effects of spatial structure on ecological dynamics (see, e.g., the excellent book of Dieckmann et al. (2004a)), and spatial structure can affect fundamental evolutionary processes such as the evolution of cooperation (e.g., Nowak & Sigmund, 2004). I will give examples of how spatial structure can affect the process of adaptive diversification in Chapter 9, but here I want to concentrate on the evolution of dispersal, which is itself one of the main determinants of spatial structure.

A common approach to incorporating spatial structure into population models is based on the concept of metapopulations (Levins, 1968), in which the total population is viewed as consisting a number of local populations living in habitat patches that are connected by dispersal (see Chap. 2, e.g., and Chap. 9 for alternative ways to model spatially structured populations). For such models, it has been known for some time that different dispersal types can coexist (Doebeli & Ruxton, 1997; Holt & McPeek, 1996; Ludwig & Levin, 1991; Mathias et al., 2001; McPeek & Holt, 1992; Olivieri et al., 1995; Parvinen, 1999). The different forces maintaining dispersal polymorphisms are nicely explained in McPeek & Holt (1992) and Mathias et al. (2001). On the one hand, if there is spatial, but no temporal variability in patch quality, and hence in local population size, a given per capita rate of dispersal generates a net flow of individuals from large local populations to small local populations. If fitness is density-dependent, this improves conditions in the source populations and worsens conditions in the sink populations, and since the majority of dispersing individuals move to the sink populations, lower dispersal rates are favored under these conditions (see also Dockery et al. (1998)). Conversely, if local population sizes undergo temporal fluctuations that are sufficiently uncorrelated, then a dispersal sink, that is, a local population receiving many migrants from high density patches, could experience better conditions in the next generations, in which case dispersal would be advantageous. Thus, with temporal fluctuations that are uncorrelated or negatively correlated between local populations, dispersal can be selectively favored as a bet-hedging strategy. Together, spatial heterogeneity, selecting for lower dispersal, and uncorrelated temporal fluctuations in patch quality, selecting for higher dispersal, can maintain coexistence between low and high dispersal strategies. This is also consistent with recent work by Schreiber & Saltzman (2009), who showed that low and high dispersal phenotypes can coexist in spatial predator-prey models with variable habitat quality and fluctuating population dynamics.

It is natural to view dispersal rates as continuously varying characters, in which case the emergence of a dispersal polymorphism can be studied in the framework of evolutionary branching (Doebeli & Ruxton, 1997; Mathias et al., 2001; Parvinen, 1999). I first consider deterministic metapopulation models in which temporal fluctuations in local population size are caused by non-linear, density-dependent competition. One-dimensional difference equations arguably yield the simplest models for (local) nonlinear population dynamics. They are typically given by an equation of the form

$$n(t + 1) = G(n(t)), \tag{7.1}$$

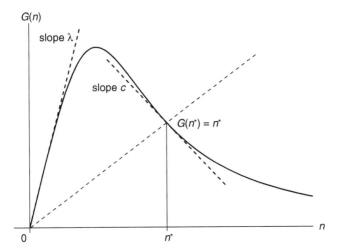

FIGURE 7.1. Schematic example of a next generation map $G(n)$. The function is approximately linear for small n, indicating exponential growth as long as population sizes are small: for small $n(t)$, $n(t+1) = G(n(t)) = \lambda n(t)$, where λ is the derivative of G at 0, and hence $n(t) = \lambda^t n(0)$, where $n(0)$ is the initial (small) population size. In the example shown, $G(n) \to 0$ as n becomes large, indicating strong density dependence (i.e., if the population size at time t is large, then the population size in the next generation will be small due to density dependence). Equilibria of the dynamical system defined by G are given as intersection of the graph of the function G with the diagonal, that is, as solution of $G(n) = n$. In the example shown, this equation has two solutions, one at 0 and one at $n^* > 0$. The derivative of G at an equilibrium determines its local stability. In particular, if $c = dG/dn(n^*)$, the equilibrium n^* is locally stable if and only if $|c| < 1$. (Note that the stability of the equilibrium 0 is determined by λ, with 0 being unstable, and hence the population viable, if and only if $\lambda > 1$, as is the case in the example shown.) See text for further explanations.

where $n(t)$ is the population density at time t, and G is some (nonlinear) function. A schematic example of such a next generation map is shown in Figure 7.1. These models are one-dimensional because the population density n, that is, the state variable, is a real number, and they are set in discrete time, that is, the population density is only known at discrete time intervals t, $t + 1$, etc. They are usually thought to describe organisms with discrete life histories, for example, organisms with a well-defined reproductive period each year (in which case the time unit would be years). More generally, they apply to populations that are censused at regular time intervals such that the population density at one point in time is a function of the density at the previous census point (Turchin, 2003). Dynamical systems given by difference equations have attracted an enormous amount of attention since the 1970s of the past century, when it became widely known that these simple systems can exhibit very complicated dynamics, called chaos (Li & Yorke, 1975; May, 1974, 1976; May & Oster, 1976). Such complicated dynamical systems have first been studied by mathematicians (e.g., Sharkovskii, 1964), but caught the attention of a wide audience when it was realized that they have many diverse applications due to the nonlinear interactions prevalent in natural systems such as the weather and the human brain. In ecology, the discovery of nonlinear dynamics and chaos as an alternative explanation for population fluctuations (May, 1974, 1976) sparked enormous interest in the prevalence of ecological chaos (e.g., Hassell et al., 1976; Turchin & Taylor, 1992), as well as the general mechanisms affecting the complexity of population dynamics.

Spatial structure is generally viewed as an important determinant of ecological dynamics, and metapopulations consisting of local populations described by difference equations can be used to study the effects of dispersal on population fluctuations. For example, one can assume that the metapopulation consists of a finite number $1, \ldots, M$ of local populations whose dynamics in isolation is given by a difference equation $\tilde{n}_i(t) = G_i(n_i(t))$, where $n_i(t)$ is the population density in patch i, $i = 1, \ldots, M$ at the start of generation t, and $\tilde{n}_i(t)$ is the population density in patch i after reproduction, but before dispersal. Dispersal could then be described by assuming that after reproduction, a fraction d of each local population enters a dispersal pool, and that all individuals surviving this dispersal pool are distributed in some specified way among the M patches, thus yielding the local populations $n_i(t + 1)$ at the start of the next generation. For example, one could assume that the local populations are arranged in a linear chain, and that dispersal only occurs to the two nearest neighboring populations, so that $n_i(t + 1) = (1 - d)\tilde{n}_i(t) + (d/2)\tilde{n}_{i-1}(t) + (d/2)\tilde{n}_{i+1}(t)$ (with some suitable specifications for what happens at the boundaries of the linear chain). Thus dispersal leads to a "coupling" of the difference equations describing local population dynamics, that is, to a system of coupled difference equations. Such "coupled map lattices" have been studied extensively by theoretical physicists (e.g., Chow & Mallet-Paret, 1995; Kaneko, 1992), who discovered phenomena such as spatio-temporal chaos and the formation of complicated spatial patterns, including traveling and spiral waves. In theoretical ecology, coupled difference equations have been used to argue that spatial structure can dampen population fluctuations and lead to persistence of locally extinction-prone ecological interactions (e.g., Comins, Hassell, & May, 1992), and to temporally less complicated or even stable population dynamics when local dynamics is chaotic in isolation (Doebeli, 1995; Doebeli & Ruxton, 1998; Gyllenberg et al., 1993; Hastings, 1993).

Here I consider such two-patch models to study evolutionary branching in dispersal rates. I assume that the "next generation" map G in eq. (7.1) can be written in the form

$$G(n) = g(n) \cdot n, \tag{7.2}$$

where the nonlinear function $g(n)$ is the per capita "reproductive output" over one time unit, that is, the expected per capita number of offspring if the parent population density is n. The so-called Ricker function

$$g(n) = \lambda \exp(-qn) \tag{7.3}$$

has been widely used in this context. Here the parameter λ is the basic reproductive output, that is, the per capita reproductive output in the absence of any density effects, and the parameter q measures the strength of density dependence. The Ricker model has been widely used, but it has certain undesirable mathematical properties, for example, regarding the position of the inflection point of the function G, which can lead to spurious and nongeneric results about the effects of dispersal on the complexity of population dynamics (Doebeli, 1995). It is therefore preferable to use the more general three-parameter function

$$g(n) = \frac{\lambda}{1 + an^b} \tag{7.4}$$

(Bellows, 1981). Perhaps the most straightforward biological interpretation of the resulting population dynamical model is that starting with a population density $n(t)$ at the beginning of generation t, each individual survives competition for resources with probability $1/(1 + an^b)$. Thus, competition acts before reproduction, and mortality due to competition is density-dependent such that survival probability is a monotonically decreasing function of density. The parameters a and b describe the type and strength of density dependence (Bellows, 1981). Individuals surviving the impact of competition each produce an expected number of λ offspring, after which the parents die, so that the density of individuals at the start of the next generation is given by eq. (7.2). An equilibrium of this dynamical system is a density n^* satisfying

$$G(n^*) = g(n^*) \cdot n^* = n^* \tag{7.5}$$

(see Fig. 7.1). Nontrivial equilibria (i.e., equilibria other than $n^* = 0$) are thus given by

$$g(n^*) = 1. \tag{7.6}$$

With $g(n)$ given by (7.4), the nontrivial equilibrium is

$$n^* = \frac{(\lambda - 1)^{1/b}}{a}. \tag{7.7}$$

For the dynamics, the most basic question is whether the system returns to the equilibrium after small perturbations, that is, whether the equilibrium is (locally) stable. It is well known that local stability is generically determined by the first derivative of the next generation function G evaluated at the equilibrium, that is, by $\frac{dG}{dn}(n^*)$ (Edelstein-Keshet, 1988; see Fig. 7.1). This can be seen by observing that for $n(t)$ close to n^*, $n(t+1) - n^* \approx \frac{dG}{dn}(n^*) \cdot (n(t) - n^*)$, as is evident from a first-order Taylor expansion of the function G at the equilibrium n^* (prove this). It follows that if $|\frac{dG}{dn}(n^*)| < 1$, then $|n(t+1) - n^*| < |n(t) - n^*|$, that is, the distance to the equilibrium n^* diminishes over time. In contrast, the distance of a perturbed population to the equilibrium n^* increases over time if $|\frac{dG}{dn}(n^*)| > 1$. Accordingly, an equilibrium n^* is called locally stable if and only if $|\frac{dG}{dn}(n^*)| < 1$.

In the present situation, $|\frac{dG}{dn}(0)| = \lambda$, hence the equilibrium $n^* = 0$ is unstable as soon as $\lambda > 1$, which simply states the obvious fact that the expected number of offspring per surviving individual must be larger than one if the population is to persist (see Fig. 7.1). According to eq. (7.7) this is also the condition for a nontrivial positive equilibrium density to exist, which, in slight abuse of notation, will be called n^* henceforth. The stability of n^* is determined by the quantity

$$c = \left.\frac{dG}{dn}\right|_{n = \frac{(\lambda-1)^{1/b}}{a}} = 1 - \frac{\lambda - 1}{b}. \tag{7.8}$$

If $0 < |c| < 1$, the population dynamics exhibits a monotonic approach to the equilibrium n^* from any starting density that is greater than zero. If $-1 < c < 0$, the system exhibits damped oscillations toward the equilibrium for any starting density greater than zero. Thus, if $|c| < 1$, the equilibrium density n^* is a global attractor. As $|c|$ increases above 1, the dynamical system exhibits the famous period-doubling route to chaos, which is illustrated in Figure 7.2. In particular, for large negative values of c, the population dynamics shows irregular, apparently random fluctuations and sensitive dependence on initial conditions, which means that despite the deterministic nature of the dynamical systems, populations whose densities are very similar at one point in time will have very different densities at some future point in time. For more details on the nature of the complicated dynamics of difference equations and other chaotic systems I refer to the extensive literature on this subject (see, e.g., Collet & Eckmann, 1980; Schuster & Just, 2005 for classical treatments).

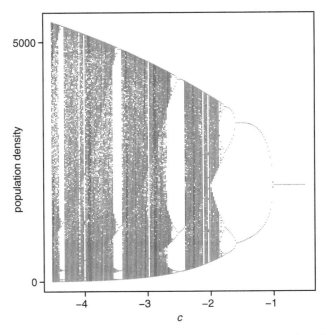

FIGURE 7.2. The period-doubling route to chaos in the difference equation $G(n+1) = g(n) \cdot n$ with $g(n)$ given by eq. (7.4). The x-axis is the parameter $c = dG/dn(n^*)$, where n^* is the nonzero solution of $G(n^*) = n^*$ given by eq. (7.7). For the plot, the parameter λ was fixed at $\lambda = 5$, and for a given value of c, eq. (7.8) was used to calculate the parameter b, and then eq. (7.7) was used to calculate the value of the parameter a under the assumption that the carrying capacity $n^* = 2000$. Using these parameter values, the dynamics was run for 1,000 transient time steps, and then the next 1,000 time steps were plotted on the y-axis. This was done for 200 equidistant c-values between -0.5 and -4.5.

Based on the local dynamics (eq. (7.2)), a two-patch metapopulation can be constructed by assuming that after reproduction has occurred in each patch, a fraction d of the offspring migrate to the other patch. In slight abuse of terminology, I call d the dispersal rate. (Strictly speaking, d is a dispersal probability, and the term "rate" is usually reserved for events occurring per unit time in continuous time models.) In addition, one can assume that dispersal has a cost, so that only a fraction s of all dispersers actually arrive at the new location. If $n_1(t)$ and $n_2(t)$ denote the population densities in the two patches at the start of generation t, and g_1 and g_2 denote the per capita reproductive output functions in the two patches, then the population densities in the two patches after reproduction, but before dispersal are given by $g_1(n_1(t)) \cdot n_1(t)$ and $g_2(n_2(t)) \cdot n_2(t)$, respectively. During the dispersal phase, a fraction $(1 - d)$ of these local populations stays in their patch, and a fraction d disperses, of which a fraction s arrives in the other patch. This leads to the following system of coupled difference equations for the discrete-time population dynamics in the two patches:

$$n_1(t + 1) = (1 - d)g_1(n_1(t)) \cdot n_1(t) + sdg_2(n_2(t)) \cdot n_2(t) \qquad (7.9)$$

$$n_2(t + 1) = (1 - d)g_2(n_2(t)) \cdot n_2(t) + sdg_1(n_1(t)) \cdot n_1(t) \qquad (7.10)$$

As mentioned, studying the dynamics of such coupled maps has been the subject of much research, which has shown, among other things, that dispersal often tends to simplify the dynamics. For example, in Doebeli (1995) I have shown that very large dispersal rates can induce stable equilibrium dynamics even when the local dynamics as determined by g_1 and g_2 are chaotic, a result that Doebeli & Ruxton (1998) have extended to metapopulations consisting of large numbers of local populations. Figure 7.3 recapitulates some results of Hastings (1993), most notably showing that intermediate dispersal rates can induce dynamically stable out-of-phase two-cycles in systems of locally chaotic maps. If, as is assumed for Figure 7.3, the local patches are equal, that is, if $g_1 = g_2$ in the dynamical system given by (7.9) and (7.10), the total density of the metapopulation remains constant on such a two-cycle, but the local populations alternate between two densities above and below the equilibrium density such that whenever one local density is high, the other one is low. Hastings (1993) also noted that larger dispersal rates tend to induce synchronous dynamics, so that the local densities in the two patches are the same at all times. As a consequence, and if there is no cost to dispersal, the dynamics of the coupled system is the same as that of an isolated patch.

To study the evolution of dispersal rates, we assume that the dispersal rate d is an evolving quantitative trait, and we use the techniques first outlined in Metz et al. (1992) to calculate, at least in principle, the invasion fitness of a rare mutant d_m in a resident population that is monomorphic for the trait value d_r. The resident is assumed to move on its population dynamical attractor, as determined by the reproductive functions g_1 and g_2, which are independent of the evolving phenotype, and by the dispersal rate d_r. This resident attractor may be very complicated, but neglecting any transient effects, it can in any case be described by the time series $(n_{r,1}(t), n_{r,2}(t))$, $t = 1, \ldots, \infty$, where $n_{r,i}(t)$ is the density of the resident in patch i at time t ($i = 1, 2$). For example, if the resident moves on an out-of-phase two-cycle as described above, we have $(n_{r,1}(t), n_{r,2}(t)) = (n_1^*, n_2^*)$ for t odd, and $(n_{r,1}(t), n_{r,2}(t)) = (n_2^*, n_1^*)$ for t even, where n_1^* and n_2^* are the two densities of the local two-cycles.

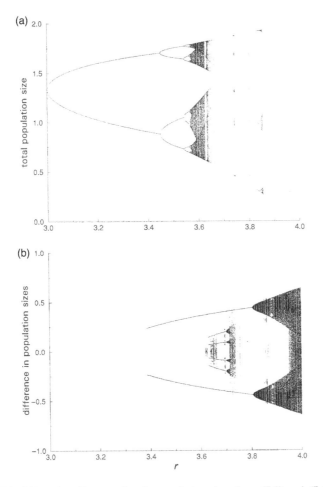

FIGURE 7.3. Bifurcation diagrams for the coupled system (eqs. (7.9) and (7.10)) with quadratic functions $g_1(x) = g_2(x) = rx(1 - x)$. The probability of dispersal is $d = 0.1$, and there is no cost to dispersal ($s = 1$). The dynamics is shown as a function of the growth parameter r (note that the quadratic functions g_1 and g_2 are biologically meaningful only for $r \leq 4$). Panel (a) shows the in-phase dynamics, in which the population density is the same in both habitat patches at all times, and in fact is the same as the dynamics in an isolated habitat patch. Note that for large values of r, the in-phase attractor tends to lose stability. Panel (b) shows the out-of-phase dynamics, which for some range of r-values coexist as alternative attractors with the in-phase dynamics. Note that there is a range of r-values for which the metapopulation has a two-cyclic out-of-phase attractor even though an isolated habitat patch with the same r-values would exhibit chaotic dynamics. Note also that different out-of-phase attractors can coexist, and that for very large r-values, the metapopulation exhibits complicated, asynchronous fluctuations in the two patches. Reproduced with permission of Ecological Society of America, from "Complex interactions between dispersal and dynamics: lessons from coupled logistic equations," Alan Hastings. *Ecology* **74**(5) 1993; permission conveyed through Copyright Clearance Center, Inc.

The rarity assumption for the mutant d_m implies that its per capita reproductive output at time t is entirely determined by the resident density, and hence equal to $g_1(n_{r,1}(t))$ in patch 1 and to $g_2(n_{r,2}(t))$ in patch 2. Therefore, during invasion of the mutant, the dynamics of its local densities $n_{m,1}$ and $n_{m,2}$ is given by

$$n_{m,1}(t+1) = (1 - d_m)g_1(n_{r,1}(t)) \cdot n_{m,1}(t) + sd_mg_2(n_{r,2}(t)) \cdot n_{m,2}(t) \quad (7.11)$$

$$n_{m,2}(t+1) = (1 - d_m)g_2(n_{r,2}(t)) \cdot n_{m,2}(t) + sd_mg_1(n_{r,1}(t)) \cdot n_{m,1}(t). \quad (7.12)$$

In matrix form, this can be written as

$$\begin{pmatrix} n_{m,1}(t+1) \\ n_{m,2}(t+1) \end{pmatrix} = G_m(t) \cdot \begin{pmatrix} n_{m,1}(t) \\ n_{m,2}(t) \end{pmatrix}, \quad (7.13)$$

where

$$G_m(t) = \begin{pmatrix} (1-d_m)g_1(n_{r,1}(t)) & sd_mg_2(n_{r,2}(t)) \\ sd_mg_1(n_{r,1}(t)) & (1-d_m)g_2(n_{r,2}(t)) \end{pmatrix}. \quad (7.14)$$

For any matrix A, let $E(A)$ be the absolute value of the dominant eigenvalue of A, that is, of the eigenvalue with the largest absolute value. Then the long-term growth rate of the rare mutant d_m in the resident d_r, that is, the invasion fitness $f(d_m, d_r)$, is given by

$$f(d_m, d_r) = \ln \left[\lim_{T \to \infty} \sqrt[T]{E \left(\prod_{t=1}^{T} G_m(t) \right)} \right] \quad (7.15)$$

(Metz et al., 1992). The adaptive dynamics of dispersal can be studied using the invasion fitness $f(d_m, d_r)$. However, analytical expressions for f, and hence for the adaptive dynamics, can generally not be obtained when the resident dynamics is complicated, and only certain simple cases can be handled analytically. For example, if the resident is on an out-of-phase two-cycles as described above, it can be proved that $\partial f(d_m, d_r)/\partial d_m(d_r, d_r) > 0$, and hence that there is selection for increased dispersal (Doebeli & Ruxton, 1997), a result that remains true as long as costs of dispersal are low enough (i.e., as long as s is close enough to one in eqs. (7.11) and (7.12)). Also, it is easy to see that if the dynamics of the two patches are synchronized, then $\partial f(d_m, d_r)/\partial d_m(d_r, d_r) \le 0$, with strict inequality if and only if there is a cost to dispersal, that is, if $s < 1$. This is intuitively obvious, for with synchronized dynamics, the density-dependent reproductive output is exactly the same in the two patches, and hence dispersal is a neutral trait, except if there is an explicit cost that selects against dispersal.

These two facts can already generate interesting evolutionary dynamics. Suppose the starting point is a metapopulation on an out-of-phase two-cycle, in which there is selection for larger dispersal rates, so that dispersal rates increase over evolutionary time. Once dispersal rates are high enough, the local dynamics of the two patches tend to become synchronized, which reverses the direction of selection, and dispersal rates start to decrease. Once they have decreased enough, the metapopulation again enters the region of stable out-of-phase attractors, and hence the direction of selection changes once again. As a result, the dispersal rate undergoes sustained evolutionary fluctuations, as was shown in Doebeli & Ruxton (1997). This is a nice example of a feedback between ecological and evolutionary dynamics: one type of ecological dynamics selects for increased dispersal, which eventually generates a different type of ecological dynamics, which in turn selects for decreased dispersal, which reverses the ecological dynamics to the original state, and so on. Thus, there are not only evolutionary fluctuations in the dispersal rate, but also in the qualitative behavior of the metapopulation dynamics.

Another case that can easily be dealt with analytically occurs when the resident population is at a stable equilibrium. If the local population densities are the same at the equilibrium, then dispersal is again neutral, except if it has a cost, in which case it is selected against. If the local population densities are different in the equilibrium state, then one can show that selection always favors lower dispersal rates in the above model. Thus, in the absence of temporal fluctuations, spatial heterogeneity selects against dispersal. Together with the fact that negatively correlated temporal fluctuations in the form of out-of-phase fluctuations favors dispersal, this result forms the basis of evolutionary branching in dispersal rates. Unfortunately, an analytical treatment of evolutionary branching in the two-patch models considered here is not feasible, and instead I resort to stochastic individual-based simulations, which will yield some noteworthy additional insights.

Challenge: Prove that $\partial f(d_m, d_r)/\partial d_m(d_r, d_r) < 0$ in resident metapopulations whose dynamics is described by eqs. (7.9) and (7.10) and exhibits a stable equilibrium with unequal local population sizes.

Based on the biological interpretation of the difference equation model (eq. (7.2)) mentioned earlier, it is straightforward to define the processes of survival, reproduction, and dispersal at an individual level. Each individual is characterized by its dispersal rate, and the numbers of individuals in the two patches at the start of generation t are denoted by N_1 and N_2, respectively.

These numbers determine the probabilities of surviving competition, which are given by $1/(1 + aN_1^b)$ in patch 1 and $1/(1 + aN_2^b)$ in patch 2. For each individual, we draw from a uniform random distribution on the interval $[0, 1]$ to determine whether the individual survives or not (it survives if the draw is smaller than the survival probability). For each surviving individual, we then draw the number of offspring from a Poisson distribution with mean λ_i, where the λ_i, $i = 1, 2$, are the basic reproductive rates in the two patches, which appear as parameter in the reproductive output functions g_1 and g_2 (see eq. (7.4)). (Note that one could use alternative probability distributions for determining the number of offspring.) Each offspring is assigned a dispersal phenotype that is equal to the parental dispersal rate with probability $1 - \mu$, where μ is the mutation probability. With probability μ, the offspring phenotype is drawn from a Gaussian distribution with mean equal to the parental dispersal rate and variance equal to σ_{mut}, a parameter describing the magnitude of mutational effects. (Draws are repeated until the draw results in a number in the feasible dispersal rate interval $[0, 1]$.) After reproduction, all parents are discarded, and each offspring stays in the patch in which it was born with probability $1 - d$, where d is the individual's dispersal phenotype. If it does not stay it moves to the other patch, but only gets there with probability s, and dies otherwise. This procedure fully determines the population sizes in the two patches at the start of generation $t + 1$, and the evolution of dispersal rates can be studied through iteration.

The first thing to note about the individual-based model concerns the eco-logical dynamics of finite populations. While the notion of a (locally) stable equilibrium is well-defined for deterministic difference equations, its signifi-cance for stochastic models is less clear. This is illustrated by the bifurcation diagram in Figure 7.4, which shows the individual-based population dynam-ics in a single habitat patch without dispersal as a function of demographic parameters. Figure 7.4 is thus the individual-based analogon of the determin-istic bifurcation diagram shown in Figure 7.2. While the stochastic model also exhibits distinct dynamic regimes as the complexity parameter is varied, the definition of the different regimes is less clear. In particular, Figure 7.4 shows that even if the deterministic difference equation exhibits a stable equilibrium, the stochastic model undergoes large fluctuations, ranging roughly between population sizes of 1,500 and 2,500 when the carrying capacity is at 2,000 in the examples shown in Figure 7.3. A similar phenomenon occurs for the eco-logical dynamics of individual-based metapopulation models with two habitat patches, in which the dispersal rate of all individuals is fixed at the same value and no mutations occur. This is illustrated by the time series shown in Figure 7.5.

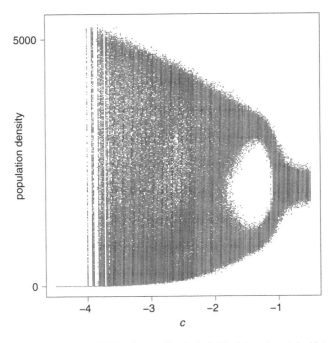

FIGURE 7.4. Same as Figure 7.2, but for stochastic individual-based models. Note that it is still possible to distinguish different dynamical regimes, with more stable population dynamics for low complexities giving way to a region with a pronounced two-cyclic component and then to more "chaotic" dynamics as the complexity is increased (i.e., as c is decreased). Also note that even in the stable regime, the population size exhibits rather large fluctuations.

It is known that demographic stochasticity can lead to persistent fluctuations of large amplitude in individual-based predator-prey models for which the corresponding deterministic model exhibits a stable equilibrium (McKane & Newman, 2005; Pineda-Krch et al., 2007; Renshaw, 1991). For difference equations of the type analyzed here, it is known that small amounts of external noise can induce a significant qualitative change in the dynamics compared to deterministic models without noise, especially near bifurcation points (Takens, 1996). However, the fact that demographic stochasticity can have similar effects in these models does not seem to be appreciated in the literature. Moreover, Figure 7.4 shows that the effects of demographic stochasticity are robust and do not depend on the deterministic model being near bifurcation points. In fact, demographic stochasticity causes persistent large fluctuations in population size even when the underlying deterministic model shows monotonic convergence to the equilibrium (see Fig. 7.5). In general, the amplitude of the fluctuations increases with increasing λ, which may not be surprising

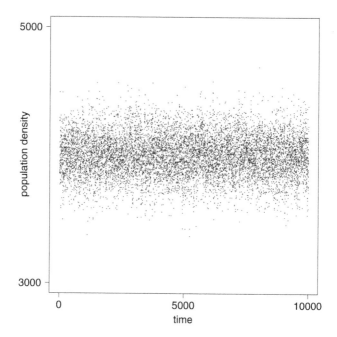

FIGURE 7.5. Time series of an individual-based model for two local population connected by dispersal. The total population size across the two patches is shown as a function of time. The two local populations are described by identical difference equations of the form used for Figures 7.2 and 7.4. The value of the complexity c was chosen to be $c = 0.5$, and the probability of dispersal was $d_m = 0.25$. Other parameter values are as in Figure 7.2. In the corresponding deterministic two-patch model, these parameters would generate monotonic convergence to the equilibrium population size (2,000 in each patch, 4,000 for the total population). However, the stochastic model exhibits large fluctuations in the total population size, which ranges roughly from 3,500 to 4,500.

considering that λ is not only the mean, but also the variance in the number of offspring produced by different individuals. Note, however, that the simulations in Figures 7.3 and 7.4 involve populations consisting of thousands of individuals, and hence individual variation in offspring number alone cannot explain the fluctuations, which are due to an interaction of this variation with nonlinear density dependence.

For the present purposes, the importance of the finding that demographic stochasticity can generate large population fluctuations lies in the fact that persistent temporal population fluctuations can generate selection for increased dispersal rates. In fact, numerical simulations of metapopulations consisting of two identical patches ($g_1 = g_2$) clearly indicate that in the individual-based model, dispersal rates can evolve away from zero even if the underlying

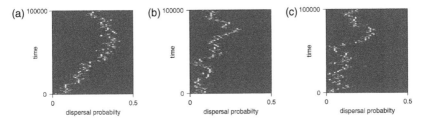

FIGURE 7.6. Evolution of dispersal in individual-based two-patch models. The plots show the dynamics of the frequency distribution of the dispersal rate, with white areas indicating high frequencies. In all panels, the underlying deterministic metapopulation model exhibits a stable equilibrium. In panel a, there is no cost to dispersal, so that dispersal is a neutral trait in the deterministic model. However, the fluctuations generated by stochasticity in the individual-based models are enough to select for high-dispersal rates. This remains true when there is an explicit cost to dispersal (panel b), or when the habitats differ in their carrying capacity (panel c). In both cases, dispersal would evolve to 0 in the underlying deterministic model. In both cases, average dispersal rates are lower than in panel a, but stochastic fluctuations can maintain dispersal at nonzero levels. Parameter values were $\lambda = 25$, $c = -0.8$, b was calculated from eq. (7.8) and a was calculated from eq. (7.7) under the assumption that the carrying capacity in the two habitat patches, given by eq. (7.7), was set to 2,000 for panels a and b, and to 2,000 and 2,500 for panel c; the dispersal survival probability was $s = 1$ in panels a and c, and $s = 0.95$ in panel b; the mutational parameters were $\mu = 0.002$ and $\sigma_{mut} = 0.02$, and initially, all individuals had a dispersal rate of 0.02 for all panels.

deterministic model has a stable equilibrium, that is, even if the selection for dispersal is neutral in the underlying deterministic model. Figure 7.6a shows an example in which the deterministic model exhibits a stable equilibrium, yet demographic stochasticity generates fluctuations in local population size that are large enough to select for high dispersal rates in the individual-based model. Even though the average rate of dispersal itself undergoes considerable fluctuations over time, it is evident from the initial 30,000 generations or so shown in Figure 7.6a that there is selection for increased dispersal over prolonged periods of time, as opposed to the neutral selection that would prevail in the underlying deterministic model.

Figure 7.6b shows an example in which nonzero dispersal rates evolve despite the fact that there is an explicit cost to dispersal, so that dispersal would be selected against in the underlying deterministic model (which again exhibits a stable ecological equilibrium). As mentioned above, selection against dispersal is also generated by differences in the habitat quality of the two patches. This is illustrated in Figure 7.6c, which is the same as Figure 7.6a, except that the patches have different carrying capacities. This difference in habitat quality would generate selection against dispersal in the underlying deterministic model (which again has stable equilibrium ecological dynamics). Indeed, in the individual-based model dispersal now evolves to lower values than in the case with equal habitats (Fig. 7.6a). Nevertheless, dispersal again evolves away from zero, as is illustrated in Figure 7.6c. These examples show that dispersal can be maintained at a level that is significantly different from zero due to demographic fluctuations alone, a fact that has been reported in a different model by (Cadet et al., 2003), but that does not seem to be widely appreciated.

To make selection for increased dispersal more robust, one can increase the dynamic complexity of the underlying deterministic model to generate larger and more complex temporal fluctuations in the individual-based simulations. If this is combined with components of selection that act against dispersal, such as habitat differences, evolutionary branching into coexisting high and low dispersal rate clusters becomes possible, as illustrated in Figure 7.7a,b. Such adaptive diversification in dispersal rates can be observed even if dispersal has substantial costs ($s < 1$ in eqs. (7.11) and (7.12)), as is illustrated in Figure 7.7c. With high-dispersal costs, the high-dispersal branch may die out some time after diversification has occurred, which constitutes another example of the phenomenon of evolution to extinction (Gyllenberg & Parvinen, 2001; Gyllenberg et al., 2002; see also Fig. 3.5 in Chapter 3). After extinction of the high dispersal branch, the low dispersal branch diversifies again, generating a cyclic evolutionary dynamics with repeated evolutionary branching and extinction (Doebeli & Ruxton, 1997). Overall, simulations of the individual-based model confirm that temporal fluctuations in local conditions can drive adaptive diversification in dispersal rates, as has previously been shown in a number of models that were based on population densities and hence implicitly assumed infinite population sizes (Doebeli & Ruxton, 1997; Holt & McPeek, 1996; Mathias et al., 2001; Parvinen, 1999, 2002). Incidentally, the individual-based simulations also show that demographic stochasticity can have rather profound effects on the evolutionary dynamics of dispersal, in particular generating selection for higher dispersal rates in scenarios in which selection in the underlying deterministic models would drive dispersal to zero.

Challenge*: Generalizing from the system given by eqs. (7.11) and (7.12), investigate the evolution of dispersal rates in metapopulations consisting of many local patches that are connected by dispersal, for example, by assuming that the patches are arranged in a square grid on which dispersal occurs between nearest neighbors. Show that such models exhibit evolutionary branching of dispersal rates.

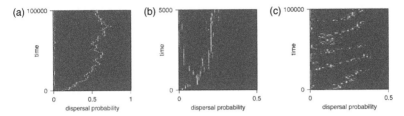

FIGURE 7.7. Evolutionary diversification of dispersal in individual-based two-patch models. The plots show the dynamics of the frequency distribution of the dispersal rate, with white areas indicating high frequencies. In all panels, the underlying deterministic metapopulation model exhibits complex dynamics, and the habitat patches have different carrying capacities. Panel a shows a case with no explicit cost to dispersal ($s = 1$), in which the opposing selective forces of (nonsynchronized) temporal fluctuations in local population size, selecting for increased dispersal, and of a difference in habitat quality, selecting for decreased dispersal, combine to generate evolutionary branching into two coexisting dispersal clusters. Panel b is a close-up of panel a during the first 5,000 generation, showing the convergence to the branching point and subsequent diversification. Panel c is the same as panel a, except that there is an explicit cost to dispersal ($s = 0.6$). With substantial costs to dispersal, the high-dispersal branch tends to evolve to extinction, leading to repeated bouts of diversification. Parameter values were $\lambda = 5$, $c = -3$, b was calculated from eq. (7.8), and a was calculated from eq. (7.7) under the assumption that the carrying capacities in the two habitat patches, given by eq. (7.7), were set to 2,000 in one patch and 5,000 in the other patch; the dispersal survival probability was $s = 1$ in panels a and b, and $s = 0.6$ in panel c; the mutational parameters were $\mu = 0.002$ and $\sigma_{mut} = 0.02$, and initially, all individuals had a dispersal rate of 0.02 for all panels.

Challenge*: Extend the individual-based simulations of systems (7.11) and (7.12) to sexual populations, using assortative mating and segregation kernels as in previous chapters. Investigate the occurrence of adaptive speciation due to evolutionary branching in dispersal rates.

Challenge*: Investigate metapopulations given by eqs. (7.11) and (7.12) under the assumption that dispersal rates may differ in the two patches. In that case, the dispersal phenotype is two-dimensional, that is, given by two numbers (d_1, d_2) describing the dispersal rate from patch 1 to patch 2 and from patch 2 to patch 1, respectively. Holt & McPeek (1996) have argued that the dispersal dimorphism consisting of high and low dispersers observed when dispersal is one-dimensional could then be replaced by a single, two-dimensional dispersal phenotype. But Parvinen (1999) has shown that evolutionary branching can occur even in two-dimensional dispersal space under suitable conditions. Both these articles have used deterministic models. Use stochastic individual-based models to confirm and extend their results.

7.2   DIVERSIFICATION IN GAMETE SIZE: EVOLUTION OF ANISOGAMY

Anisogamy refers to the differentiation in gamete size between males (the sex producing small gametes) and females (the sex producing large gametes), which is a fundamental aspect of most multicellular sexual organisms (Maynard Smith & Szathmáry, 1997). Anisogamy is at the origin of many other types of differentiation between the genders in their physiology, morphology and behavior, and of important evolutionary mechanisms such as sexual conflict and sexual selection. The prevalent explanation for the evolution of anisogamy is due to Parker et al. (1972) and is based on a trade-off between size and number of gametes produced. In conjunction with the assumption that zygote survival probabilities increase with zygote size, such a trade-off can generate frequency-dependent selection that allows for coexistence between types producing many small gametes and types producing few large gametes. More precisely, when everybody is producing many small gametes, a type producing few large gametes has an advantage, because all its offspring zygotes are large and hence have high survival probabilities. Conversely, when everybody is producing few large gametes, a type producing many small gametes has an advantage, because it will have many zygote offspring that are large enough to ensure high survival probabilities. Thus, the two types can coexist because each type has an advantage when rare.

There is some empirical support for this theory from comparative data (Bell, 1978; Randerson & Hurst, 2001a). Alternative hypotheses for the evolution of anisogamy have also been proposed (Bonsall, 2006; Hurst, 1990; Hurst & Hamilton, 1992; Randerson & Hurst, 2001b), but here I will reexamine the basic mechanisms proposed by Parker et al. (1972) in the light of adaptive dynamics and evolutionary branching. Even though it is natural to view gamete size as a continuously varying trait, only few of the existing models have looked at the evolution of anisogamy as a dynamical system in continuous trait space (e.g., Matsuda & Abrams, 1999). In fact, most of the models based on Parker et al. (1972) invoke, implicitly or explicitly, mutations of large effects to explain the evolution of anisogamy from an isogamous ancestral state. Such models essentially investigate the evolutionary stability of isogamous and anisogamous populations and can be distinguished by whether they incorporate different mating types (e.g., Bell, 1978; Bulmer & Parker, 2002; Charlesworth, 1978; Hoekstra, 1980; Matsuda & Abrams, 1999; Maynard Smith, 1982; Parker et al., 1972) or not (e.g., Bulmer & Parker, 2002; Matsuda & Abrams, 1999; Maynard Smith, 1978). The origin of two (or more) types of gametes that can only produce zygotes with gametes from the other type(s) may predate the origin of anisogamy (Hoekstra, 1987; Wiese et al., 1979), and the existence of mating types can facilitate the evolution of anisogamy under certain conditions (see Matsuda & Abrams, 1999, and later this chapter). However, when loci controlling mating types and loci controlling gamete size are unlinked, which was likely the case initially (Maynard Smith, 1982, page 49), models with mating types become mathematically equivalent to models without mating types (Matsuda & Abrams, 1999). Therefore, and to highlight the disruptive force of the size-number trade-off in gamete production without enhancing mechanisms, I consider here mainly models without mating types. The analysis essentially follows Maire et al. (2001), a paper that appeared in a short-lived and now-defunct journal.

I envisage a life cycle with discrete generations in which haploid parent individuals produce gametes that fuse with other, randomly chosen gametes to form diploid zygotes, which in turn divide meiotically to produce new parent individuals. Each parent individual is given by its trait value x, which can be thought of as an allele determining the size x of the individual's gametes. Each individual has a fixed amount of resources available to produce gametes, and the resources required to produce a gamete of size x are proportional to x. Hence the number of gametes produced by an x-individual is proportional to $1/x$, and without loss of generality we assume equality. Thus, the smaller the gametes produced, the more numerous they are. If a gamete produced by an x-individual fuses with a gamete produced by a y-individual, it forms a zygote of size $x + y$, which has a survival probability that is proportional to some positive function $S(x + y)$. If the zygote survives, it produces two parent individuals, one with trait value (allele) x and one with trait value (allele) y. Assuming that nothing else impinges on individual fitness, the reproductive success of an x-individual in a population consisting entirely of x-individuals is therefore proportional to $S(2x)/x$. On the other hand, the reproductive success of a rare mutant y in a resident population that is monomorphic for x is proportional to $S(x + y)/y$ (with the same constant of proportionality), because the mutant produces $1/y$ gametes, which all fuse with x-gametes because the mutant is rare. Assuming that population size is constant, the growth rate of a rare mutant type y relative to the resident type x, that is, the invasion fitness, is thus

$$f(x, y) = \frac{\dfrac{S(x + y)}{y}}{\dfrac{S(2x)}{x}} = \frac{xS(x + y)}{y(S(2x))}. \tag{7.16}$$

Note that $f(x, x) = 1$ as it should be, because 1 is the "neutral" growth rate in discrete time. Given the invasion fitness, the analysis follows standard adaptive dynamics theory. First, the invasion fitness determines the selection gradient $D(x)$ as

$$D(x) = \left.\frac{\partial f(x, y)}{\partial y}\right|_{y=x} = \frac{S'(2x)}{S(2x)} - \frac{1}{x}, \tag{7.17}$$

where $S'(2x)$ is the first derivative of S evaluated at $2x$. Thus, the adaptive dynamics of the trait x are given by

$$\frac{dx}{dt} = D(x) = \frac{S'(2x)}{S(2x)} - \frac{1}{x}. \tag{7.18}$$

(As usual, strictly speaking, the right-hand side of this expression needs to be multiplied by a quantity describing the mutational process in the resident x, which depends on the genetics of the trait x as well as on the population size of the resident. However, since population size is constant, we can assume that this quantity is a constant as well, which we can set equal to one by rescaling time.)

One then finds the equilibrium points of the adaptive dynamics as solutions x^* of $D(x^*) = 0$, that is, of the implicit equation

$$x^* = \frac{S(2x^*)}{S'(2x^*)}. \tag{7.19}$$

An equilibrium x^* is convergence stable if and only if

$$\frac{dD}{dx}(x^*) = \frac{2S''(2x^*)}{S(2x^*)} - \frac{1}{x^{*2}} < 0 \tag{7.20}$$

(where eq. (7.19) is used to evaluate the second derivatives in dD/dx). Finally, an equilibrium x^* is evolutionarily stable if and only if

$$\left.\frac{\partial^2 f}{\partial y^2}\right|_{y=x=x^*} = \frac{S''(2x^*)}{S(2x^*)} < 0. \tag{7.21}$$

Since S is a positive function, $S(2x^*) > 0$ in any case. Therefore, a convergent stable equilibrium x^* satisfying $S''(2x^*) > 0$ is an evolutionary branching point, and hence a potential starting point for evolutionary diversification in gamete size, that is, for the evolution of anisogamy.

If the survival function S is linear, $S(z) = a + bz$, then we must assume $b > 0$ (so that survival increases as a function of zygote size), and it easily follows that the adaptive dynamics (7.18) has a unique equilibrium $x^* = -a/b$. If $a > 0$ this equilibrium is < 0 and hence not in the range of biologically feasible gamete sizes. In this case $D(x) < 0$ for all feasible gamete sizes x, and hence x will evolve to the smallest value that is biologically possible. If $a \leq 0$, the equilibrium x^* is also ≥ 0, and condition (7.20) immediately shows that it is an attractor for the adaptive dynamics. However, condition (7.21) then shows that the linear case is degenerate in the sense that the second derivative of the invasion fitness at x^* is zero. In particular, the equilibrium x^* is evolutionarily neutral (i.e., neither stable nor unstable, so that in a resident at x^*, all mutants have the same relative growth rate of one as the resident).

This degeneracy disappears with quadratic survival functions $S(z) = a + bz + cz^2$. In this case, eq. (7.19) yields $x^* = -a/b$ as unique equilibrium of the adaptive dynamics, which is biologically feasible only if a and b have different signs. Evaluating condition (7.20) shows that convergence stability of the equilibrium requires $ab^2 < 0$, which is equivalent to $a < 0$. Thus, convergence stable and biologically feasible attractors of the adaptive dynamics exist whenever $a < 0$ and $b > 0$. Note that if $a > 0$ and $b < 0$, a biologically feasible equilibrium still exists, but instead of an attractor it is a repellor. In this case, if the evolutionary dynamics is initialized at resident values smaller than x^*, gamete size would decrease over evolutionary time, whereas gamete size would increase (to the maximal biologically feasible value) if the adaptive dynamics was started above x^*.

Condition (7.21) immediately shows that a biologically feasible equilibrium x^* is evolutionarily unstable if $c > 0$, that is, if survival is an accelerating function of zygote size. Thus, with quadratic zygote survival functions $S(z) = a + bz + cz^2$, the adaptive dynamics has an evolutionary branching point whenever $a < 0$, $b > 0$ and $c > 0$. Assuming these conditions to be satisfied, two phenotypes that are close enough and lie on either side of the evolutionary branching point can coexist (see Appendix). To investigate evolution after the system has reached the branching point, one therefore needs to consider the two-dimensional adaptive dynamics of two coexisting lineages. Let x_1 and x_2 be two coexisting resident phenotypes, where we can assume $x_2 < x^* < x_1$. To derive the two-dimensional adaptive dynamics, we need to determine the invasion fitness functions $f(x_1, x_2, y)$ of rare mutants y appearing in the residents (x_1, x_2). For this, we first need to determine the coexistence equilibrium of x_1 and x_2.

Let p be the frequency of x_1 in a population containing only the phenotypes x_1 and x_2. Since the numbers of gametes produced by the two types are $1/x_1$

and $1/x_2$, the frequency q of x_1 among gametes is given by

$$q = \frac{p\dfrac{1}{x_1}}{p\dfrac{1}{x_1} + (1-p)\dfrac{1}{x_2}} = \frac{p}{p + (1-p)\dfrac{x_1}{x_2}}. \tag{7.22}$$

A gamete produced by x_1 forms a zygote with another x_1-gamete with probability q and a zygote with an x_2-gamete with probability $1 - q$. Therefore, the reproductive output of an x_1-individual is

$$(1/x_1)\,(qS(2x_1) + (1-q)S(x_1 + x_2)). \tag{7.23}$$

Similarly, the reproductive output of an x_2-individual is

$$(1/x_2)\,(qS(x_1 + x_2) + (1-q)S(2x_2)). \tag{7.24}$$

At the (ecological) coexistence equilibrium these two reproductive outputs must be the same, which, after some calculation, yields an equilibrium frequency of

$$\tilde{p} = \frac{x_1\,(-a - x_2(b + c(x_2 - x_1)))}{(x_1 - x_2)(2cx_1x_2 - a)}. \tag{7.25}$$

(Note that it is easy to see that $0 < \tilde{p} < 1$ for $x_2 < x^* < x_1$ and x_1 and x_2 close enough to branching point $x^* = -a/b$.)

Now consider a mutant y appearing in a resident population consisting of x_1 and x_2 that is at its (ecological) equilibrium $\tilde{p}$. When the mutant is rare, it only forms zygotes with resident gametes, and by similar arguments as above, the reproductive output of a mutant individual is

$$\frac{1}{y}\,(\tilde{q}S(x_1 + y) + (1 - \tilde{q})S(x_2 + y)), \tag{7.26}$$

where $\tilde{q} = \tilde{p}/\,(\tilde{p} + (1 - \tilde{p})(x_1/x_2))$ is the proportion of x_1-gametes produced at the equilibrium $\tilde{p}$. Therefore, the invasion fitness for mutants appearing in the resident (x_1, x_2) is

$$f(x_1, x_2, y) = \frac{\dfrac{1}{y}\,(\tilde{q}S(x_1 + y) + (1 - \tilde{q})S(x_2 + y))}{\dfrac{1}{x_1}\,(\tilde{q}S(2x_1) + (1 - \tilde{q})S(x_1 + x_2))}$$

$$= \frac{\dfrac{1}{y}\,(\tilde{q}S(x_1 + y) + (1 - \tilde{q})S(x_2 + y))}{\dfrac{1}{x_2}\,(\tilde{q}S(x_1 + x_2) + (1 - \tilde{q})S(2x_2))}. \tag{7.27}$$

(The denominators in these expressions are the same because the resident is at its equilibrium $\tilde{p}$.) The adaptive dynamics in the resident strains x_1 and x_2 are then given by

$$\frac{dx_1}{dt} = D_1(x_1, x_2) = \left.\frac{\partial f}{\partial y}\right|_{y=x_1} \tag{7.28}$$

$$\frac{dx_2}{dt} = D_2(x_1, x_2) = \left.\frac{\partial f}{\partial y}\right|_{y=x_2} \tag{7.29}$$

where D_1 and D_2 are the selection gradients in the two resident strains. We again assume that the mutational processes in the two resident strains are the same, and hence can be scaled to one in both strains simultaneously. With quadratic zygote survival functions, elementary calculations yield

$$D_1(x_1, x_2) = \frac{c(x_1 - x_2)(b + c(x_1 + x_2))}{x_1(-2ac + (b + cx_1)^2 + 2cx_2(b + 3cx_1) + c^2 x_2^2)} \tag{7.30}$$

$$D_2(x_1, x_2) = -\frac{c(x_2 - x_1)(b + c(x_1 + x_2))}{x_2(-2ac + (b + cx_1)^2 + 2cx_2(b + 3cx_1) + c^2 x_2^2)} \tag{7.31}$$

Note that $D_2 = -D_1$. Using that $x_1 > x_2 > 0$, as well as $a < 0$ and $b, c > 0$, it follows that $D_1(x_1, x_2) > 0$ and $D_2(x_1, x_2) < 0$ for all biologically feasible x_1 and x_2. Thus, x_1 will increase evolutionarily to become as large as possible, whereas x_2 will decrease to become as small as possible. The result of this process is obviously anisogamy. Note that even with strong anisogamy, there are in principle three types of zygotes. However, two of these should be rare. On the one hand, with random zygote formation, zygotes formed by two large gametes will be rare because large gametes will be rare compared to small gametes. On the other hand, random zygote formation will often result in zygotes being formed by two small gametes, but these zygotes will be very small and hence have a very low chance of surviving. Thus, most of the surviving zygotes will be formed by one small gamete and one large gamete. Incidentally, this situation sets the stage for the evolution of mating types in a reinforcement-like process, resulting in each mating type producing one type of gamete. This has been discussed in Parker (1978) and Charlesworth (1978) and will not be pursued here.

*Challenge**: Construct models for the evolution of mating types based on evolutionary branching in gamete size.

In this context, it is interesting to note that if one assumes the a priori existence of two mating types, then the evolution of anisogamy due to the size-number trade-off in gamete production becomes inevitable if the mating type loci are tightly linked to the loci determining gamete size. This has been pointed out by Matsuda & Abrams (1999) and can be seen by noting that under these assumptions, evolution of gamete size in the two mating types is described by two different equations, one for each mating type. In fact, a slight modification of the approach outlined before yields the adaptive dynamics of the gamete sizes x_1 and x_2 in the two mating types as

$$\frac{dx_1}{dt} = \frac{S'(x_1 + x_2)}{S(x_1 + x_2)} - \frac{1}{x_1} \tag{7.32}$$

$$\frac{dx_2}{dt} = \frac{S'(x_1 + x_2)}{S(x_1 + x_2)} - \frac{1}{x_2}. \tag{7.33}$$

This dynamical system has the isogamous equilibrium (x^*, x^*) given implicitly by $x^* = S(2x^*)/S'(2x^*)$, that is, by the same equation as the singular point of the one-dimensional adaptive dynamics (7.18). While this equilibrium is often convergent stable for the adaptive dynamics without mating types (see earlier), it is never stable for the adaptive dynamics (7.32) and (7.33) of the system with two mating types. Indeed, the Jacobian matrix of this system at the equilibrium (x^*, x^*) is

$$\begin{pmatrix} \dfrac{S''(2x^*)}{S(2x^*)} & \dfrac{S''(2x^*)}{S(2x^*)} - \dfrac{1}{x*^2} \\ \dfrac{S''(2x^*)}{S(2x^*)} - \dfrac{1}{x*^2} & \dfrac{S''(2x^*)}{S(2x^*)} \end{pmatrix} \tag{7.34}$$

This matrix has an eigenvector $(-1, 1)$ along the anti-diagonal with positive eigenvalue $1/x*^2$, independent of the zygote survival function S. Therefore, whenever the isogamous equilibrium (x^*, x^*) exists, it is convergence unstable in the direction of the antidiagonal. This implies that any small asymmetry in gamete size between the two mating types is reinforced and will lead to larger asymmetries, that is, anisogamy (Matsuda & Abrams, 1999). Thus, with the basic ecological setup of Parker et al. (1972), isogamy is impossible with two mating types when mating type and gamete size are tightly linked.

Going back to models without mating types, it is of course possible to analyze the adaptive dynamics of gamete size based on survival functions S other than the quadratic ones used before. For example, assume $S(z) = z^n$, $n > 0$, as in Parker et al. (1972). In the degenerate case $n = 2$, every phenotype $x > 0$ is singular (i.e., the selection gradient is zero everywhere), and every point is evolutionarily unstable. In general, with such survival functions, eq. (7.17) yields the selection gradient as $D(x) = (1/x)(n/2 - 1)$. Hence $D(x)$ is

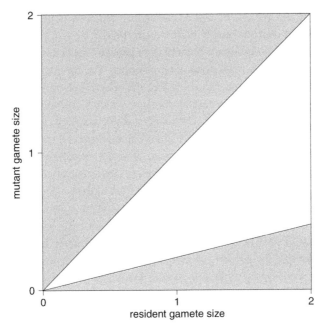

FIGURE 7.8. Pairwise invasibility plot for gamete size based on the invasion fitness function $f(x, y) = (x + y)^3 / 8x^2 y$ (eq. (7.16)) resulting from the size survival function $S(z) = z^3$. Gray and white areas indicate resident-mutant pairs for which the mutant has an invasion fitness of >1 and <1, respectively. Since the area above the diagonal is gray, there is global directional selection for increased gamete size. However, for any given resident, large mutations leading to substantially smaller gamete size than the resident can also invade, as indicated by the gray area in the lower right corner (which is bounded by the line $y = (\sqrt{5} - 2)x$). It follows that dimorphic coexistence of large and small gamete types is possible, but requires large mutations.

always negative if $n < 2$, in which case gamete size evolves to the smallest possible size. Conversely, if $n > 2$ the selection gradient is always greater than zero and gamete size evolves to the largest possible value. In particular, the survival functions used by Parker et al. (1972) cannot generate evolutionary branching. Note, however, that the selection gradient (eq. (7.17)) only predicts the direction of selection for small mutations, and that coexistence between types that are sufficiently different is possible. In fact, if $n > 2$, so that there is always directional selection for larger gamete size and mutant types that are smaller and sufficiently close to the resident cannot invade, invasion is possible for mutants that are smaller than, but sufficiently different from the resident. This is illustrated by the pairwise invasibility plot in Figure 7.8. Indeed, this is the basis for anisogamy in the original model of Parker et al. (1972), in which anisogamy required large mutations of a magnitude that depends on n (Charlesworth, 1978).

Nevertheless, evolutionary branching in gamete size is possible with many other zygote survival functions. For example, for the saturating function $S(z) = a - \exp(-z^2)$, one can show that the adaptive dynamics (7.18) has a convergent stable singular point if $0 < a < 1$, which is an evolutionary branching point if $a > 1/2$, and an evolutionarily stable strategy (ESS) if $a < 1/2$. More complicated evolutionary dynamics can be observed for zygote survival functions that have an intermediate maximum, for example, because zygotes that are too large suffer some disadvantages in terms of maintenance. For example, with a Gaussian survival function $S(z) = \exp[-(z - z_0)^2/2\sigma^2]$, one can show that there is no singular gamete size if $z_0 < 8\sigma^2$, but two singular points exist simultaneously if $z_0 > 8\sigma^2$. In that case one singular point is a repellor located below z_0, and the other is an attractor located above z_0. This implies that the adaptive dynamics of gamete size depends on the initial conditions, converging to the lowest possible gamete size if started below the repellor, and converging to the attractor if started above the repellor. Moreover, the attractor is an evolutionary branching point if $z_0 > 9\sigma^2$. I note that an analytical treatment of the adaptive dynamics after evolutionary branching, as carried out above for the case of quadratic S, is not feasible in general, because the ecological equilibrium of two coexisting gamete size types can generally not be obtained analytically.

Challenge: Prove the claims made above for the zygote survival functions $S(z) = a - \exp(-z^2)$ and $S(z) = \exp[-(z - z_0)^2/2\sigma^2]$. Investigate the adaptive dynamics resulting from these survival functions after evolutionary branching. Is it possible that there are further branching points, so that more than two gamete size types can coexist?

Overall, it is clear that the size-number trade-off in gamete production, in conjunction with increasing zygote survival over at least some range of increasing zygote size, can readily give rise to evolutionary branching in gamete size, that is, the evolution of anisogamy. This seems to be an attractive alternative explanation to the models based on Parker et al. (1972), which require mutations of large effects. If true, the evolution of anisogamy through evolutionary branching would also refute John Maynard Smith's claim that "males were the first sex" (Maynard Smith, 1982). His remark was based on the observation that in the models he studied, "primitive" conditions would generate selection for ever-smaller gametes. Upon a change in conditions, that is, in the zygote survival function, populations of "males" producing small gametes could then be invaded by "females" producing sufficiently large gametes. With evolutionary branching, initial populations, whether consisting of "females" producing large gametes or of "males" producing small gametes, would first evolve to the branching point, which has some intermediate gamete size and hence represents a "neutral sex," and would only subsequently split into coexisting "males" and "females."

7.3 DIVERSIFICATION IN TROPHIC PREFERENCE: EVOLUTION OF COMPLEXITY IN ECOSYSTEMS

Ecosystems typically consist of many different trophic levels, with the lowest level being inorganic matter, and species at higher levels engaging in complex networks of trophic interactions. Understanding the evolution of such ecological networks is a fundamental problem. In particular, it is interesting to ask how more complex ecosystems can emerge evolutionarily from simpler ones. Troost et al. (2005b) have addressed this problem by studying the evolution of specialization in mixotrophic plankton. Mixotrophs are characterized by their ability to use both inorganic and organic resources for growth, that is, for the production of biomass. In contrast, autotrophic species only use inorganic matter for growth, while heterotrophs only use organic resources. Mixotrophy appears to be common in planktonic protists, which play important roles in aquatic food webs (Troost et al., 2005b). Because mixotrophy allows complete recycling of nutrients, a mixotrophic species together with its inorganic nutrients form the simplest possible ecosystem. Troost et al. (2005b) investigated how evolutionary branching in the trophic preference of the mixotroph could generate a more complicated ecosystem consisting of an autotrophic and a heterotrophic species (as well as inorganic matter), in which the heterotrophic species represents a new trophic level whose members do not feed at the lowest (inorganic) level.

The ecological framework for mixotrophy in which the evolutionary analysis of Troost et al. (2005b) unfolds is rather complicated and based on the physiological mechanisms for the use of energy and nutrients. Without going into the detailed derivation, which can be found in Troost et al. (2005b) and Kooijman (2000), the ecological model can be described as follows. It contains four state variables: X_C is the concentration of inorganic carbon, X_N is the concentration of inorganic nitrogen, X_D is the biomass of organic carbon and nitrogen contained in detritus (i.e., in dead plankton), and X_V is the biomass of organic carbon and nitrogen contained in live plankton. X_V is a proxy for the population density of plankton. The ecological dynamics of these quantities are determined by the affinities of the plankton for inorganic and organic nutrients, denoted by j_A and j_H, where the subscripts stand for autotrophy (A) and heterotrophy (H). The evolving plankton trait is trophic preference, denoted by ρ. This trait can vary in the interval [0, 1] and affects the affinities j_A and j_H. For a given resident plankton strain with preference ρ, these affinities are given by

$$j_A = \rho f(X_C) \tag{7.35}$$

$$j_H = (1 - \rho)g(X_D), \tag{7.36}$$

where f and g are monotonously increasing and saturating positive functions of their argument. These functions essentially represent the plankton's functional response describing consumption of carbon in inorganic and organic form. The analytical expression for these functions is rather complicated and given in Troost et al. (2005b), but not needed for the present purposes. Note that the functional responses f and g do not depend on ρ. Thus, the trait ρ only influences the relative preference for carbon in organic and inorganic form: the more autotrophic a plankton strain is (i.e., the higher its ρ-value), the higher its affinity for inorganic carbon and the lower its affinity for organic, and vice versa for more heterotrophic plankton strains (i.e., for lower ρ-values). Thus, eqs. (7.35) and (7.36) incorporate a trade-off between autotrophy and heterotrophy.

The ecological dynamics is then given by the four equations

$$\frac{dX_D}{dt} = X_V \left(-c_1 j_H + d\right) \tag{7.37}$$

$$\frac{dX_C}{dt} = X_V \left(c_1 j_H - y(\rho)(j_A + j_H) + k\right) \tag{7.38}$$

$$\frac{dX_N}{dt} = X_V \left(c_1 j_H - y(\rho)(j_A + j_H) + k\right) n \tag{7.39}$$

$$\frac{dX_V}{dt} = X_V \left(y(\rho)(j_A + j_H) - k - d\right) \tag{7.40}$$

Here c_1, k, d, and n are parameters, and $y(\rho)$ is a function whose meaning will become clear shortly.

These equations have the following explanations. First, organic detritus biomass is consumed by the plankton and converted into inorganic carbon and nitrogen at a rate $X_V c_1 j_H$, where the parameter c_1 is a conversion coefficient, so that $c_1 j_H$ is the per capita rate of heterotrophic consumption by the plankton. On the other hand, detritus is generated by death of plankton at a total rate $d X_V$, where d is the per capita death rate. These two processes define the dynamics of X_D as described by eq. (7.37).

Second, inorganic carbon is produced by heterotrophy at a total rate $X_V c_1 j_H$, and then consumed and converted into plankton biomass at a total rate of $X_V y(\rho)(j_A + j_H)$, which incorporates both autotrophic and heterotrophic growth. Thus, the model describes heterotrophic consumption as first decomposing detritus into inorganic carbon, which is then converted into plankton biomass. Autotrophic consumption is an additional term describing conversion of inorganic carbon into biomass. Both these conversions occur with an efficiency given by the function $y(\rho)$, and it is assumed that this efficiency depends on the trait ρ in such a way that it incorporates a cost to mixotrophy. In other words, it is assumed that $y(\rho)$ has a (unique) minimum in the interval $[0, 1]$. In addition, it is also assumed that plankton individuals produce inorganic carbon at a constant per capita rate k due to maintenance. These three processes, production of inorganic carbon through heterotrophy, consumption of inorganic carbon through heterotrophy and autotrophy, and production of inorganic carbon due to planktonic maintenance, generate eq. (7.38).

Equation (7.39), describing the dynamics of the density of inorganic nitrogen, is redundant: it is the same as the second equation, multiplied by a constant n, which is the number of nitrogen atoms per carbon atom in biomass (Troost et al., 2005b). Thus, the equation is redundant because the model assumes that the ratio of nitrogen to carbon is the same at all times.

Finally, eq. (7.40) describing the dynamics of the plankton biomass X_V simply follows by bookkeeping from the first two equations: inorganic carbon is converted into plankton biomass at a per plankton capita rate of $y(\rho)(j_A + j_H)$, and biomass is lost at per capita rates k and d due to maintenance and death. In fact, the fourth equation is also redundant, because it is assumed that the ecosystem is closed for mass, so that the per capita rate of change in plankton biomass is simply equal to minus the sum of the per capita rates of change in detritus and inorganic carbon. However, the per capita rate of change of plankton biomass, $y(\rho)(j_A + j_H) - k - d$, will come in handy as the invasion fitness to be used to derive the adaptive dynamics of the trait ρ.

To determine the evolutionary dynamics of ρ one first needs to find the equilibrium $(\hat{X}_D(\rho), \hat{X}_C(\rho), \hat{X}_N(\rho), \hat{X}_V(\rho))$ of the ecological dynamics (7.37)–(7.40). As mentioned, only the first two equations, and the equilibrium is actually determined by a solution $(\hat{X}_D(\rho), \hat{X}_C(\rho))$ of the two equations

$$-c_1 j_H(\hat{X}_D(\rho)) + d = 0 \qquad (7.41)$$

$$c_1 j_H - y(\rho)\left(j_A(\hat{X}_C(\rho)) + j_H(\hat{X}_D(\rho))\right) + k = 0. \qquad (7.42)$$

Obtaining an analytical expression for the equilibrium $(\hat{X}_D(\rho), \hat{X}_C(\rho))$ is not possible in general. However, it is intuitively clear that $\hat{X}_D(\rho)$ should be an increasing function of ρ, that is, $\hat{X}'_D(\rho) > 0$, because higher degrees of autotrophy (higher ρ) should increase the equilibrium biomass of organic detritus. Similarly, $\hat{X}_C(\rho)$ should be an decreasing function of ρ, that is, $\hat{X}'_D(\rho) < 0$, because higher degrees of autotrophy (higher ρ) should decrease the equilibrium amount of inorganic carbon.

Given a monomorphic resident plankton population that is at the ecological equilibrium $(\hat{X}_D(\rho), \hat{X}_C(\rho))$, the per capita growth rate of a rare mutant type ρ_{mut}, that is, the invasion fitness $F(\rho, \rho_{mut})$, is obtained from eq. (7.40) as

$$F(\rho, \rho_{mut}) = y(\rho_{mut})(j_{A,mut}(X_C) + j_{H,mut}(X_D)) - k - d. \qquad (7.43)$$

Here the affinities of the mutant types $j_{A,mut}$ and $j_{H,mut}$ are defined in analogy to eqs. (7.35) and (7.36):

$$j_A = \rho_{mut} f(X_C) \qquad (7.44)$$

$$j_H = (1 - \rho_{mut}) g(X_D), \qquad (7.45)$$

where f and g are the functional responses. As long as the mutant ρ_{mut} is rare, the arguments for these functions in (7.43) are $X_C = \hat{X}_C(\rho)$ and $X_D = \hat{X}_D(\rho)$, where ρ is the resident trait value.

To proceed with analyzing the adaptive, it is convenient to specify the function $y(\rho)$ explicitly. In the simplest case considered in Troost et al. (2005b), this function is of the form $y(\rho) = 1/\tilde{y}(\rho)$, where $\tilde{y}(\rho)$ is a quadratic polynomial, and for the present purposes, we assume that

$$y(\rho) = \frac{c_2}{\rho(1 - \rho)}, \qquad (7.46)$$

where c_2 is a parameter. This is not exactly the form derived in Troost et al. (2005b), but this function has the desired property of having an intermediate minimum and will serve the purpose of showing that evolutionary branching is possible in the type of models defined above. With this choice of the function y, the invasion fitness becomes

$$F(\rho, \rho_{mut}) = \frac{c_2}{1 - \rho_{mut}} f(\hat{X}_C(\rho)) + \frac{c_2}{\rho_{mut}} g(\hat{X}_D(\rho)) - k - d. \qquad (7.47)$$

The selection gradient derived from this invasion fitness is

$$D(\rho) = \left. \frac{\partial F}{\partial \rho_{mut}} \right|_{\rho_{mut}=\rho} = \frac{c_2}{(1 - \rho)^2} f(\hat{X}_C(\rho)) - \frac{c_2}{\rho^2} g(\hat{X}_D(\rho)), \qquad (7.48)$$

and the adaptive dynamics of the trait ρ is

$$\frac{d\rho}{dt} = mD(\rho), \tag{7.49}$$

where m describes the mutational process. For simplicity, we assume that $m = 1$ at all times.

As usual, singular points of the adaptive dynamics (eq. (7.47)) are solutions ρ^* of $D(\rho^*) = 0$, that is, of $(1 - \rho^*)^2 g(\hat{X}_D(\rho^*)) = \rho^{*2} f(\hat{X}_C(\rho^*))$. This equation has solutions $\rho^* \in (0, 1)$ under general conditions (Troost et al., 2005b), and assuming that ρ^* is such a solution, its convergence stability is determined by

$$\left.\frac{dD}{d\rho}\right|_{\rho=\rho^*} = \frac{2c_2}{(1 - \rho^*)^3} f(\hat{X}_C(\rho^*)) + \frac{2c_2}{\rho^{*3}} g(\hat{X}_D(\rho^*))$$

$$+ \frac{c_2}{(1 - \rho^*)^2} f'(X_C(\rho^*))X_C'(\rho^*) - \frac{c_2}{\rho^{*2}} g'(X_D(\rho^*))X_D'(\rho^*). \tag{7.50}$$

On the other hand, the evolutionary stability of a singular point ρ^* is determined by

$$\left.\frac{\partial^2 F(\rho, \rho_{mut})}{\partial \rho_{mut}^2}\right|_{\rho_{mut}=\rho=\rho^*} = \frac{2c_2}{(1 - \rho^*)^3} f(\hat{X}_C(\rho^*)) + \frac{2c_2}{\rho^{*3}} g(\hat{X}_D(\rho^*)). \tag{7.51}$$

Clearly, this second derivative of the invasion fitness function is always >0, and so any singular point is always evolutionarily unstable. Essentially, this is a consequence of the assumption of costs for mixotrophy, as implemented by the function $y(\rho)$, which has an intermediate minimum and increases toward both margins of the trait interval $(0, 1)$.

To see that singular points can in principle be convergence stable, first note that the expression for convergence stability can be written as

$$\left.\frac{dD}{d\rho}\right|_{\rho=\rho^*} = \left.\frac{\partial^2 F(\rho, \rho_{mut})}{\partial \rho_{mut}^2}\right|_{\rho_{mut}=\rho=\rho^*} + \frac{c_2}{(1 - \rho^*)^2} f'(X_C(\rho^*))X_C'(\rho^*)$$

$$- \frac{c_2}{\rho^{*2}} g'(X_D(\rho^*))X_D'(\rho^*). \tag{7.52}$$

In addition, note that $X_C'(\rho^*) < 0$ and $X_D'(\rho^*) > 0$, as mentioned above. Together with the fact that $f'(X_C(\rho^*)) > 0$ and $g'(X_D(\rho^*)) > 0$ by assumption and construction (Troost et al., 2005b), it follows that

$$\left.\frac{dD}{d\rho}\right|_{\rho=\rho^*} = \left.\frac{\partial^2 F(\rho, \rho_{mut})}{\partial \rho_{mut}^2}\right|_{\rho_{mut}=\rho=\rho^*} - Q, \tag{7.53}$$

where Q is a positive quantity. Therefore, in principle, it is possible that $\frac{dD}{d\rho}$ is negative at the singular point, that is, that the singular point is convergent stable.

The exact model equations of Troost et al. (2005b), and in particular the exact equations for the functions f, g, and y, are actually rather complicated, and we leave it as an exercise to the reader to investigate whether the model formulated before exhibits evolutionary branching for the functions used in Troost et al. (2005b).

> *Challenge*: Using the functions $f(X_C)$, $g(X_D)$, and $y(\rho)$ specified in Troost et al. (2005b), show that the adaptive dynamics of the trait ρ can exhibit evolutionary branching.

Another difference between the model formulated earlier and the model in Troost et al. (2005b) is that in their model, phenotype space was actually two-dimensional, with two preferences ρ_A for autotrophy and ρ_H for heterotrophy. Thus, Troost et al. (2005b) did not assume a trade-off of the form $\rho_A = \rho$ and $\rho_H = 1 - \rho$. Nevertheless, they showed that evolutionary branching in trophic preferences is possible in their model, leading to coexistence of autotrophs and heterotrophs, and in particular to the establishment of a new trophic level. The same authors have extended their model to spatially structured populations, in which there is a gradient of light intensity, and hence of energy along a water column, and in which there is transport of biomass and nutrients along the water column (Troost et al., 2005a). In this spatial model the phenotype space is assumed to be one-dimensional as in the model outlined here, and the authors concluded that spatial structure actually facilitates evolutionary branching into autotrophic and heterotrophic plankton species. Interestingly, they found that high concentrations of inorganic nutrients made evolutionary branching in trophic preference more likely, whereas mixotrophy is a more likely outcome when inorganic nutrients are scarce. This conforms with the intuition that complex ecosystems should be more likely to evolve in nutrient-rich environments.

CHAPTER EIGHT

Cultural Evolution: Adaptive Diversification in Language and Religion

Evolution can occur whenever there are units of reproduction that produce other such units that inherit some characteristics of the parent units. If the units of reproduction vary in their reproductive output, there will be evolutionary change. "Intellectual content" can satisfy these simple requirements. An idea or a theory can be viewed as a unit living in the brain of an individual human (or animal). It can mutate within that brain, and it can be passed on to the brains of other individuals, thereby reproducing itself (typically with modification). For a multitude of potential reasons, some ideas and theories are more successful at such reproduction through transmission than others, hence there is typically differential reproductive success. As a consequence, there is cultural evolution of intellectual content such as ideas and theories.

Based on the notion of "meme," this perspective has been very lucidly advocated by Richard Dawkins (Dawkins, 1976). Since then, the study of cultural evolution has entered main stream science, and a number of excellent books have been written on the subject (e.g., Boyd & Richerson, 1985; Cavalli-Sforza & Feldman, 1981; Richerson & Boyd, 2005). Nevertheless, the notion of "meme" as the unit of cultural or religious content remains somewhat controversial. For example, the use of memes in cultural evolution is sometimes criticized because memes are supposedly discrete entities that replicate faithfully, but it is often hard to identify such discrete units in real cultural contexts, and a more flexible interpretation of cultural memes would be appropriate, for example, by considering continuously varying memes (such as the length of an arrow, Henrich et al., 2008). Also, such quantitative memes are not necessarily replicated faithfully, since the copy of a meme x that is transmitted from one human (e.g., the teacher) to another (e.g., the pupil) might have type $y \neq x$. In fact, transmission of cultural content may be a complicated process, in which the meme colonizing a human brain may be some weighted average of memes from other humans, and in which the acquisition of new memes by humans is biased and constrained by cognitive processes.

In traditional evolutionary terms, the issue of whether memes are categorical and discrete, or quantitative, is similar to the issue of whether one should use evolutionary models with discrete or continuous trait values, and of course, depending on the problem studied, either approach is feasible. Similarly, the issue of the modes of cultural transmission would, in classical evolutionary biology, roughly correspond to questions about the modes of reproduction and inheritance, and in principle, evolutionary models are not restricted to certain classes of reproduction and inheritance. For example, if cultural reproduction, that is, transmission, occurs through an averaging process, this would correspond to a process of blending inheritance (Cavalli-Sforza & Feldman, 1981). Henrich et al. (2008) have recently clarified a number of misconceptions about the theory of cultural evolution, and I think that using a generalized notion of memes as quantitative cultural content is a promising approach to studying cultural content.

In fact, and even though there is a rather large literature on cultural evolution, my impression is that this perspective has so far not been exploited to the fullest extent. Most often, when cultural evolution is conceptualized, the reproducing units are not the units of cultural content themselves, but instead the (typically human) units of reproduction that carry the cultural content. For example, in the models for the evolution of language studied by Martin Nowak and his colleagues (e.g., Mitchener & Nowak, 2003; Nowak & Krakauer, 1999), the evolutionary dynamics of language is determined by the reproductive success of individuals speaking the language. This is a very interesting and valid approach that nevertheless does not treat the language itself as the reproducing unit that is transmitted among suitable "host" individuals. With regard to the evolution of cultural diversity, the perspective of using the hosts of cultural content, rather than the cultural content itself, as the evolutionary units, naturally tends to generate the perspective of determining the "winners" among a preexisting set of different cultures (e.g., Diamond, 1997; Lim et al., 2007). This approach is roughly equivalent to studying "species selection" and forgoes the question of how diversity arises within a single, homogeneous culture. In contrast, the perspective of viewing the cultural memes themselves as evolutionary units naturally lends itself to studying adaptive cultural diversification within a population of cultural memes if selection on these memes, for example, in terms of transmission, is frequency-dependent. A rare example of a model for cultural diversification based on selection on individual memes is Dercole et al. (2008), who describe evolutionary innovation in an economical context (see also Dercole & Rinaldi, 2008). However, their models are essentially logistic competition models of the type studied in Chapter 4, but with the evolutionary variable simply interpreted as a trait characterizing a technological product. In this chapter, I go a step further by taking the analogy of memes and their hosts literally and adopting an epidemiological perspective of cultural content colonizing human hosts. This perspective has been advocated by a number of researchers, most notably perhaps by Sperber (1996). However, despite some early attempts (e.g., Cavalli-Sforza & Feldman, 1981), it seems that to date this epidemiological perspective has not been rigorously adopted as a basis for the mathematical modeling of cultural evolution. There is a large body of theoretical and mathematical literature on the epidemiological dynamics of pathogens in humans and other animals. I think that if suitably adapted, such epidemiological models could be very well suited to address many questions in cultural evolution. In particular, investigating the epidemiology of cultural content colonizing human hosts may shed new light on cultural diversification, a process of fundamental importance in human societies.

8.1   DIVERSIFICATION OF LANGUAGES

I first consider a very schematic epidemiological model for the evolution and diversification of language. It is well known that languages can be very diverse over small spatial scales. For example, in Switzerland, a wealth of local Swiss German dialects have been maintained over the centuries despite ever-increasing cultural exchange and mixing between regions. (I grew up in a small town approximately 20 km from the city of Basel, yet everyone who grew up in the city itself knows, upon hearing me talk, that I do not hail from the city proper, and can moreover tell that I learned my language in the countryside to the east rather than to the south of Basel [while the languages spoken to the west and and north of Basel are Alsatian (French) and Alemannian (German), respectively].)

The nature and origin of language are topics at the core of the humanities (Chomsky, 1972; Pinker, 2000) that have fascinated people throughout the centuries. That language evolves according to the laws of natural selection appears to be a relatively recent concept (Szathmáry & Maynard Smith, 1995). One method to formalize this concept mathematically is through evolutionary game theory, as suggested for example by Nowak & Krakauer (1999). Martin Nowak and his coworkers have used game theory to study the evolution of language based on the notion of "Universal Grammar" (Chomsky, 1972). In their models, a universal grammar allows acquisition of a finite set of different component grammars. Each individual (human) speaks one of these component grammars. The fitness of an individual is determined by the success of communication with other individuals in the population, which is in turn determined by a matrix describing the pairwise affinity between different grammars. Offspring inherit their parent's Universal Grammar, and learn their parent's component grammar, except if a learning error occurs, in which case they learn another one from the finite set of component grammars admitted by the parental Universal Grammar. Mitchener & Nowak (2003) studied competition between specialist and generalist Universal Grammars, that is, between Universal Grammars admitting only one component grammar and Universal Grammars admitting two or more component grammars. However, they did not address the question of how a given Universal Grammar may evolve to have more than one component grammars in the first place.

This question could be addressed as follows. Let us assume that rather than admitting a finite set of component grammars, a given Universal Grammar admits a continuum of component grammars, characterized by the real variable x. The individuals under selection are not humans, but instead language memes that represent entire component grammars and are characterized by their trait value x. These language memes colonize human hosts and reproduce when a colonized host transmits the meme to other hosts, which could be the host's offspring or other host individuals with which the colonized host is interacting. I assume that a meme's reproductive rate increases with similarity to other memes in the population. This corresponds to the commonly made assumption that a host has improved survival and/or reproduction if it shares the same language meme with many other hosts, thus increasing the host's ability to communicate. Such an increase in a meme's reproductive rate could also occur if its transmission to "naive" hosts is increased through being a common meme (e.g., because there is more logistic support for teaching). Thus, let $\alpha(x - y)$ be some function measuring similarity between memes x and y, so that α has a maximum at 0. If $\phi(x)$ is the current frequency distribution of the language memes, then the fitness of meme x contains a component that is proportional to

$$r\left(\int_y \alpha(x, y)\phi(y)\, dy\right), \qquad\qquad (8.1)$$

where $r(z)$ is a monotonically increasing function of its argument.

On the other hand, it is quite easy to imagine that rare language memes also have an advantage by conferring a status of originality or novelty to its host. For example, it is known that slang words in urban languages evolve faster than other words, presumably because using novel slang words confers a certain status (e.g., of being "cool" or "hip," etc.) to the speaker. Presumably, such an advantage in status due to novelty decreases as the new expression, or meme, becomes more common. Consequently, I assume for my admittedly rather artificial model that the fitness of a language meme also contains a component that is proportional to

$$s\left(\int_y \alpha(x, y)\phi(y)\, dy\right), \tag{8.2}$$

where $s(z)$ is a monotonically decreasing function of its argument.

Finally, I assume that there is some frequency-independent component $b(x)$ of meme fitness, which could, for example, be due to cognitive constraints in the hosts (such as ease of learning), so that there is a language meme x_0 for which $b(x)$ is maximal. Together, these assumptions imply that the reproductive output of meme x is proportional to $b(x) + r(A(x)) + s(A(x))$, where $A(x) = \int_y \alpha(x, y)\phi(y)\, dy$ and $\phi(y)$ is the frequency distribution of the different memes. For simplicity, I assume that $\alpha(x - y)$ and $b(x)$ have Gaussian form:

$$\alpha(x - y) = \alpha_0 \exp\left(\frac{-(x - y)^2}{2\sigma_\alpha^2}\right) \tag{8.3}$$

and

$$b(x) = \exp\left(\frac{-(x - x_0)^2}{2\sigma_b^2}\right). \tag{8.4}$$

Two cases are of interest when deriving the adaptive dynamics of the meme trait x. First, in a population that is monomorphic for meme x, the distribution $\phi(y)$ is a delta function centered at x. Hence $A(x) = 1$, and the reproductive output of an individual meme in such a population is proportional to $b(x) + r(1) + s(1)$. Second, in a resident population that is monomorphic for meme x, we have $A(y) = \alpha(x - y)$ for a rare mutant meme y, and hence the reproductive output of the mutant y is proportional to $b(y) + r(\alpha(x - y)) + s(\alpha(x - y))$. Therefore, the relative fitness of the rare mutant y in a resident x, that is, its invasion fitness, is

$$f(x, y) = b(y) + r(\alpha(x - y)) + s(\alpha(x - y)) - (b(x) + r(1) + s(1)). \tag{8.5}$$

The adaptive dynamics of the language trait x is now given as usual by

$$\frac{dx}{dt} = m \left.\frac{\partial f(x, y)}{\partial y}\right|_{y=x}$$

$$= b'(x) + r'(1) \left.\frac{\partial \alpha(x - y)}{\partial y}\right|_{y=x} + s'(1) \left.\frac{\partial \alpha(x - y)}{\partial y}\right|_{y=x} = b'(x). \tag{8.6}$$

Here m is a parameter that describes the mutational process in language memes, for which I assume that memes undergo small mutations when they are transmitted between human hosts. (Note that the model assumes that the host population size, and hence the meme population size, are constant, so that m is independent of x.) From the definition of the function $b(x)$, it easily follows that the trait x_0 is a unique, convergence stable singular point of the adaptive dynamics. The evolutionary stability of this singular point is determined by

$$\left.\frac{\partial^2 f(x, y)}{\partial y^2}\right|_{y=x=x_0} = -m\frac{1}{\sigma_b^2} - \frac{1}{\sigma_\alpha^2}(r'(1) + s'(1)). \tag{8.7}$$

By assumption, $r'(1) > 0$ and $s'(1) < 0$. It is therefore possible that $r'(1) + s'(1) < 0$, and hence that $\partial^2 f(x, y)/\partial y^2(x_0, x_0) > 0$, that is, that x_0 is an evolutionary branching point. In this case, frequency-dependent selection on language memes leads to cultural diversification. Note that the existence of evolutionary branching points in this model is simply a consequence of the assumption that rarity of a language meme confers some transmission advantage, and that evolutionary branching is also possible if no advantage to being common is assumed (e.g., if $r(z) = 0$ for all z). Also note that it is possible to construct an essentially identical model based on reproductive success of the host individuals carrying certain language memes (rather than on the reproductive success of the memes themselves). This would correspond to the models of Mitchener & Nowak (2003) mentioned earlier, with one Universal Grammar, a one-dimensional continuum of component grammars, and the additional assumption of an advantage to speaking a rare grammar. Thus, in this example it is not necessary to adopt the perspective that the memes themselves are the reproducing units that are transmitted between hosts. Nevertheless, I find this perspective useful, partly because it can, at least in principle, easily be generalized to other areas of cultural evolution, such as the evolution of religions.

8.2   DIVERSIFICATION OF RELIGIONS

Religions are sets of ideas, dogmas, and laws of whose validity individual humans can become convinced. Thus, human minds are the hosts of religious memes, where memes again need to be understood in a generalized sense as potentially comprising whole sets of ideas. These memes can in turn exert considerable influence on the behavior of their human hosts. In principle, it seems possible that an understanding of the cultural dynamics of religion can be achieved by investigating the interaction between religious memes and their hosts, that is, by investigating how religious memes affect not only the behavior of individual hosts, but also the social structure of host populations, and how behavior and social structure in turn affect the transmission of religious memes among host individuals. It is probably fair to say, particularly for widespread religions, that most host populations of a given religion are hierarchically structured, with relatively few hosts enjoying many material or status-related benefits, and many hosts enjoying fewer such benefits from adopting the given religion. As the number of host individuals of a given religion grows, this social structure may give rise to unrest, particularly in the lower social ranks. As a consequence, individuals may turn to alternative religions that offer less repression, and in which they can attain improved social status. For example, it has been suggested that social unrest led to the split of the Protestant Church from the Catholic Church at the beginning of the sixteenth century (Tuchman, 1985). At that time, political and economic developments led to ever-increasing financial needs of the Catholic Church, which burdened its followers through taxation and other means, for example, the sale of indulgences. This in turn led to hardship and spiritual decay, which may have contributed to the secession of a more egalitarian, less repressive and spiritually more attractive religion. Thus, some hosts of Catholic memes may have lost that meme due to effects that the Catholic memes themselves generated in their host society. Moreover, it is likely that hosts that lost faith in the Catholic Church were nevertheless susceptible to alternative religious meme, especially if those memes were similar, but promised to improve the conditions of its followers. Alternative memes may have seemed particularly attractive as long as they were not very common, and hence did not have the same detrimental effects on its hosts as the Catholic meme. Of course, many other forces impinge on the well-being of hosts of a particular religious meme, and hence on the probabilities of transmission and loss of such memes. For example, common memes may offer social and economic protection, whereas hosts of rare memes may suffer persecution. Nevertheless, we have proposed in Doebeli & Ispolatov (2010c) that mechanisms such as the ones alluded to above could cause negative frequency-dependent selection on religious memes, and that such frequency dependence could cause adaptive diversification in evolving meme populations. In the following, I recall the models of Doebeli & Ispolatov (2010c) and discuss possible extensions.

I again take the perspective that human individuals are potential hosts to religious memes (or, more generally, to cultural memes of any kind), and that it is therefore natural to base a formal analysis of the dynamics of religion on epidemiological models. Such models have been extensively studied in the context of the dynamics of pathogens and their plant or animal hosts (e.g., Anderson & May, 1992; Brauer & Castillo-Chavez, 2000; Otto & Day, 2007). We have encountered epidemiological models in Chapter 5, and the simplest form of a large class of such models have two dynamic variables, each describing a subpopulation of the total host population: S denotes the density of susceptible individuals in the host populations, that is, individuals that are not yet hosts to any religious memes, and C denotes the density of infected, or colonized, hosts, that is, host individuals whose minds have adopted a religion. The analysis in Doebeli & Ispolatov (2010c) is then based on the following model for the ecological dynamics of susceptible and infected hosts:

$$\frac{dS}{dt} = r_S S \left(1 - \frac{S + C}{K_S}\right) - \tau SC + lC \tag{8.8}$$

$$\frac{dC}{dt} = r_C C \left(1 - \frac{S + C}{K_C}\right) + \tau SC - lC \tag{8.9}$$

Here it is assumed that both susceptible and infected hosts grow logistically. For simplicity, I assume that offspring of susceptible hosts are also suscepti-ble (i.e., not yet colonized), and offspring of religious (i.e., infected) hosts are also infected. Thus, in the absence of religious memes, susceptible hosts grow logistically with intrinsic growth rate r_S and carrying capacity K_S, and in the absence of susceptibles, religious hosts grow logistically with intrinsic growth rate r_C and carrying capacity K_C. The two growth terms are coupled by assum-ing that growth depends on the sum of the two densities S and I. In addition, susceptible hosts adopt religion, that is, become infected, at a per capita rate τ that is proportional to the number of religious hosts. However, religious hosts also lose their beliefs at a per capita rate l and become susceptible once again, leading to a decrease in C at a rate lC and a corresponding increase in S.

The simplest way to introduce religious variability is to assume that reli-gious memes are characterized by a one-dimensional trait x, and that $C(x)$ describes the density distribution of the various religious types. Of course, in reality one can use many different variables to characterize religions, that is, in reality the phenotype space of religion is multidimensional. However, it has been suggested that for some basic distinctions a low-dimensional description might suffice. For example, Whitehouse (1995) argued that religions can be characterized by either having a high frequency of low-arousal rituals, or a low frequency of high-arousal rituals. If this is true, one could describe religions simply by the (one-dimensional) frequency of their rituals. Be that as it may, here we restrict our attention to a one-dimensional religious meme phenotype x. To introduce frequency-dependent selection on the trait x, I use a measure of rarity similar to the language example above by defining, for any given meme type x, the "overcrowding" function

$$A(x) = \int_y \alpha(x - y)C(y)\,dy, \tag{8.10}$$

where $\alpha(x - y)$ is a unimodal function of the form

$$\alpha(x - y) = \alpha_0 \exp\left(-\frac{1}{2}\left[\frac{|x - y|}{\sigma_\alpha}\right]^{b_\alpha}\right). \tag{8.11}$$

The exponent b_α in $\alpha(x - y)$ is a positive real number. For example, if $b_\alpha = 2$, $\alpha(x - y)$ is a Gaussian function. The parameter σ_α is a measure for the strength of frequency dependence, with lower σ_α corresponding to stronger frequency dependence, that is, to less influence of more distant phenotype for overcrowding. We further assume that the per capita rate of loss of the religious meme is some function $l(A(x))$, where $l(z)$ increases monotonically with increasing z. For simplicity, I will assume $l(z) = z$. This implies that hosts are more likely to lose common religious memes than rare religious memes. As explained above, one rationale for this assumption is that once a religion becomes common, the social structure may change such that the benefits gained from adopting the religion decrease for the majority of hosts, so that, on average, hosts of such memes become more likely to lose faith. In the last part of this chapter I will consider a model in which the social structure is described explicitly, but for now I only model these effects implicitly using the function $l(z)$. With religious variability, the differential equation for C must be replaced by a partial differential equation for the dynamics of the distribution $C(x)$. I refer to Chapter 9 for more examples of partial differential equation models. This yields the following epidemiological model:

$$\frac{dS}{dt} = r_S S\left(1 - \frac{S + \int_x C(x)\,dx}{K_S}\right) - S\int_x \tau(x)C(x)\,dx + \int_x A(x)C(x)\,dx \tag{8.12}$$

$$\frac{\partial C}{\partial t} = r_C \int_y M_{y,\sigma_m}C(y)\,dy - \frac{r_C C(x)\left(S + \int_x C(x)\,dx\right)}{K_C} + \tau(x)SC(x)$$
$$- A(x)C(x) + D_x\frac{\partial^2 C(x)}{\partial x^2}. \tag{8.13}$$

Note that the dynamics of susceptible hosts is still given by an ordinary differential equation, as susceptible hosts are not differentiated phenotypically. The per capita rate of loss of religion in colonized hosts with religious type x is $A(x)$ as described above, so that the total rate of loss is $\int_x A(x)C(x)\,dx$, which appears as a positive term on the right-hand side of the dynamic equation (8.12) for susceptibles. To describe transmission, I have made the additional assumption that the transmission parameter $\tau(x)$ is a function of the religious trait x. This function is assumed to reflect some intrinsic properties of religious ideas that determine their likelihood of transmission to susceptible hosts. For example, some ideas might entice their carriers to proselytize more than others, but such activities might come at certain costs. The function $\tau(x)$ is then assumed to reflect the balance of such costs and benefits. Specifically, we assume that this function is unimodal, so that there is a unique "optimal" religious type in terms of transmissibility. This introduces a stabilizing component of selection on the trait x. We assume that this function has the following form:

$$\tau(x) = \tau_0 \exp\left(-\frac{1}{2}\left[\frac{|x - x_0|}{\sigma_\tau}\right]^{b_\tau}\right), \tag{8.14}$$

where the exponent b_τ is a positive real number. (Note again that for $b_\tau = 2$, $\tau(x)$ is a Gaussian function.) In analogy to traditional epidemiological models, we assume that the rate at which hosts colonized by religious meme x convince susceptible individuals of their religion is $\tau(x)C(x)$, so that the total per capita rate of transmission for susceptible hosts is $\int_x \tau(x)C(x)\,dx$. Assuming mass action, the total rate if loss of susceptibles due to transmission is then $S\int_x \tau(x)C(x)\,dx$, which appears above on the right-hand side of the dynamic equation (8.12) for susceptibles. This is of course a rather simplified assumption, and in reality the process of transmission is much more complicated. However, it seems reasonable to assume that the rate at which susceptible hosts become colonized by religious memes x, for example, through teachings and propaganda, increases as the amount of people that already host this meme increases. Also, even if a susceptible person accepts a new religious meme, it is likely that in this newly colonized person the religious meme is slightly different from the meme from which it was copied. For example, a susceptible person might adopt a religion only after examining various meme models, so that the newly colonized brain would host a meme that represents some average from the memes already present in other hosts. Incorporating such effects would appear to be a promising area for future research.

The first term on the right-hand side of the dynamic equation (8.13) for $C(x)$, $r_C \int_y M_{y,\sigma_m}(x)C(y)\,dy$, includes mutation and describes all offspring that are born to infected parents with all possible religious types y and whose type mutated to x. As usual, the mutation kernel M_{y,σ_m} has normal form with mean y and variance σ_m^2. The second term on the right-hand side of the partial differential equation (8.13) for $C(x)$ is the total death rate of hosts colonized by religious meme x, and $\int_x C(x)\,dx$ is the total population size of colonized hosts, so that $S + \int_x C(x)\,dx$ is the total host population size. The third and fourth terms correspond to transmission of religious trait x from individuals colonized by x to susceptibles, and to loss of religion in individuals colonized by x, respectively. Finally, the last term, $D_x \partial^2 C(x)/\partial x^2$, is a diffusion term along the religious trait axis that reflects (small) changes of religious belief, that is (small) mutations in religious memes that have already colonized the host.

To keep the notation simple, below I assume $r_S = r_C = r$ and $K_S = K_C = K$, so that susceptible and infected hosts do not differ in their logistic growth parameters. Again, it would be straightforward to incorporate more complicated scenarios by assuming that these growth parameters depend on whether hosts are infected or not, and on the type of religion that they host. However, with the assumptions made here, the religious trait x only affects rates of loss and transmission, but it does not affect the birth and death rates of colonized hosts. Thus, selection on religion is not mediated by differential viability and/or reproductive success of the hosts, but instead by differential loss and gain of cultural memes by colonized and susceptible hosts.

By restricting the dynamics given by the partial differential equation (8.13) to atomic distributions given by delta functions, one can extract the adaptive dynamics of the religious trait x. We then first consider the ecological dynamics of monomorphic populations. (I refer to the Appendix and to previous chapters for more detailed explanations of the adaptive dynamics approach.) If the population of memes is monomorphic for trait x, the distribution $C(z)$ is a delta function with total weight $C(x)$ centered at x, so that $A(x) = C(x)$. Equations (8.12) and (8.13) then become a system of ordinary differential equations:

$$\frac{dS}{dt} = rS\left(1 - \frac{S + C(x)}{K}\right) - \tau(x)SC(x) + \alpha_0 C(x)C(x) \qquad (8.15)$$

$$\frac{dC(x)}{dt} = rC(x)\left(1 - \frac{S + C(x)}{K}\right) + \tau(x)SC(x) - \alpha_0 C(x)C(x). \qquad (8.16)$$

It is easy to see that this ecological dynamics has a unique equilibrium

$$(S^*(x), C^*(x)) = \left(\frac{K\alpha_0}{\alpha_0 + \tau(x)}, \frac{K\tau(x)}{\alpha_0 + \tau(x)}\right) \qquad (8.17)$$

at which both $S^*(x) > 0$ and $C^*(x) > 0$. Moreover, the Jacobian matrix of this system has two negative eigenvalues at the equilibrium $(S^*(x), C^*(x))$, and the equilibrium is globally stable in the sense that the system will converge to this equilibrium from any initial condition with both densities greater than zero.

Challenge: Prove these assertions.

Assuming that system (8.15) and (8.16) is at its equilibrium $(S^*(x), C^*(x))$ for some resident x, the per capita growth rate of a rare mutant y, that is, the invasion fitness $f(x, y)$, is given by

$$f(x, y) = r \left(1 - \frac{S^*(x) + C^*(x)}{K} \right) + \tau(y)S^*(x) - \alpha(y - x)C^*(x). \quad (8.18)$$

As a consequence, the adaptive dynamics of the religious trait x is determined by the selection gradient

$$D(x) = \frac{\partial f(x, y)}{\partial y} \bigg|_{y=x} = \tau'(x)S^*(x). \quad (8.19)$$

It easily follows that x_0, the maximum of the transmission function $\tau(x)$, is the unique convergent stable singular point of this adaptive dynamics. Moreover, for the second derivative of the invasion fitness function at the singular point we find

$$\frac{\partial^2 f(x, y)}{\partial y^2} \bigg|_{y=x=x_0} = \tau''(x_0) \frac{K\alpha_0}{(\alpha_0 + \tau_0)} - \alpha''(0) \frac{K\tau_0}{(\alpha_0 + \tau_0)}. \quad (8.20)$$

For example, in the Gaussian case where the exponents appearing in the functions τ and α satisfy $b_\tau = b_\alpha = 2$, it follows that the singular point x_0 is an evolutionary branching point if and only if

$$\sigma_\alpha < \sigma_\tau. \quad (8.21)$$

In general, adaptive religious diversification can occur if frequency dependence is strong enough, that is, if the disadvantage resulting from being a common meme is large enough. Because σ_α is a measure for how fast religious types can gain an advantage by being different from common types, and σ_τ measures how fast transmissibility decreases with increasing distance from the optimum x_0, this can be interpreted as diversification in religion occurring if the advantage gained from rarity outweighs the disadvantage due to having

BOX 8.1

INDIVIDUAL-BASED EPIDEMIOLOGICAL MODELS OF CULTURAL EVOLUTION

This box is essentially the same as the Appendix in Doebeli & Ispolatov (2010c) and explains how to extend individual-based models based on birth and death rates to individual-based epidemiological models incorporating transmission and loss of cultural memes. To construct such models based on the deterministic model given by eqs. (8.12) and (8.13), we have to distinguish the different types of events that can occur at the level of individuals: birth, death, loss of faith, and transmission of faith. All these events occur at certain rates. For example, all host individuals have a per capita birth rate $r_S = r_C$, so that the total birth rate of susceptible hosts, B_S, is $r_S S$, and the total birth rate of colonized hosts, B_C, is $r_C C$, where S and C are the (integer) numbers of susceptible and colonized hosts present in the population at a given time. For both susceptible and colonized host individuals, the per capita death rate is $r_S SC/K_S = r_I SC/K_C$, and total death rates D_S and D_C for susceptible and colonized individuals are $r_S S^2 C/K_S$ and $r_I SC^2/K_C$, respectively. For a host colonized by religious type x, the per capita rate at which this type is transmitted to susceptible hosts is $\tau(x)S$, where $\tau(x)$ is the transmission function (8.14). The total rate of transmission, T, is therefore $\sum_i \tau(x_i)S$, where the sum runs over all colonized hosts. The per capita rate of loss of religion of host individuals colonized by religious type x is given by $l(A(x)) = A(x)$ (eq. (8.10)), so that the total rate of loss, L, is $\sum_i A(x_i)$. Finally, the per capita rate of diffusion along the trait axis is given by a parameter m, so that the total rate of diffusion is $M = mC(x)$.

The individual-based model is implemented using the Gillespie algorithm (Erban et al., 2007; Gillespie, 1977). At any given time t, all individual rates as well as the total rates B_S, B_C, D_S, D_C, T, L, and M are calculated as described above. Then the type of event that occurs next, birth or death of a susceptible or a colonized host, transmission of religious content, loss of religious content, or diffusion, is chosen with probabilities proportional to the total rates for these events, that is, with probabilities B_S/E, etc., where $E = B_S + B_C + D_S + D_C + T + L + M$ is the total event rate. For the event chosen, the individual to perform this event is chosen with probabilities proportional to the individual rates for the chosen event. For example, if loss of faith is the chosen event, individual i is chosen to lose faith with probability $A(x_i)/L$, where x_i is the religious type of individual i. This individual is then removed from the population of colonized hosts, and the number of susceptible hosts is augmented by one. Similarly, if the chosen event is transmission, individual i among the colonized hosts is chosen for transmission with probability $\tau(x_i)S/T$, where x_i is again the religious type

BOX 8.1 (*continued*)

of individual i. Individuals for birth and death events are chosen analogously. If an individual dies it is removed from the population (and the numbers S or C are updated accordingly). If a susceptible individual gives birth the number S is augmented by one. If an individual colonized with type x_i gives birth, a new colonized host is added to the population carrying a type x' that is chosen from a normal distribution with mean the parental type x and a certain (small) width σ_{mut}. This reflects "mutation" during transmission of religious content from parent to offspring. Finally, if diffusion occurs, a colonized host individual is chosen randomly and the phenotype of its colonizing meme changed by drawing the new phenotype from a Gaussian distribution with mean the old phenotype and variance given by a parameter σ_m.

Performing one individual event in the manner described before completes one computational step in the individual-based model, which advances the system from time t to time $t + \Delta t$ in real time. To make the translation from discrete computational steps to continuous real time, Δt is drawn from an exponential probability distribution with mean $1/E$, where E is the total event rate. Starting from some initial population containing S_0 susceptible hosts and C_0 hosts colonized by religious types $x_1^0, \ldots, x_{N_0}^0$ at time zero, iteration of the computational steps described above generates the stochastic cultural evolutionary dynamics of a finite population in continuous time.

■

lower transmissibility. After evolutionary branching, different types of diversification dynamics are possible, depending on the choice of the functions τ and α. These dynamics can for example be studied using individual-based models that implement the epidemiological assumptions made for the meme-host models introduced above. As in the individual-based models of earlier chapters, the corresponding models for cultural epidemiology again have individual birth and death rates for both susceptible and colonized hosts, which are defined based on the logistic growth terms in (8.12) and (8.13). However, we now also need rates for transmission and loss of religious memes, as well as for diffusion along the trait axis. Box 8.1 explains the setup of these individual-based models.

Salient examples of the evolutionary dynamics resulting from the individual-based models are shown in Figure 8.1. In the first example, both the overcrowding kernel α and the transmission function τ are Gaussian. That is, $b_\alpha = b_\tau = 2$ in eqs. (8.11) and (8.14). Figure 8.1(a) shows that when the branching condition (8.21) is not satisfied, no diversification occurs. In

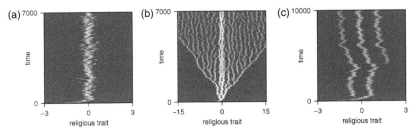

FIGURE 8.1. Evolution of the religious distribution $C(x)$ obtained from individual-based models. In panel (a), no diversification occurs. In panel (b), multiple sequential branching in the Gaussian case results in a multitude of religious types (with an essentially unimodal total distribution). In panel (c), cultural adaptive diversification stops after two branching events and results in the coexistence of three distinct religious clusters. Parameter values were $K_C = K_S = 10^4$, $r_C = r_S = 1$, $\sigma_{mut} = 0.02$, $\tau_0 = 0.006$, and $\alpha_0 = 0.003$, $m = 0.1$ and $\sigma_m = 0.05$ for all panels, $\sigma_\tau = 1$, $\sigma_\alpha = 0.5$ and $b_\tau = b_\alpha = 2$ in panel (a); $\sigma_\tau = 1$, $\sigma_\alpha = 0.5$ and $b_\tau = b_\alpha = 2$ in panel (b); $\sigma_\tau = 1$, $\sigma_\alpha = 0.5$, $b_\tau = b_\alpha = 3$ in panel (c). Populations were initialized with equal numbers of susceptibles and infecteds, which had a Gaussian distribution with mean -2 and a small variance.

contrast, Figure 8.1(b) shows that when this condition is satisfied, evolutionary branching occurs after convergence to the optimal transmission phenotype x_0. After the first branching event, the two emerging phenotypic branches first diverge and then undergo secondary evolutionary branching. This process repeats itself until the phenotype space fills up with a multitude of different meme types whose total distribution approximates a unimodal distribution. Such a scenario could describe the adaptive diversification of an ancestral religion into a multitude of related denominations (such as the Protestant denominations in North America today).

It should be noted that the outcome of the cultural evolution model for Gaussian overcrowding and transmission functions bears some similarity to the results from the asexual competition models studied in Chapter 3. There we have seen that when the branching condition is satisfied in competition models with Gaussian competition kernels and carrying capacity functions, the (deterministic) adaptive dynamics typically results in the coexistence of many different phenotypic branches whose total distribution approaches a unimodal distribution. As I have pointed out in Chapter 3, and as will be described in more detail in the next chapter, this result is reflected in the fact that the corresponding partial differential equation models for competition in asexual populations with Gaussian kernels always have a Gaussian equilibrium distribution. In fact, the same is true here: with Gaussian overcrowding and

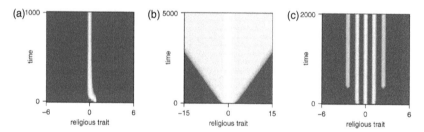

FIGURE 8.2. Evolution of the religious distribution $C(x)$ obtained from numerical solution of eqs. (8.12) and (8.13). No diversification occurs in panel (a), and the equilibrium consist of a narrow, unimodal phenotype distribution. Panel (b) shows diversification in the form of a broad and unimodal distribution when the transmission and loss functions are Gaussian. In this case, the cultural phenotype distribution converges to a Gaussian equilibrium distribution. In contrast, panel (c) shows diversification in the form of multimodal equilibrium distributions, with each mode representing a separate religious lineage. Parameter values were the same as for Figure 8.1, except that the diffusion parameters m and σ_m in the individual-based model were replaced by the diffusion coefficient $D_x = 0.0001$ in the partial differential equation model.

transmission functions, it is easy to see that the deterministic partial differential equation model given by (8.12) and (8.13) has a Gaussian equilibrium distribution of colonized hosts (see also Figure 8.2).

As we know by now, the Gaussian case has to be treated with caution, because it can be structurally unstable. Indeed, with non-Gaussian overcrowding and competition functions, the individual-based model shows qualitatively different evolutionary dynamics, and can result, if the branching condition is satisfied, in the coexistence of but a few clearly distinct religious memes. This scenario is illustrated in Figure 8.1(c) and corresponds to the cultural evolution of clearly distinct religions (such as the main religious branches emanating from Judaism). Overall, it is clear that these individual-based epidemiological models for cultural evolution can generate adaptive diversification in cultural content.

Corresponding results hold for the dynamics of the partial differential equation model given by (8.12) and (8.13). In general, this system is analytically intractable, but it can be solved numerically (Doebeli & Ispolatov, 2010c). Such simulations again reveal two basic dynamic regimes of adaptive cultural diversification. In the first, the equilibrium distribution of colonized hosts is a unimodal function with a large variance, as shown in Figure 8.2(b). This corresponds to the individual-based scenario shown in Figure 8.1(b), that is, to "denominational" diversification. In the second, the equilibrium distribution is multimodal, as shown in Figure 8.2(c). This corresponds to the individual-based scenario shown in Figure 8.1(c) and reflects the adaptive emergence of distinctly different religions. Note that the deterministic model shows the emergence of five modes for the same parameter values that yield only three modes in the individual-based model (Figure 8.1(c)). This is because the marginal modes resulting from the partial differential equation model have very low densities, and hence cannot be established in the individual-based model with small population size (the quasi-stationary population size of the individual-based model shown in Figure 8.1(c) is approximately 600).

Overall, whether diversification occurs in either the individual-based models and the partial differential equation model (8.12) and (8.13), and whether diversity manifests itself in wide unimodal or in multimodal meme distributions, depends on the parameters of the models, as illustrated in Figures 8.1 and 8.2. First, diversification occurs when frequency dependence is strong enough, that is, when the parameter σ_α is small enough compared to the parameter σ_τ, which corresponds well with the condition for evolutionary branching (eq. (8.20)) derived from the adaptive dynamics. Second, whether diversification results in unimodal or multimodal equilibrium distributions depends on the exponents b_α and b_τ, that is, on the nature of the overcrowding function α and the transmission function τ. Generally speaking, larger exponents, and hence "platykurtic" functions α and τ that fall off less sharply from their maximum than Gaussian functions, tend to generate multimodal diversification. This is in agreement with results from the individual-based competition models described in Chapter 3, as well as with results from competition models based on partial differential equations, which will be studied in Chapter 9.

To conclude the present chapter, I will extend the model (8.12) and (8.13) for the cultural dynamics of religion to stage-structured populations, in which susceptible hosts belong to different social classes, and in which the assumption that repression of lower social classes can lead to loss of faith is described in terms of these social classes. For this I assume that social class or status is described by a continuous real variable $u > 0$. The subpopulation of susceptible hosts is then described by a density distribution $S(u)$, while the subpopulation of infected hosts is described by a two-dimensional density distribution $C(u, x)$. For the logistic growth of both susceptible and religious hosts, I assume that each social class has a class-specific carrying capacity $K(u)$ such that $K(u)$ decreases as u increases, reflecting the assumption that higher social classes are occupied by fewer individuals. For example, one could assume a negative exponential function $K(u) = K_0 \exp(-u/K_1)$ with some parameters K_0 and K_1.

I further assume that susceptibles belonging to social class u can only be infected by religious hosts belonging to social classes $u' \geq u$. Thus, the per capita rate at which susceptibles of class u become religious is

$$\int_x \left(\tau(x) \int_{u' \geq u} C(u', x)\, du' \right) dx, \qquad (8.22)$$

where $\tau(x)$ is the transmission function as before. I also assume that when a susceptible of social class u becomes religious, it will remain in that social class, independent of the social class of the religious individual that infected the susceptible individual. Thus, the total rate at which individuals of social class u acquire the religious meme x is

$$\tau(x)S(u) \int_{u' > u} C(u', x)\, du'. \qquad (8.23)$$

To determine the rate of loss of faith in religious hosts due to repression in common memes, I assume that repression only occurs from other religious hosts of higher social classes, and that repression diminishes with increasing distance between religious memes. As before, this latter assumption leads to frequency-dependent selection on religious memes, with sharper decreases in repression with religious distance corresponding to stronger frequency dependence. Here I will assume that the degree of frequency dependence experienced may depend on an individual's social class u. Thus, I assume that the parameter describing the decrease in repression with increasing religious distance is a function of the social status u.

Accordingly, the per capita rate of loss of faith of hosts of social class u and with religious meme x is

$$\int_y \alpha(x - y, u) \left(\int_{u' > u} C(u', y)\, du' \right) dy, \qquad (8.24)$$

where the overcrowding function $\alpha(x - y, u)$ is now of the form

$$\alpha(x - y, u) = \alpha_0 \exp\left(-\frac{1}{2} \left[\frac{|x - y|}{\sigma_\alpha(u)} \right]^{b_\alpha} \right). \qquad (8.25)$$

As a function of social status u, the width $\sigma_\alpha(u)$ of the overcrowding function may be increasing or decreasing. If it is increasing, then individuals with higher social status experience a lower degree of frequency dependence, which means that for individuals of higher social status it is less advantageous to be different from the mainstream than for individuals with lower social status. The converse is true if $\sigma_\alpha(u)$ is a decreasing function of u.

Finally, I assume that religious hosts of social class u that lose faith become susceptibles in the same social class u. This leads to the following partial differential equations describing the dynamics of the distributions $S(u)$ and $C(u, x)$:

$$
\frac{\partial S(u)}{\partial t} = r_S S(u) \left(1 - \frac{\int_{u'} S(u') \, du' + \int_{u',x'} C(u', x') \, du' dx'}{K(u)} \right)
$$

$$
- S(u) \int_x \left(\tau(x) \int_{u'>u} C(u', x) \, du' \right) dx
$$

$$
+ \int_x C(u, x) \left(\int_y \alpha(x - y, u) \left(\int_{u'>u} C(u', y) \, du' \right) dy \right) dx
$$

$$
+ D_u \frac{\partial^2 S(u)}{\partial u^2}, \tag{8.26}
$$

$$
\frac{\partial C(u, x)}{\partial t} = r_C C(u, x) \left(1 - \frac{\int_{u'} S(u') \, du' + \int_{u',x'} C(u', x') \, du' dx'}{K(u)} \right)
$$

$$
+ \tau(x) S(u) \int_{u'>u} C(u', x) \, du'
$$

$$
- C(u, x) \int_y \alpha(x - y, u) \left(\int_{u'>u} C(u', y) \, du' \right) dy
$$

$$
+ D_u \frac{\partial^2 C(u, x)}{\partial u^2} + D_x \frac{\partial^2 C(u, x)}{\partial x^2}. \tag{8.27}
$$

The first term on the right-hand sides of these equations corresponds to logistic growth, and $\int_{u'} S(u') \, du'$ and $\int_{u',x'} I(u', x') \, du' dx'$ are the total densities of susceptible and religious hosts. The transmission terms are based on the rates given by expressions (8.21) and (8.22), while the terms describing loss of faith are based on the per capita rate given by expression (8.23). In addition, the dynamic equations for $S(u)$ and $C(u, x)$ contain diffusion terms. $D_u \partial^2 S(u)/\partial u^2$ and $D_u \partial^2 C(u, x)/\partial u^2$ describe (nondirectional) movement between social classes, reflecting gain or loss of social status. The social diffusion constant D_u is assumed to be the same in susceptible and religious hosts. Finally, $D_x \partial^2 C(u, x)/\partial x^2$ reflects (small) mutations in the religious meme, which can occur when memes are not inherited faithfully from religious parents. This diffusion term in the x-direction essentially replaces the mutation kernel in the model (eqs. (8.12) and (8.13)) (more precisely, the diffusion term

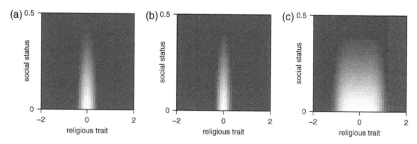

FIGURE 8.3. Examples of equilibrium distributions of religious traits (*x*-axis) and social status (*y*-axis). These distributions were obtained from the numerically simulating model (eqs. (8.25) and (8.26)). In panel (a), the width $\sigma_\alpha(u)$ of the loss kernel α is independent of social status, and the parameters were chosen so that no diversification is observed in this case (for the same parameters, no diversification would be observed in the simpler model (eqs. (8.12) and (8.13)) without social status). Panel (b) shows that still no diversification occurs if $\sigma_\alpha(u)$ is a decreasing function of social status u, so that frequency dependence is weak for low u. Panel (c) shows that diversification occurs in the opposite case, in which $\sigma_\alpha(u)$ is an increasing function of social status u, so that frequency dependence is strong for low u. Parameter values were $K_0 = K_1 = 1$ for the carrying capacity function $K(u) = K_0 \exp(-u/K_1)$, $r_S = r_C = 1$, $D_u = 0.001$, $D_x = 0.00001$, $\tau_0 = 1$, $\sigma_\tau = 0.75$, $\alpha_0 = 1$, $b_\tau = b_\alpha = 2$ for all panels; $\sigma_\alpha(u) = 1$ for panel (a), $\sigma_\alpha(u) = 2(1 - 1.25u)$ for panel (b), and $\sigma_\alpha(u) = 2.5u$ for panel (c). In all cases, the equilibrium distribution appears to be independent of the initial conditions.

represents a second order approximation of the mutational process for narrow, symmetric mutation kernels). The religious diffusion constant D_x reflects the faithfulness of religious imprinting in the offspring.

Figure 8.3 shows results of numerical integrations of the dynamical system (eqs. (8.25) and (8.26)) for different functions $\sigma_\alpha(u)$, and hence for different assumptions about how the strength of frequency-dependent selection on culture changes with social status of the colonized individuals. Figure 8.3(a) shows the control case, in which $\sigma_\alpha(u)$ is constant and does not generate strong enough frequency dependence for diversification. In Figure 8.3(b), $\sigma_\alpha(u)$ is a decreasing function of u, such that the degree of frequency dependence is the same as in Figure 8.3(a) for cultural memes colonizing some intermediate social class u. This means that for memes colonizing lower social classes, frequency dependence is weak (high $\sigma_\alpha(u)$), and in higher social classes frequency-dependence is strong (low $\sigma_\alpha(u)$). In this case, diversification does not occur, despite the fact that if only cultural evolution in high social classes were considered (i.e., if values in the u-direction were constrained to be higher than a certain threshold), then cultural diversification would indeed occur due to strong frequency dependence. Thus, in this example the fact that frequency dependence is weak in the more abundant lower classes, and hence overall selection remains stabilizing for cultural content colonizing these lower classes, prevents cultural diversification altogether. Conversely, if $\sigma_\alpha(u)$ is an increasing function of social class u, so that frequency dependence is strong on cultural memes colonizing lower classes, diversification does indeed occur, despite the fact that frequency dependence on memes colonizing higher social classes is weak. This is shown in Figure 8.3(c). These examples show that whether cultural diversification occurs may depend on the social classes in which cultural memes experience strong frequency-dependent selection. They also illustrate the potential usefulness of incorporating more realism and complexity, such as social stratification, into epidemiological models of cultural evolution.

   In this chapter, I have attempted to apply the theory of adaptive diversification to cultural evolution. The historic record contains many examples of the types of diversification occurring in these models. Diversification in languages has been rampant throughout history, and must have often occurred under substantial contact between hosts of diverging language memes. Similarly, it seems clear that religious diversification has often occurred, and continuous to occur, under conditions of substantial contact. The emergence of partially overlapping sects that differ from each other, for example, in the details of the interpretation of holy texts, such as the proliferation of protestant denominations in the United States, and major splits that lead to the emergence of separate religious hierarchies, such as between the Catholic, Protestant, and Eastern Orthodox churches, may serve as examples of the type of diversification dynamics resulting from the models described here. The models illustrate that diversifying processes should be expected to operate whenever the likelihood of secession from a dominant culture increases with increasing dominance of the mainstream culture. Intuitively, it is not hard to imagine that the attractiveness of a culture diminishes as the culture becomes more dominant, dogmatic, and perhaps oppressive, and that the desire to stand out and be different increases in increasingly conformist cultures.

   Following Daniel Dennett (Dennett, 2009), I think it is important to model cultural evolution by acknowledging that cultural memes have their own fitness, that is, by considering the cultural memes themselves as the units of selection, rather than the humans carrying the cultural memes. Cultural memes, such as languages and ideologies, clearly exhibit reproduction and heredity through their transmission between human hosts. Of course, these memes ultimately need their human hosts for survival and reproduction (e.g., a book's content only comes "alive" once the book is read). Just as the survival and reproduction of symbionts and pathogens is tied to their effects on their hosts, the evolutionary fate of cultural memes is tied to their impact on human individuals. And just as viewing individual organisms as hosts of evolving symbionts or pathogens offers the appropriate perspective for studying the evolution of those symbionts and pathogens, viewing human individuals as hosts of evolving memes offers a useful perspective. Of course, humans and culture may also coevolve, just as hosts and symbionts or pathogens may coevolve. For example, human brains may change evolutionarily due to differential capacity for absorbing or generating cultural memes.

The perspective of cultural memes as the evolutionary unit for cultural evolution also serves to objectify the significance of cultural content, such as religion. Cultural content is best viewed not as fixed and preexisting, but as evolving due to its effect on human individuals, who ultimately decide whether to accept or reject such content. Regarding processes of cultural diversification, a more traditional approach consists of viewing different human populations as carrying different cultural ideas, and of investigating competition between such human populations. In the language of host-pathogen models, this would correspond to considering different host populations carrying different pathogens and asking which of the host populations can outcompete the other. Because in this perspective success is based on characteristics imparted or imposed by the pathogen on a group of hosts, this perspective is akin to group selection. In contrast, studying the evolution of pathogens or the evolution of cultural content in a single host population is based on individual selection on the pathogens or on the cultural content. In the models discussed here, this difference to traditional approaches is reflected in the fact that the religious trait x does not affect survival and reproduction in the host. Thus, selection on culture is not mediated by differential viability or reproductive success in the host population, but by the fact that different religious ideas have different rates of being transmitted to susceptible hosts, and different rates of being lost from colonized hosts. Selection is frequency-dependent, because the rate of loss of faith is assumed to depend on overcrowding. In principle, one could envisage many other sources of frequency-dependent selection in the epidemiology of cultural memes. Many different facets of human culture appear to be incredibly diverse, and I would not be surprised if frequency-dependent selection on cultural memes turns out to be the major driving force for many processes of cultural diversification.

Adaptive Diversification and Speciation as Pattern Formation in Partial Differential Equation Models

The previous chapters have discussed adaptive diversification mostly based on the theory of adaptive dynamics on the one hand, and on individual-based models on the other hand. Adaptive dynamics models offer the possibility of mathematical tractability, but they are based on the assumption of essential monomorphy of resident populations. Individual-based models do not assume monomorphic populations, but they appear to be too complicated for mathematical analysis. In this chapter, we will consider a class of models that we have already encountered in Chapter 8, and that lie somewhere in between these two approaches. Partial differential equations (pde's) can describe the dynamics of phenotype distributions of polymorphic populations, and they allow for a mathematically concise formulation from which some analytical insights can be obtained. It has been argued (e.g., Polechova & Barton, 2005) that because partial differential equations describe polymorphic populations, results from such models are fundamentally different from those obtained using adaptive dynamics. However, this contention is unfounded, as we have shown in Doebeli et al. (2007), and as we will see in this chapter. In partial differential equation models, diversification manifests itself as pattern formation in phenotype distribution. More precisely, diversification occurs when phenotype distributions become multimodal, with the different modes corresponding to phenotypic clusters, or to species in sexual models. We will see that such pattern formation occurs in partial differential equation models for competitive as well as for predator-prey interactions. Partial differential equations are also very well suited to study the dynamics of spatially structured populations, and the chapter concludes with a discussion of diversification due to competition in populations occupying a continuous spatial arena.

9.1 PARTIAL DIFFERENTIAL EQUATION MODELS FOR ADAPTIVE DIVERSIFICATION DUE TO RESOURCE COMPETITION

Consider populations in which individuals are characterized by a quantitative trait x, and let $\phi(x)$ be the density distribution of the population. Thus, $\phi(x)dx$ is the density of individuals having trait values in the interval $(x, x + dx)$. To describe the temporal change of the distribution $\phi(x)$, we need to have an expression for the rate of change of $\phi(x)$ at each phenotype x, that is, a partial differential equation for $\partial\phi(x)/\partial t$. This is done by determining the per capita birth and death rates of individuals of type x. Let us assume that the trait x affects competitive interactions as described in Chapter 3, that reproduction is asexual and faithful, and that all individuals have a constant per capita growth rate r. The per capita death rate is $rN_{\text{eff}}(x)/K(x)$, where $N_{\text{eff}}(x)$ is the effective density experienced by individuals of phenotype x, and where the carrying capacity function $K(x)$ implements a frequency-independent component of selection on x. Because $\phi(x)$ is a continuous distribution, the effective density $N_{\text{eff}}(x)$ is given by an integral

$$N_{\text{eff}} = \int \alpha(x, y)\phi(y)dy, \tag{9.1}$$

where $\alpha(x, y)$ is the competition kernel (see Chapter 3) describing the competitive impact that individuals of phenotype y have on x-individuals (the integral runs over all of phenotype space). The partial differential equation describing the dynamics of $\phi(x)$ is then simply obtained by multiplying the local density $\phi(x)$ by the difference between per capita birth and death rates:

$$\frac{\partial\phi(x)}{\partial t} = \left(r - \frac{rN_{\text{eff}}(x)}{K(x)}\right)\phi(x) = r\phi(x)\left(1 - \frac{\int \alpha(x, y)\phi(y)dy}{K(x)}\right). \tag{9.2}$$

As discussed in Chapter 3, Gaussian functions are a common, if only phenomenological choice for the carrying capacity function K and the competition kernel α. Here we considerably widen the class of possible functions by considering the following functional forms for K and α:

$$K(x) = K_0 \exp\left(-\frac{|x|^{2+\epsilon_K}}{2\sigma_K^{2+\epsilon_K}}\right) \tag{9.3}$$

$$\alpha(x, y) = \exp\left(-\frac{|x - y|^{2+\epsilon_\alpha}}{2\sigma_\alpha^{2+\epsilon_\alpha}}\right) \tag{9.4}$$

where ϵ_K and ϵ_α are real numbers > -2. All carrying capacity functions of type (9.3) imply frequency-independent, stabilizing selection for $x = 0$, because this phenotype minimizes the frequency-independent component of the per capita death rate (the optimal phenotype $x = 0$ is chosen arbitrarily). Similarly, all competition kernels of type (9.4) imply that competitive impacts decrease with phenotypic distance $|x - y|$. Note that for $\epsilon_K = \epsilon_\alpha = 0$, K and α both have Gaussian form. In this case, if $\sigma_\alpha < \sigma_K$, then system (9.2) has an equilibrium distribution $\tilde{\phi}(x)$ that is itself Gaussian. More precisely, for

$$\tilde{\phi}(x) = \frac{K_0 \sigma_K}{\sqrt{2\pi(\sigma_K^2 - \sigma_\alpha^2)\sigma_\alpha^2}} \exp\left(-\frac{x^2}{2(\sigma_K^2 - \sigma_\alpha^2)}\right), \tag{9.5}$$

the integral $N_{\text{eff}}(x) = \int \alpha(x, y)\tilde{\phi}(y)dy = K(x)$ for all x, and hence the right-hand side of eq. (9.2) is identically zero.

In general, it is of course well known that equilibrium states of dynamical systems need not be stable. Stability analysis of partial integro-differential equations such as (9.2) is generally difficult, and it is often convenient to test local stability, as well as other dynamical properties such as dependence on initial conditions, using numerical simulations of the model. All the simulations presented in this chapter were run by my colleague Slava Ispolatov, who implemented a simple adaptive Euler method (or, equivalently, a first-order Runge-Kutta algorithm, see, e.g., Press et al. (2007)) to obtain numerical solutions of the various partial differential equations. Phenotype space was discretized into a finite number of equidistant bins over a range that ensured that marginal phenotypes had a very low carrying capacity, so that using absorbing boundary conditions was feasible and did not affect the outcome of the simulations. The discretization of phenotype space results in a system of couple differential equations (with the number of equations equal to the number of bins), and for each (small) time step in the simulation, population densities in each bin were updated using the linear Euler method for integrating differential equations. The integrals occurring in the partial differential equation (9.2), as well as in other partial differential equations used in this chapter, were calculated using the trapezoid rule (Press et al., 2007). If the updated population densities were negative, the time step was reduced until the updated densities were positive, and the failed step repeated. To ensure the independence of the simulation results of both the scale of the spatial discretization (number of bins) and the scale of the temporal discretization (time step), individual simulations were checked by increasing the number of bins, and by using smaller time steps. These procedures were used for all numerical integrations of partial differential equations reported in this chapter.

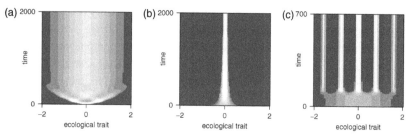

FIGURE 9.1. Numerical solutions of the evolutionary dynamics of phenotype distributions given by the partial differential equation (9.2). In panels (a) and (b), the carrying capacity function and the competition kernel are Gaussian ($\epsilon_K = \epsilon_\alpha = 0$). In panel (a), $\sigma_\alpha < \sigma_K$, which leads to the maintenance of variation, but not to multimodality. In panel (b), $\sigma_\alpha > \sigma_K$, which leads to the loss of variation. The distribution converges to a delta peak centered at 0 in the limit of large times. In panel (c), the carrying capacity is still Gaussian, but the competition kernel is platykurtic ($\epsilon_\alpha = 2$). $\sigma_\alpha < \sigma_K$ now leads to diversification in the form of multimodal equilibrium distributions. Parameter values were $K_0 = 1$, $\sigma_K = 1$, and $\sigma_\alpha = 0.5$ for panel (a), $K_0 = 1$, $\sigma_K = 1$, and $\sigma_\alpha = 1.2$ for panel (b), and $K_0 = 1$, $\sigma_K = 1$, and $\sigma_\alpha = 0.5$ for panel (c). The simulations were initialized with Gaussian distributions with mean 0 and variance 0.5.

For the competition model (9.2), the numerical simulations for $\epsilon_K = \epsilon_\alpha = 0$ indicate that the phenotype distribution indeed converges to the equilibrium $\tilde{\phi}$ given by (9.5) from arbitrary initial conditions (strictly speaking, this is only true for initial conditions that are >0 everywhere, because the model (9.2) does not include mutations; however, mutations could easily be incorporated, as we will see shortly, in which case any initial condition would converge to (9.5)). Thus, the distribution $\tilde{\phi}$ is globally stable (Figure 9.1(a)). Note that the equilibrium distribution $\tilde{\phi}$ corresponds to a polymorphic population with phenotypic variance $\sigma_K^2 - \sigma_\alpha^2$.

On the other hand, if $\sigma_\alpha > \sigma_K$ (and still with $\epsilon_K = \epsilon_\alpha = 0$), then simulations indicate that the system always converges to a solution consisting of a Dirac delta function with all its weight centered at $x = 0$ (Figure 9.1(b)). In other words, in this case the population becomes monomorphic for the "optimal" phenotype. Thus, the partial differential equation model for Gaussian carrying capacities and competition kernels exhibits a transition from a polymorphic to a monomorphic evolutionary regime as σ_α decreases below the value of σ_K. At this point, it is important to note that in general, the location of this transition between monomorphic and polymorphic dynamic regimes changes as the dimensionality of the phenotype space changes. While the region of monomorphic dynamics, $\sigma_\alpha > \sigma_K$, is relatively large in the models for one-dimensional phenotype spaces discussed above, the corresponding region of parameter space in models for the evolutionary dynamics of multidimensional phenotype distributions becomes smaller as the dimension of the phenotype space increases. Thus, diversification due to frequency-dependent competition generally tends to occur much more easily in high-dimensional phenotype spaces. This is explained in Box 9.1.

BOX 9.1

EVOLUTION OF DIVERSITY IN HIGH-DIMENSIONAL PHENOTYPE SPACES

This box describes extensions of the basic model (9.2) for evolutionary dynamics under frequency-dependent competition to high-dimensional phenotype spaces. We consider the Gaussian case, in which the carrying capacity, representing the stabilizing component of selection, and the competition kernel, representing the frequency-dependent component of selection, are given by expressions (9.3) and (9.4) with $\epsilon_K = \epsilon_\alpha = 0$. In this case model (9.2) for one-dimensional phenotype spaces always admits a stable equilibrium solution (Hernandez-García et al., 2009). If $\sigma_\alpha > \sigma_K$, the stabilizing component of selection dominates, and the equilibrium distribution consists of a Delta peak centered at the maximum of the carrying capacity, that is, at $x = 0$. If $\sigma_\alpha < \sigma_K$, frequency-dependent selection dominates, and the equilibrium distribution is given by (9.3). This distribution has a positive variance, and hence in this case phenotypic variation is maintained by frequency-dependent competition.

Model (9.2) can be extended to m-dimensional phenotype spaces by generalizing the exponents of the ecological functions (9.3) and (9.4) using quadratic forms. Quadratic forms are determined by two $m \times m$-matrices K and A, and if $x = (x_1, \ldots, x_m)$ and $y = (y_1, \ldots, y_m)$ are two m-dimensional phenotype vectors, the relevant quadratic forms are $x^T K x$ and $(x - y)^T A (x - y)$, where ^{T} denotes the transposed vector (and where $(x - y) = (x_1 - y_1, \ldots, x_m - y_m)$). The m-dimensional competition kernel and the carrying capacity then become

$$K(x) = K_0 \exp\left(-x^T K x\right)$$
$$\alpha(x, y) = \exp\left(-(x - y)^T A (x - y)\right). \tag{B9.1}$$

The evolutionary dynamics of m-dimensional phenotype distribution $\phi(x)$ is obtained using expressions (B9.1) as in model (9.2). The simplest case occurs when the two matrices K and AA are diagonal with elements $k_{11}, \ldots, k_{mm}$ and

BOX 9.1 (*continued*)

$a_{11}, \ldots, a_{mm}$, respectively. Then the functions (B9.1) are separable:

$$K(x) = K_0 \prod_{i=1}^{m} \exp\left(-x_{ii}^2 k_{ii}\right)$$

(B9.2)

$$\alpha(x, y) = \prod_{i=1}^{m} \exp\left(-(x_{ii} - y_{ii})^2 a_{ii}\right).$$

Note that comparison to the one-dimensional case can be made by the reparametrization $k_{ii} = 1/(2\sigma_{K,ii}^2)$ and $a_{ii} = 1/(2\sigma_{\alpha,ii}^2)$. Extrapolating from the one-dimensional case, it then follows that the m-dimensional equilibrium distributions of the evolutionary dynamics are given by multivariate Gaussian distributions having a positive variance in the direction of those phenotypic components i for which $a_{ii} > k_{ii}$, and having zero variance in the other phenotypic directions. In particular, the conditions for maintenance of diversity are essentially the same as in the one-dimensional case.

This changes when the quadratic forms K and A are not diagonal, and the difference can be most easily appreciated when $m = 2$. For argument's sake, let us assume that in each phenotypic direction the system is poised on the verge of diversification, which means that the diagonal elements of the two matrices K and A are equal, $k_{11} = a_{11}$ and $k_{22} = a_{22}$, that is, when considered alone each phenotypic dimensions corresponds to the case $\sigma_\alpha = \sigma_K$ in the one-dimensional model (eq. (9.2)). By rescaling the phenotypic axes we can further assume that $k_{ii} = a_{ii} = 1$ for $i = 1, 2$. Instead of assuming that the two-dimensional ecological functions are given by the products (B9.2), we now assume that these products also contain a cross term involving both phenotypic dimensions:

$$K(x) = K_0 \exp\left(-x_{11}^2\right) \exp\left(-x_{22}^2\right) \exp(-2x_{11}x_{22}k_{12})$$

$$\alpha(x, y) = \exp\left(-(x_{11} - y_{11})^2\right) \cdot \exp\left(-(x_{22} - y_{22})^2\right)$$

(B9.3)

$$\times \exp(-2(x_{11} - y_{11})(x_{22} - y_{22})a_{12}).$$

Note that this is a natural assumption, for it is generally unlikely that ecological properties of organisms will be determined multiplicatively by the various phenotypic components of high-dimensional phenotypes as in eq. (B9.2). Instead, generally such ecological properties will be determined by interactions between different phenotypic components, as reflected by the cross terms in (B9.3). Also note that the coefficients appearing in these cross terms have been denoted $2k_{12}$ and $2a_{12}$ for convenience, because with this notation, the quadratic forms K and A determining the ecological functions in (B9.1) are given by

$$K = \begin{pmatrix} 1 & k_{12} \\ k_{12} & 1 \end{pmatrix}$$

$$A = \begin{pmatrix} 1 & a_{12} \\ a_{12} & 1 \end{pmatrix}.$$

(B9.4)

Note that we are not making any assumptions about the sign of the off-diagonal elements k_{12} and a_{12}. The coordinate change $z_1 = (x_1 + x_2)/\sqrt{2}$ and $z_2 = (x_1 - x_2)/\sqrt{2}$ diagonalizes both matrices K and A, which in the new coordinates become

$$\hat{K} = \begin{pmatrix} 1 + k_{12} & 0 \\ 0 & 1 - k_{12} \end{pmatrix}$$

$$\hat{A} = \begin{pmatrix} 1 + a_{12} & 0 \\ 0 & 1 - a_{12} \end{pmatrix},$$

(B9.5)

where $1 \pm k_{12}$ and $1 \pm a_{12}$ are the eigenvalues of the matrices K and A. Thus, in the composite coordinates (z_1, z_2) the ecological functions are separable as in eq. (B9.2), and hence the evolutionary dynamics are determined by the eigenvalues $1 \pm k_{12}$ and $1 \pm a_{12}$. Generically, that is, unless $k_{12} = a_{12}$, it immediately follows that the condition for diversification is satisfied in one of the two composite directions: either $1 + a_{12} > 1 + k_{12}$, or $1 - a_{12} > 1 + k_{12}$ must hold. Therefore, adding generic interactions in the form of nonzero, off-diagonal elements in the quadratic forms K and/or A generates robust diversification in systems that are otherwise poised on the brink of diversification. In much the same way, interactions between phenotypic components can also generate diversification in systems lying robustly in parameter regions leading to the loss of diversity without interactions (Doebeli & Ispolatov, 2010a).

It can be shown analytically that the same effect occurs for dimensions $m > 2$ (Doebeli & Ispolatov, 2010a). In fact, the diversifying effect of interactions between components of high-dimensional phenotypes becomes stronger with increasing dimension of the phenotype space. This is illustrated in Figure 9.B1, which shows the probability of diversification as a function of the dimension m and of the magnitude of the off-diagonal elements in the quadratic forms K and A. The figure illustrates that the diversifying effect of phenotypic complexity in the form of interactions between components of high-dimensional phenotypes in determining ecological properties is a very strong and general effect.

BOX 9.1 (*continued*)

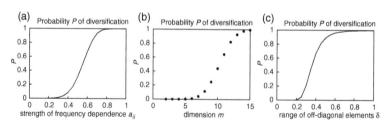

FIGURE 9.B1. Effects of interactions between phenotypic components on the probability of diversification. Starting with diagonal matrices K with identical diagonal elements $k_{ii} = 1$, and A with identical diagonal elements $a_{ii} < 1$, in the ecological functions (B9.2), the condition $a_{ii} > k_{ii}$ for the maintenance of diversity is not satisfied in any phenotypic component. The panels show the probability that addition of uniformly and randomly chosen off-diagonal elements in the range $[-\delta, \delta]$ at each off-diagonal position in both matrices results in the condition for diversification $a > k$ satisfied in at least one (composite) direction after simultaneous diagonalization of the resulting quadratic forms (see Doebeli & Ispolatov (2010a) for details). This probability is shown as a function of the diagonal elements a_{ii} in the starting matrix A (panel (a)), as a function of the dimension of phenotype space m (panel (b)), and as a function of the parameter δ determining the range from which the off-diagonal elements are drawn (panel (c)). Results were obtained from 10^5 random samplings of off-diagonal elements k_{ij} and a_{ij}, with standard diagonalization techniques applied to the resulting quadratic forms. The parameters that are fixed are the dimension of phenotype space $m = 5$ and the range $\delta = 0.2$ for panel (a), $a_{ii} = 0.2$ and $\delta = 0.2$ for panel (b), and $m = 5$ and $a_{ii} = 0.2$ for panel (c). Panel (a) shows that the probability of diversification becomes large even if $k_{ii} - a_{ii}$ is substantially larger than δ, and panel (b) shows that even for k_{ii} much larger than a_{ii}, that is, even when stabilizing selection is overwhelming frequency dependence in each component direction, the probability of diversification due to off-diagonal elements approaches one for high dimensions m. Note also the saturating effect of increasing the range from which off-diagonal elements are drawn (panel (c)). After Doebeli & Ispolatov (2010a).

In particular, this effect does not depend on the sign of the interactions, and it also occurs for non-Gaussian models based on multidimensional extensions of the general ecological functions (eqs. (9.2) and (9.3)) (Doebeli & Ispolatov, 2010a). Because ecological properties of many organisms are likely to be determined nonmultiplicatively by many different phenotypic components, these results imply that the conditions for frequency-dependent competition to maintain genetic and phenotypic diversity are likely to be satisfied in many natural ecosystems.

Returning to competition models with one-dimensional phenotypes of the form (9.2), we have seen in Chapter 3 that the corresponding adaptive dynamics model for resource competition undergoes a similar transition from monomorphic to polymorphic dynamic regimes. For $\sigma_\alpha > \sigma_K$, the adaptive dynamics model exhibits an evolutionarily stable strategy (ESS) at $x = 0$ (see Chapter 3), which corresponds well with model (9.2) showing convergence to a monomorphic population centered at $x = 0$ (or a Gaussian equilibrium distribution with a small variance in the presence of mutations). As σ_α becomes smaller than σ_K, the adaptive dynamics generates polymorphisms. However, the nature of these polymorphisms appears to be very different in the two approaches. While the adaptive dynamics model exhibits evolutionary branching for $\sigma_\alpha < \sigma_K$, and hence the emergence of two distinct phenotypic clusters, the pde model (9.2) converges to the unimodal phenotype distribution $\tilde{\phi}$ with a large variance. This equilibrium distribution describes a polymorphic population, but it does not show diversification in the form of pattern formation into multiple modes that would correspond to distinct phenotypic clusters.

Nevertheless, upon closer inspection the differences between the two approaches seem to disappear. In fact, as we have seen in Chapter 3, the adaptive dynamics model with Gaussian ecological functions shows repeated evolutionary branching events, and this phenomenon seems to be independent of parameters as long as $\sigma_\alpha < \sigma_K$. Thus, it seems reasonable to conjecture that in theory, for any $\sigma_\alpha < \sigma_K$, an infinite sequence of evolutionary branchings in the adaptive dynamics model would eventually generate a discretized version of the continuous phenotype distribution $\tilde{\phi}$. This would essentially reconcile the adaptive dynamics models with the pde models, and show that diversification in the form of coexistence of clearly distinct phenotypic clusters does not occur in either model.

However, we have also seen in Chapter 3 that the occurrence of repeated branching events in the adaptive dynamics models critically depends on the choice of functions for the carrying capacity and the competition kernel. In particular, for some choices of these functions evolutionary branching only occurs once, resulting in the coexistence of two distinct phenotypic clusters that do not further diversify. Analogous phenomena can be observed in pde models. For example, if we choose $\epsilon_K = \epsilon_\alpha = 2$ in expressions (9.3) and (9.4) for K and α, numerical simulation of the pde models shows that the phenotype dynamics converges to an equilibrium distribution that exhibits pattern formation, that is, clearly distinct modes that correspond to different phenotypic clusters. An example is shown in Figure 9.1(c).

That evolution can result in multimodal phenotypic clustering rather than in continuous, unimodal phenotype distribution has already been observed by Sasaki & Ellner (1995) for frequency-independent evolution in fluctuating environments. For frequency-dependent competition, the conditions under which models of the type (9.2) exhibit multimodal pattern formation have been studied extensively in a number of recent papers (Barabas & Meszéna, 2009; Fort et al., 2009; Gyllenberg & Meszéna, 2005; Hernandez-García et al., 2009; Pigolotti et al., 2007; Pigolotti et al., 2010). For competition kernels that only depend on the phenotypic distance $z = x - y$, such as the ones used here, these conditions can be expressed in terms of the Fourier transform

$$\hat{\alpha}(k) = \int_{-\infty}^{\infty} \alpha(z) \exp(-ikz) dz. \tag{9.6}$$

A competition kernel α is called positive if its Fourier transform is a positive function, that is, if $\hat{\alpha}(k) > 0$ for all k. For example, competition kernels of the form (9.4) are positive if and only if $\epsilon_\alpha \leq 2$ (Hernandez-García et al., 2009). In particular, Gaussian competition kernels are positive, and in fact lie exactly at the margin of positivity in terms of the parameter ϵ_α, which is essentially the reason why (deterministic) Gaussian competition models are structurally unstable. Hernandez-García et al. (2009) have shown that if the partial differential equation (9.2) has an equilibrium solution $\tilde{\phi}(x)$ that is strictly positive (i.e., $\tilde{\phi}(x) > 0$ for all x), then this solution is dynamically stable if and only if the Fourier transform of the competition kernel is positive. As a consequence, the dynamics given by (9.2) may not converge to unimodal equilibrium distributions, and instead exhibit multimodal pattern formation, if either a strictly positive unimodal equilibrium does not exist, or if the competition does not have a positive Fourier transform. In fact, this leaves ample opportunity for pattern formation, as is illustrated in Figure 9.2, which shows that pattern formation is a generic phenomenon in pde models of the type (9.2) that can occur for a wide range of the parameters ϵ_K and ϵ_α.

It is apparent from Figure 9.2 that the Gaussian case (i.e., the case in which both K and α have Gaussian form) is a special case that should be treated with caution. For many modeling purposes it is tempting to use Gaussian functions, because they have some nice analytical properties. In particular, a convolution of two Gaussian functions is again a Gaussian function, which makes it, for example, possible to find the analytical expression (9.5) for the equilibrium distribution of the pde model (9.2). However, this analytical tractability may come at the cost of structural instability, meaning that models that are "nearby" in modeling space exhibit qualitatively different behavior (such as when models with ϵ_K and ϵ_α small, but nonzero, exhibit pattern formation, as shown in Fig. 9.2). Mathematical proofs of various aspects of the structural

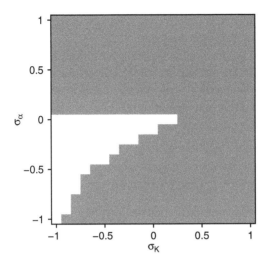

FIGURE 9.2. Multimodal pattern formation as a function of the parameters ϵ_K and ϵ_α. For the figure, the widths σ_K and σ_α of the carrying capacity and the competition kernel were chosen such that the ecological conditions for the maintenance of phenotypic variation are satisfied in the Gaussian case (i.e., $\sigma_\alpha < \sigma_K$). The parameters ϵ_K and ϵ_α in (9.3) and (9.4) were varied between -1 and 1 in steps of 0.1 (so that the exponents in the carrying capacity and the competition kernels ranged from 1 to 3), and for each parameter combination, model (9.2) was solved numerically and run to equilibrium, which was then checked for multimodality. Gray areas represent parameter combinations that led to multimodal pattern formation. The Gaussian case corresponds to $\epsilon_\alpha = \epsilon_K = 0$ and does not lead to pattern formation. Note that values of ϵ_α that are >0 generate nonpositive competition kernels (Hernandez-García et al., 2009). In the example shown, pattern formation occurs for all such competition kernels. But pattern formation can also be observed for competition kernels with positive Fourier transform ($\sigma_\alpha \leq 0$, lower right of the panel), which indicates the lack of strictly positive unimodal equilibrium distribution for those parameter values. Parameter values were $r = 1$, $K_0 = 1$, $\sigma_K = 2$, and $\sigma_\alpha = 1$. Results appear to be independent of initial conditions.

instability of Gaussian competition models have for example been given by Gyllenberg & Meszéna (2005) and by Hernandez-García et al. (2009). In fact, as we will see shortly, even models that are based on Gaussian functions sometimes exhibit dynamic behavior that is very different from that expected based on the analytical properties of Gaussian functions.

It is also worth pointing out that non-Gaussian models not only often exhibit qualitatively different types of diversification, but the conditions for diversification can also be qualitatively different. For example, the non-Gaussian model with $\epsilon_\alpha = \epsilon_K = 2$ in (9.3) and (9.4) exhibits diversification for

$\sigma_\alpha < \sqrt{2}\sigma_K$, rather than only for $\sigma_\alpha < \sigma_K$, as in the Gaussian case. This is explained in Box A.1 in the Appendix.

System (9.2) is the simplest possible pde model for frequency-dependent competition and only includes the most basic genetics. A straightforward way to incorporate more genetic realism is to assume that reproduction, while still asexual, is not faithful and instead results in mutations that are described by a mutation kernel

$$M(x, y) = \frac{1}{\sqrt{2\pi}\,\sigma_{mut}} \exp\left(-\frac{(x-y)^2}{2\sigma_{mut}^2}\right). \tag{9.7}$$

Here $M(x, y)$ is the probability that the offspring of a mother with phenotype y has a phenotype in the interval $(x, x + dx)$. (Note that $M(x, y)$ is normalized so that its integral over x is equal to 1, corresponding to the fact that each offspring has some phenotype.) Incorporating mutation into the birth term of system (9.2) results in the pde

$$\frac{\partial \phi(x)}{\partial t} = r \int M(x, y)\phi(y)dy - \frac{r\phi(x)\int \alpha(x, y)\phi(y)dy}{K(x)}. \tag{9.8}$$

In general, as long as the width σ_{mut} of the mutational kernel is narrow enough, introducing mutation does not change the qualitative behavior of the model, and instead only has the expected blurring effect. For example, with Gaussian K and α, the model still has a Gaussian equilibrium (Fig. 9.3(a)), and with $\sigma_\alpha > \sigma_K$, numerical simulations show that the phenotype dynamics now always converges to a narrow unimodal distribution centered at $x = 0$, rather than to the delta peak with 0 width seen in the absence of mutation (Fig. 9.3(b)). This new equilibrium distribution essentially reflects a mutation-selection balance. For non-Gaussian models, pattern formation in the form of multimodality can be readily observed despite the peak-eroding action of mutation, as illustrated in Figure 9.3(c). Thus, as expected, mutation alone does not prevent diversification as long as the mutation kernel is sufficiently narrow.

Challenge: Prove that model (9.8) with Gaussian carrying capacity and Gaussian competition kernel always has a Gaussian equilibrium distribution.

To take the genetics one step further and describe the effects of mating in sexual populations, I follow the approach based on segregation kernels used for individual-based models in Chapter 4. Thus, I assume that a mating between

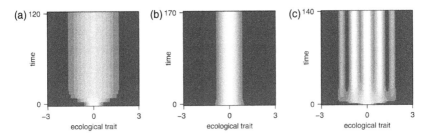

FIGURE 9.3. Numerical solutions of the evolutionary dynamics of phenotype distributions given by the partial differential model (9.8). In panels (a) and (b), the carrying capacity function and the competition kernel are Gaussian ($\epsilon_K = \epsilon_\alpha = 0$). In panel (a), $\sigma_\alpha < \sigma_K$, which leads to the maintenance of variation, but not to multimodality. In panel (b), $\sigma_\alpha > \sigma_K$, which leads to a much narrower equilibrium distribution, in which some variation is nevertheless maintained due to mutations. In panel (c), the carrying capacity is still Gaussian, but the competition kernel is platykurtic ($\epsilon_\alpha = 2$). Despite a fairly large width of the mutation kernel, $\sigma_\alpha < \sigma_K$ can again lead to diversification in the form of multimodal equilibrium distributions. Parameter values were the same as in Figure 9.1, and $\sigma_{mut} = 0.1$ in panels (a) and (b), and $\sigma_{mut} = 0.2$ in panel (c).

parents of phenotypes u and v results in a distribution of offspring $S_{\frac{u+v}{2}}(x)$, which is a Gaussian (normal) distribution with variance σ_S^2 and centered at the midparent value $(u + v)/2$, where σ_S is a system parameter. As discussed in Chapter 4, the segregation kernel is essentially an empirical quantity, and is not derived from any underlying genetic assumptions. In principle, it is of course possible to use different segregation kernels, and to accommodate different genetic scenarios by using segregation kernels whose shape parameters may depend on the parent values x and y, as well as on the current phenotype distribution ϕ.

I further assume that mate choice is based on the phenotypes of potential mating partners. Again I assume that assortative mate choice is unilateral and based on the (normalized) choice function

$$A(u, v) = \frac{1}{\sqrt{2\pi}\,\sigma_A} \exp\left[-\frac{(x - y)^2}{2\sigma_A^2}\right], \tag{9.9}$$

which describes the relative preference an individual of a given phenotype u has for mates with phenotype v. For a given phenotype distribution $\phi(v)$, the integral

$$N(u) = \int A(u, v)\phi(v)\,dv \tag{9.10}$$

is a measure for the total amount of mates available to an u-individual. As a baseline, I want to control for costs of assortment due to being rare and choosy by letting all individuals having the same total per capita birth rate. This is done by using the quantity $N(u)$ for normalization. Then, the probability $P(u, v)$ that a u-individual mates with an individual of phenotype in the interval $(v, v + dv)$ is

$$P(u, v) = \frac{A(u, v)\phi(v)dv}{N(u)}. \tag{9.11}$$

Note that normalizing by $N(u)$ implies $\int P(u, v)dv = 1$, which in turn implies that there is no direct cost to being choosy and/or rare. (However, note that, as in Chap. 4, rare individuals are chosen less often as mating partners when mating is assortative.) We will consider direct costs to assortment later.

If a u-individual mates with a v-individual, it produces x-offspring at rate $S_{\frac{u+v}{2}}(x)$, where S is the segregation kernel. Since the per capita birth rate is r and there are $\phi(u)$ u-individuals, the rate at which u-individuals produce x-offspring through matings with individuals with phenotype in $(v, v + dv)$ is $r\phi(u)P(u, v)S_{\frac{u+v}{2}}(x)$. Finally, the total rate of change in $\phi(x)$ due to birth is obtained by integrating this over u:

$$\left.\frac{\partial\phi(x)}{\partial t}\right|_{birth} = \int r\phi(u)P(u, v)S_{\frac{u+v}{2}}(x)dvdu$$

$$= r\int \frac{\phi(u)}{N(u)}\left(\int A(u, v)\phi(v)S_{\frac{u+v}{2}}(x)dv\right)du. \tag{9.12}$$

The rate of change due to competition (i.e., death) remains the same as in the asexual model, so that the full pde model for sexual populations with assortative mating takes the form

$$\frac{\partial\phi(x)}{\partial t} = r\int \frac{\phi(u)}{N(u)}\left(\int A(u, v)\phi(v)S_{\frac{u+v}{2}}(x)dv\right)du$$

$$- \frac{r\phi(x)\int \alpha(x, y)\phi(y)dy}{K(x)}. \tag{9.13}$$

We always assume that the segregation kernel S and the mate choice function A are Gaussian, but the carrying capacity function K and the competition kernel α may or may not be Gaussian. If they are Gaussian, it is straightforward to see, using the nice mathematical properties of Gaussian functions,

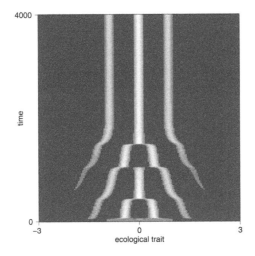

FIGURE 9.4. Pattern formation in the evolutionary dynamics of phenotype distributions in sexual population with Gaussian carrying capacity and competition kernel. The figure shows that adaptive speciation can occur even in the purely Gaussian case if mating is assortative. Parameter values were $r = 1$, $K_0 = 1$, $\sigma_K = 1$, $\sigma_\alpha = 0.5$, $\sigma_s = 0.05$, $\sigma_A = 0.2$, and the population was initialized with a Gaussian distribution centered at zero and with a variance of one. Introducing some perturbation to this initial condition to make it asymmetric does not change the equilibrium distribution, which appears to be attracting all initial conditions as long as the initial distribution is not too narrow.

that the sexual system (9.13) has again a Gaussian equilibrium distribution whose variance σ_{eq} can be calculated from the following expression:

$$\frac{2(\sigma_A^2 + \sigma_{eq}^2)^2}{4\sigma_{eq}^4(\sigma_S^2 + \sigma_{eq}^2) + 2\sigma_A^4(2\sigma_S^2 + \sigma_{eq}^2) + \sigma_A^2\sigma_{eq}^2(8\sigma_S^2 + 5\sigma_{eq}^2)}$$
$$= \frac{1}{2\sigma_{eq}^2} + \frac{1}{2(\sigma_\alpha^2 + \sigma_{eq}^2)} - \frac{1}{2\sigma_K^2}. \tag{9.14}$$

Challenge: Prove the previous statement.

However, unlike in the asexual model, this Gaussian equilibrium distribution need not be dynamically stable. Instead, for intermediate strengths of assortment (i.e., intermediate values of σ_A), the pde dynamics can converge to a multimodal pattern and hence exhibit diversification, as is illustrated in Figure 9.4. This again emphasizes that the Gaussian case can be tricky: not only may such equilibrium distribution not exist for models that are arbitrarily close to the Gaussian case (Gyllenberg & Meszéna, 2005), but even if they do exist, as, for example, in case of Gaussian K and α, they may be dynamically unstable.

The following simple argument, which is due to Slava Ispolatov, gives some indication of possible causes of pattern formation in the sexual system. It is based on considering harmonic distributions ϕ of the form

$$\phi(x, t) = \phi_0(t) \exp(ikx), \tag{9.15}$$

that is, distributions that oscillate with frequency k in the x-direction. Evaluating the Gaussian integrals on the right-hand side of the birth term, eq. (9.12), yields

$$\left.\frac{\partial \phi(x, t)}{\partial t}\right|_{birth} = \phi(x, t) \left(k^2(3\sigma_A^2/8 - \sigma_S^2/2)\right). \tag{9.16}$$

If $\sigma_A^2 > 4\sigma_S^2/3$, the solution to this equation grows indefinitely, making the amplitude of the oscillations in x larger and larger. This shows that if σ_A is increased from lower to higher values (i.e., if the strength of assortment is decreased from high values), oscillations in x can grow indefinitely, which supports the observation of dynamic instability of unimodal distributions occurring as σ_A is increased. (Note that these arguments essentially only hold as long as assortment is not too weak, for it is intuitively obvious that very high σ_A, which essentially correspond to random mating, should always generate a unimodal equilibrium solution.)

We already know from the asexual systems (9.2) and (9.8) that pattern formation can occur due to purely ecological interactions if the carrying capacity function K or the competition kernel α are non-Gaussian. It thus comes as no surprise that pattern formation can also be seen in the sexual model for non-Gaussian K and α, as illustrated in Figure 9.5. In fact, Doebeli et al. (2007) have shown that pattern formation occurs in the sexual model for a wide range of assortment strengths and functional forms of K and α. Note that pattern formation in sexual species is tantamount to adaptive speciation, for when the equilibrium of the sexual model has multiple modes, the different modes effectively correspond to different incipient species that are reproductively isolated due to assortative mating. Overall, these results show that adaptive diversification and speciation not only occur in models based on adaptive dynamics, but also in the "maximally polymorphic" pde models studied in this chapter.

In fact, the sexual pde models exhibit another class of very interesting dynamics, which occurs when not only the unimodal distributions are unstable, but essentially any distribution is dynamically unstable. In principle, this

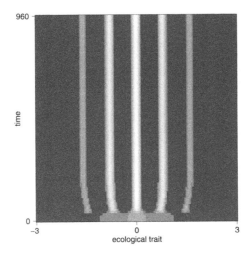

FIGURE 9.5. Pattern formation in the evolutionary dynamics of phenotype distributions in sexual population with assortative mating. The carrying capacity is Gaussian, but the competition kernel is not ($\epsilon_\alpha = 2$). Other parameters are as in Figure 9.4. The equilibrium distribution appears to be independent of initial conditions.

could happen if the stability analysis outlined before reveals instability of all possible distributions. In this case, the pde model will exhibit nonstationary behavior, that is, fluctuating dynamics. Two such scenarios are illustrated in Figure 9.6. In the first, the populations fluctuate between different multimodal states, and in the second the population seems to be poised in a unimodal distribution, but is intermittently subject to large phenotypic fluctuations. The first behavior could be classified as an irregular limit cycle in the space of distributions, and it is interesting to note that the dynamics shown do not appear to settle on a true cycle, despite the symmetry of the model (9.13). In fact, the dynamics shown in Figure 9.6(a) are the deterministic equivalent of the dynamics of individual-based sexual competition models shown in Figure 4.3. As in the case of the individual-based models, it seems important to note that the fluctuating dynamics in the deterministic model (9.13) seems to be ultimately caused by the birth term, that is, by the genetics. In fact, I have never seen fluctuating dynamics in the corresponding asexual models (9.2) and (9.8). Thus, it is not the ecological interactions per se that cause the temporal instability, which would, in any case, be surprising, because the ecology is based on the simple logistic competition model, which can generally be expected to exhibit very stable dynamics. Instead, it is the mixing of phenotypes through mating that, in conjunction

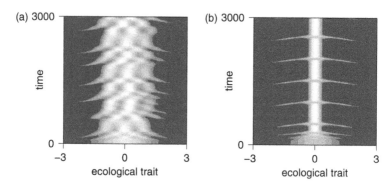

FIGURE 9.6. Nonequilibrium dynamics of phenotype distributions in sexual competition models. Panel (a) shows an example of an irregular limit cycle. The simulation was started with asymmetric initial conditions, and despite the symmetric nature of the model, asymmetries in the phenotype distribution appear to persist indefinitely, leading to complicated evolutionary dynamics. Panel (b) shows an example of intermittent dynamics. This simulation was started with symmetric initial conditions, and the dynamics converges on a limit cycle, in which the system is close to a unimodal equilibrium for most of the time, with short intermittent bursts of bimodality that seem to originate at the edges of the phenotype distribution where densities are very low. Parameter values were $r = 1$, $K_0 = 1$, $\sigma_K = 1$, $\sigma_\alpha = 0.5$, $\sigma_s = 0.1$, $\sigma_A = 0.2$, and the population was initialized with a Gaussian distribution with variance 1 and centered at 0.4 for panel (a), and at 0 for panel (b).

with the ecological interactions, causes the fluctuations. More precisely, it seems that instabilities tend to be due the fact that rare phenotypes have the same birth rate as common phenotypes despite mating being assortative. This makes it possible for the phenotype distribution to repeatedly grow in regions where it has very low densities, as can be seen in the examples shown in Figure 9.6.

One way to see that the no-cost assumption of assortment indeed tends to facilitate fluctuations is to introduce costs in the form of Allee effects, which can be done along the same lines as for the individual-based models in Chapter 4. Recall that $N(u)$, given by eq. (9.10), is a measure for the total amount of available mates in the sense that when mating is assortative, $N(u)$ will be smaller for rare phenotypes than for common ones. Thus, if we multiply the birth rate of u-individuals by a monotonically increasing function $f(N(u))$, we introduce a cost to being rare when mating is assortative. Following Noest (1997) we use the function $f(z) = z/(\eta + z)$, where η is a parameter that determines the severity of the costs, with $\eta = 0$ corresponding to the no-cost case. If we multiply the density of u-individuals, $\phi(u)$, by $f(N(u))$ in the birth term

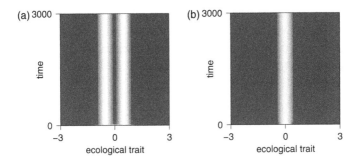

FIGURE 9.7. Phenotype dynamics of sexual models with Allee effect. The figure is the same as Figure 9.6(a), but with an Allee effect. The Allee effect stabilizes the dynamics and generates multimodal (panel (a)) or unimodal (panel (b)) equilibrium distributions, depending on initial conditions. Parameter values were the same as in Figure 9.6(a), with an Allee effect introduced by setting $\eta = 0.1$. The populations were initialized with a Gaussian distribution centered at 0 with variance 1 for panel (a), and with variance 0.1 for panel (b).

of the sexual system (9.13), we obtain the pde

$$\frac{\partial \phi(x)}{\partial t} = r \int \frac{\phi(u)}{c + N(u)} \left(\int A(u, v)\phi(v)S_{\frac{u+v}{2}}(x)dv \right) du$$
$$- \frac{r\phi(x) \int \alpha(x, y)\phi(y)dy}{K(x)}. \tag{9.17}$$

Using similar arguments as given earlier for the destabilizing effects of assortment, it can be shown that the Allee effect introduced by adding the parameter η does indeed have a stabilizing effect, as it makes the conditions for instability more restrictive. Figure 9.7 illustrates this stabilizing effect by showing that an Allee effect can lead to stable equilibrium distribution dynamics in systems showing fluctuating dynamics in the absence of Allee effects. Note that in the presence of Allee effects, the evolutionary dynamics may depend on initial conditions. This is also illustrated in Figure 9.7 and is in accordance with the results for individual-based models with Allee effects described in Chapter 4. Nevertheless, Figure 9.7(a) shows that multimodal phenotype distributions can emerge even if there are costs to assortative mating in the form of Allee effects. This again supports the general contention that pde models for sexual populations can generate adaptive speciation due to pattern formation in phenotype space under a wide range of conditions.

Challenge*: Develop partial differential equation models that incorporate the evolution of assortative mating along the lines described in Chapter 4 for individual-based models. For this, one needs to consider two-dimensional phenotype spaces, in which one dimension corresponds to the ecological trait, and the other to the degree of assortment. As a consequence, numerical solutions of the partial differential equation models will become computationally more expensive.

9.2 PARTIAL DIFFERENTIAL EQUATION MODELS FOR PREDATOR-PREY INTERACTIONS

Here I extend the predator-prey models studied using adaptive dynamics in Chapter 5 to polymorphic populations and investigate coupled partial differential equations for the dynamics of phenotype distributions in both the prey and the predator. As in Chapter 5 I assume that prey and predator individuals are characterized by quantitative traits x and y, respectively, which determine predator attack rates. The prey trait x also determines the carrying capacity. After suitable rescaling, we assume that attack rates are highest when prey and predator phenotypes coincide, and decline according to the Gaussian function $\beta(x, y) = \beta_0 \exp\left(-\frac{(x-y)^2}{2\sigma_\beta^2}\right)$ with increasing phenotypic distance between prey and predator. Let $\phi(x)$ and $\psi(y)$ be the density distributions of the prey and predator phenotypes. The partial differential equation describing the dynamics of the prey distribution takes the following form:

$$\frac{\partial \phi(x)}{\partial t} = r \int B_\phi(x, z)\phi(z)dz - \phi(x)\frac{r\phi(x)\int \phi(z)dz}{K(x)}$$

$$- \phi(x)\int \beta(x, y)\psi(y)dy. \tag{9.18}$$

Here $B_\phi(x, z)$ is a prey birth term to be explained shortly. The second term on the right-hand side describes density dependence in the prey: $\int \phi(z)dz$ is the total prey density, and $K(x)$ is, as before, the carrying capacity function given by eq. (9.3). Note that there is no frequency dependence assumed for competition in the prey. The third term on the right-hand side describes the effects of predation: $\int \beta(x, y)\psi(y)dy$ is the total attack rate on prey of type x, and we assume a linear functional response.

For asexual reproduction, the function B_ϕ is simply the mutation kernel of the prey, that is, the probability that the offspring of a mother with phenotype z has a phenotype in the interval $(x, x + dx)$:

$$B_\phi(x, z) = M_{prey}(x, z) = \frac{1}{\sqrt{2\pi}\,\sigma_{mut,prey}} \exp\left(-\frac{(x-z)^2}{2\sigma_{mut,prey}^2}\right). \tag{9.19}$$

If reproduction is sexual, the function B_ϕ is defined as in the competition models in the previous section, using a segregation kernel S_{prey} and assuming that assortative mating is based on a (unilateral) choice function A_{prey}. As before, these functions are assumed to be normal functions with variances $\sigma^2_{S,prey}$ and $\sigma^2_{A,prey}$. Thus, B_ϕ takes the form

$$B_\phi(x, z) = \frac{1}{N(z)} \int A_{prey}(z, u)\phi(u)S_{prey, \frac{z+u}{2}}(x)du. \qquad (9.20)$$

Here $N(z) = \int A_{prey}(z, u)\phi(u)du$ is again the total amount of available mates.

The partial differential equation describing the dynamics of the predator distribution takes the following form:

$$\frac{\partial \psi(y)}{\partial t} = c \int B_\psi(y, z)\Psi(z)dz - d\psi(y) \qquad (9.21)$$

Here $\Psi(z)$ is the amount of prey captured by predators of type z:

$$\Psi(z) = \psi(z) \int \beta(z, x)\phi(x)dx, \qquad (9.22)$$

and c is the rate at which captured prey is converted into predator offspring. d is the per capita predator death rate, and $B_\psi(y, z)$ is again a birth term. For asexual reproduction, the function B_ψ is simply the mutation kernel of the predator:

$$B_\psi(y, z) = M_{pred}(y, z) = \frac{1}{\sqrt{2\pi}\sigma_{mut,pred}} \exp\left(-\frac{(y - z)^2}{2\sigma^2_{mut,pred}}\right). \qquad (9.23)$$

If reproduction is sexual, the function B_ψ is defined using a normal segregation kernel S_{pred} with variance $\sigma^2_{S,pred}$, and assuming that assortative mating is based on a normal function A_{pred} with variance $\sigma^2_{A,pred}$:

$$B_\phi(y, z) = \frac{1}{N(z)} \int A_{pred}(z, v)\psi(v)S_{pred, \frac{z+v}{2}}(y)dv, \qquad (9.24)$$

where $N(z) = \int A_{pred}(z, v)\psi(v)dv$ is the total amount of available mates.

The coupled partial differential equations (9.18) and (9.21) can be integrated numerically to study the dynamics of the prey and predator phenotype distributions. It is interesting to note that in contrast to the asexual competition model (9.2), the asexual predator-prey model given by (9.18) and (9.21) without mutation, that is, with delta functions for the mutation kernels, never admits solutions consisting of Gaussian equilibrium distribution with positive variance. Therefore, in both the prey and the predator diversification is always expected to correspond to multimodal pattern formation. If a population is sexual, multimodality is only possible if mating is assortative, in which case different phenotypic modes represent different incipient species. In the corresponding adaptive dynamics models of Chapter 5, we have seen that diversification tends to occur when the parameter σ_β, which determines how fast attack rates decrease with phenotypic distance between prey and predator, is decreased compared to the parameter σ_K, which determines the width of the prey carrying capacity. Qualitatively similar phenomena can be seen in both the asexual and sexual pde models, in which pattern formation readily occurs when σ_β is small enough. Examples of pattern formation in asexual models are shown in Figure 9.8. Note that as expected from the model setup, diversification can occur either only in the prey, or in both predator and prey, but not in the predator alone.

Challenge: Prove that the coupled partial differential equation model given by (9.18) and (9.21) for asexual reproduction without mutation, that is, with delta functions for the mutation kernels in both the prey and the predator, does not admit Gaussian equilibrium distributions with positive variance for any choice of parameters.

The pde models can also exhibit nonstationary dynamics, as shown in Figure 9.9. In the case shown, phenotype distributions in both species are essentially bimodal. In each species, the two modes are stationary in phenotype space, but they undergo temporal fluctuations in abundance in such a way that when one of the modes in a species is at its maximum abundance, the other mode of that species is at its minimum, and vice versa. That is, in each species the two modes fluctuate out of phase and continually replace each other. Thus, diversification essentially results in two temporally fluctuating predator-prey species pairs, and in each pair, the fluctuations in the predator lag behind the fluctuations in the prey, exhibiting classical ecological predator-prey cycles. Moreover, in both the prey and the predator the average phenotype undergoes fluctuations due to the temporal fluctuations of the different modes of the phenotype distributions. Because of the ecological lag between the predator and the prey, these fluctuations in the mean phenotype exhibit a cyclic evolutionary arms race, as shown in Figure 9.9(e) and (f). This arms race of the mean phenotypes closely corresponds to similar types of dynamics observed in adaptive dynamics models (Dieckmann et al., 1995), and in multilocus and quantitative genetic models (Gavrilets, 1997; Nuismer et al., 2005).

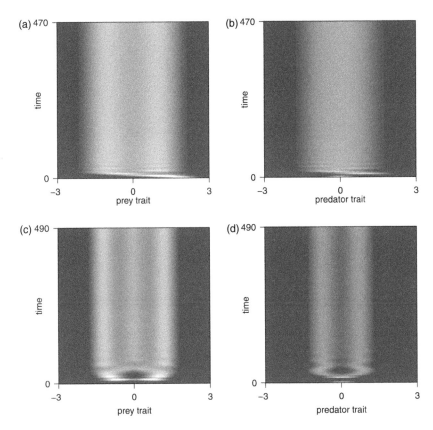

FIGURE 9.8. Pattern formation in the evolutionary dynamics of phenotype distributions in asexual predator-prey models. Panels (a) and (b) show an example in which the prey becomes bimodal and the predator remains unimodal at equilibrium. Panels (c) and (d) show an example where both the prey and the predator show multimodal pattern formation. The two scenarios only differ in the width of the mutation kernel and thus illustrate that this parameter can have qualitative effects on the evolutionary dynamics. Parameter values were $K_0 = 1$, $\sigma_K = 1$, $\beta_0 = 10$, $\sigma_\beta = 0.5$, $\sigma_{mut,prey} = \sigma_{mut,pred} = 0.2$, $c = 1$, and $d = 1$ for panels (a) and (b). Panels (c) and (d) are the same, except that $\sigma_{mut,prey} = \sigma_{mut,pred} = 0.05$. The equilibrium distributions appear to be independent of initial conditions.

Figure 9.10 shows examples of the dynamics of phenotype distributions in sexual predator-prey models. The examples illustrate that adaptive speciation in the form of multimodal pattern formation is possible if mating is assortative. The figure also illustrates the potential effects of the strength of assortment, with weaker assortment (Figure 9.10(a)–(d)) leading to more complicated dynamics, which can be stabilized by decreasing the width of the

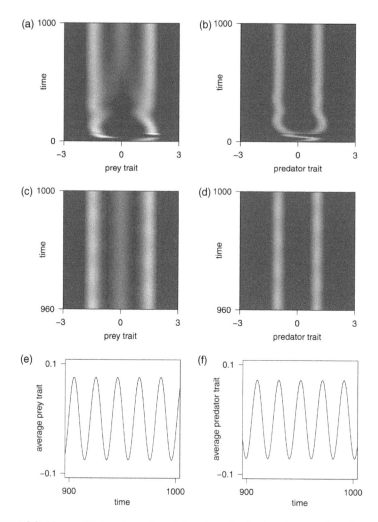

FIGURE 9.9. Nonequilibrium dynamics of phenotype distributions in asexual predator-prey models. Panels (a) and (b) show an example in which both prey and predator distributions become multimodal, with the different modes undergoing temporal fluctuations in abundance while remaining stationary in phenotype space. The temporal fluctuations occur on short time scales and can be seen when the time series is magnified (panels (c) and (d)). Despite stationary locations of the various modes in phenotype space, the out-of-phase temporal fluctuations in abundance generate fluctuations of the average phenotype, as shown in panels (e) and (f). Note that the amplitude of the fluctuations of the average is much smaller than the distance between the modes of the phenotype distributions (which is simply a consequence of taking the average). Nevertheless, it is clear that the mean predator and prey phenotypes undergo a cyclic arms race. Parameter values were $K_0 = 1$, $\sigma_K = 1$, $\beta_0 = 10$, $\sigma_\beta = 0.7$, $\sigma_{mut,prey} = \sigma_{mut,pred} = 0.015$, $c = 1$, and $d = 1$. The long-term dynamics appear to be independent of initial conditions.

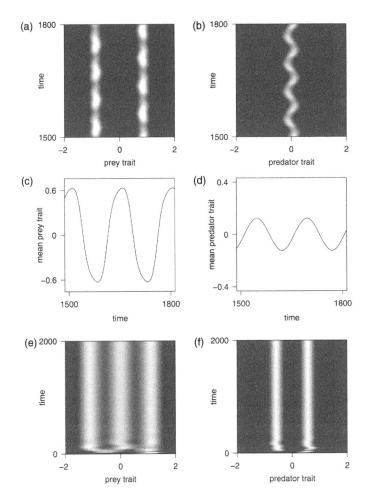

FIGURE 9.10. Evolutionary dynamics of phenotype distributions in sexual predator-prey models. Panels (a) and (b) show an example with nonequilibrium dynamics, similar to the case shown in Figure 9.9. Note, however, that in the sexual example shown both the prey and the predator not only undergo temporal fluctuations in abundance in each mode of the distribution, but the modes themselves also undergo fluctuations in their locations in phenotype space. This is particularly clear for the dynamics of the (unimodal) predator distribution, which shifts cyclically toward the more abundant of the two modes of the prey distribution. This again generates an evolutionary arms race, which is reflected in the dynamics of the mean phenotypes (panels (c) and (d)). Panels (e) and (f) illustrate the effect of increasing the strength of assortment in the prey, which stabilizes the dynamics and generates equilibrium distributions in which both the prey and the predator are multimodal. Parameter values were $K_0 = 1$, $\sigma_K = 1$, $\beta_0 = 1$, $\sigma_\beta = 0.5$, $\sigma_{s,prey} = \sigma_{s,pred} = 0.05$, $c = 10$, and $d = 1$ for all panels, and $\sigma_{A,prey} = \sigma_{A,pred} = 0.15$ for panels (a)–(d), and $\sigma_{A,prey} = 0.1$ and $\sigma_{A,pred} = 0.05$ for panels (e) and (f). The long-term dynamics appear to be independent of initial conditions, and for purposes of clarity, only a few hundred generations of the time series after the systems have reached their dynamic attractors are shown.

preference function (Figure 9.10(e) and (f)). Thus, just as in the sexual competition models increased assortment appears to have a stabilizing effect on the evolutionary predator-prey dynamics.

It is apparent from the results reported before that pattern formation and diversification can readily occur in both asexual and sexual pde models for coevolving predator and prey populations. To further illustrate the potential of antagonistic interactions for promoting speciation, Box 9.2 illustrates a sexual plant-pathogen model in which pattern formation is not due to assortative mating, but to hybrid inviability. In this model, pathogen pressure leads to divergent evolution in the plant immune system, which in turn leads to hybrid incompatibility due to autoimmune responses. The model thus establishes an interesting connection between properties of the immune system and speciation, a theme that is receiving increasing attention in the literature (Eizaguirre et al., 2009).

Challenge: Investigate how introducing Allee effects changes the dynamics of the predator-prey pde models for sexual populations, and in particular whether an Allee effect can stabilize otherwise nonstationary dynamics.

*Challenge**: Use partial differential equation models to investigate the evolution of assortative mating in predator-prey models.

*Challenge**: Construct partial differential equation models for interspecific mutualistic interactions, for example, along the lines of the models for interspecific mutualism described in Chapter 6. Show that the resulting pde models can generate diversification in the form of multimodal pattern formation in one or both of the interacting species.

9.3   PARTIAL DIFFERENTIAL EQUATION MODELS FOR ADAPTIVE DIVERSIFICATION IN SPATIALLY STRUCTURED POPULATIONS

Partial differential equations are also very useful to study evolutionary dynamics in spatially structured populations in which geographical location is a continuous variable. Populations are then not only described by their density distribution in phenotype space, but also in geographical space. As exemplified by John Endler's seminal work (Endler, 1977), one theme that has occupied both theoreticians and empirical biologists for a long time is the question of how spatial gradients in the abiotic environment affect the origin and

BOX 9.2

SPECIATION DUE TO HYBRID NECROSIS IN PLANT-PATHOGEN MODELS

The type of reproductive isolation considered in most models of adaptive speciation is isolation due to assortative mating, that is, prezygotic isolation. The role of postzygotic isolation due to hybrid inviability or infertility has received less attention in this context. In plants, a well-known example of postzygotic isolation is hybrid necrosis, defined as a set of highly deleterious and often lethal phenotypic characteristics in hybrids. In hybrid necrosis, a mixture of genes from different strains becomes deleterious even though the contributing genes were harmless, or even beneficial, in the parents. Bomblies et al. (2007) have suggested that inappropriate activation of the plant immune system can lead to hybrid necrosis in *Arabidopsis thaliana* caused by an epistatic interaction of loci controlling plant immune responses. In Ispolatov & Doebeli (2009b) we have introduced the model presented below, in which diversification in genes controlling immune responses occurs due to pathogen pressure, and postzygotic reproductive isolation between different resistance phenotypes is due to autoimmune reactions in individuals containing resistance genes from different clusters.

   When a pathogen attacks a cell of a host plant, it often injects effector proteins that manipulate target proteins in the host cell and thereby contribute to the success of the pathogen (Jones & Dangl, 2006). These host targets, the "guardees," are guarded by other host proteins (the "guards") that monitor the guardee's molecular structure. When a pathogen effector induces changes in the molecular structure of the guardee, they are recognized by the guard proteins, which then activate the immune response. Successful pathogen attack requires fine-tuning of the effector to the guardee, and efficient immune response requires fine-tuning of the guardee and the guard. In particular, if hybridization between different host strains leads to hybrids in which guard and guardees come from different lineages, the guardee might recognize the guard as modified even in the absence of pathogen attack, leading to immune response and subsequent necrosis of the hybrid even in the absence of pathogens.

   In the model, guardee and guard proteins are described by two phenotypic coordinates, g and r. Thus, the host phenotype space is two-dimensional. The pathogen is characterized by a single coordinate, e, describing the genetic makeup of the parasite's effector proteins. The effector proteins interact with the guardee protein of a host plant plant, and frequency-dependent selection is again generated through the assumption that the pathogen attack is most effective for small distances $|e - g|$ between the pathogen phenotype e and the host's guardee phenotype g.

BOX 9.2 (*continued*)

In the host, the r and g coordinates determine the immune reaction of the plant to a parasite effector, as well as autoimmune reactions. We assume that there is a trade-off between mounting an efficient immune response in the presence of pathogens and being prone to deleterious autoimmune reactions. If $g >> r$, the probability that the guard binds to the guardee protein and triggers an immune response is small, independent of whether pathogen effectors are present or not. In this case, the plant immune system is less sensitive, making the plant more susceptible to pathogen attack, but less prone to autoimmune reactions. Conversely, if $r >> g$, the guard protein has a high propensity to bind to the guardee, again irrespective of whether the pathogen is present or absent. In this case, the plant immune system is more sensitive, making the plant less susceptible to pathogen attack, but more prone to autoimmune reactions. Specifically, the probability for an (r, g)-plant to die from parasitic infection once attacked by a pathogen is proportional to $\exp[(g - r)/\sigma_I]$, while the probability to die from autoimmune reactions is proportional to $\exp[(r - g)/\sigma_{AI}]$, where σ_I and σ_{AI} are system parameters. As a consequence of this trade-off, plants tend to survive best if their phenotypes satisfy $r \approx g$.

To model the dynamics of phenotype distributions in both the host plant and the pathogen, let $h(r, g)$ be the host density distribution and $p(e)$ the pathogen density distribution. Then

- For a given plant phenotype (r, g), the rate of attack from the pathogen is proportional to a weighted sum over all pathogen phenotypes,

$$\int p(e)a(e - g)de, \qquad (B9.6)$$

where the function $a(g - e)$, which is assumed to be Gaussian with variance σ_a^2, reflects how well a pathogen phenotype e can attack a plant with guardee phenotype g. Once attacked, the probability that the plant individual dies is proportional to $\exp[(g - r)/\sigma_I]$, as described above. This leads to a total death rate of plants with phenotypes (r, g) due to pathogen attack given by

$$-\delta_I h(r, g) \exp\left(\frac{g - r}{\sigma_I}\right) \int p(e)G_a(e - g)de. \qquad (B9.7)$$

where δ_I is a parameter.

- In the plant population, death also occurs due to autoimmune reactions, whose magnitude in plants of phenotype (r, g) is proportional to $\exp[(r - g)/\sigma_{AI}]$ (see earlier). The resulting death rate is given by

$$-\delta_I h(r, g) \exp\left(\frac{r - g}{\sigma_{AI}}\right). \tag{B9.8}$$

For simplicity, it is assumed that the rate coefficient δ_I is the same for the autoimmune and pathogen-induced death terms.

- To complete the death terms for the plant, we assume that density dependent competition in the absence of the pathogen results in a logistic death term of the form

$$-\delta_C \frac{h(r, g)H(t)}{K(r, g)}. \tag{B9.9}$$

where δ_C is a parameter, and $H(t)$ is the total density of the plant population:

$$H(t) = \int_{r,g} h(r, g)\,dr\,dg. \tag{B9.10}$$

$K(r, g)$ is the carrying capacity function, which is assumed to be of the form

$$K(r, g) = K_0 \exp\left[-\frac{(r - r_0)^2}{2\sigma_r^2}\right] \exp\left[-\frac{(g - g_0)^2}{2\sigma_g^2}\right]. \tag{B9.11}$$

$K(r, g)$ reflects stabilizing selection on r and g that is unrelated to the immune response.

- For the birth term we assume that individuals are haploid and have two loci with continuously varying alleles encoding the two phenotypes r and g. An offspring with phenotype (r', g') either inherits the two alleles r' and g' from different parent, or from the same parent. In the first case, the offspring comes from a mating between (r', g'') and (r'', g'). Such matings occur with probability $\frac{h(r', g'')h(r'', g')}{2H(t)}$, where $H(t)$ is the total plant density as before. In the second case, the probability that such an offspring is produced is simply $h(r', g')$. Adding the two cases together, the total probability that an (r', g') offspring is the result of mating is thus

$$P(r', g') = \int \frac{h(r', g'')h(r'', g')}{2H(t)}\,dr''\,dg'' + \frac{h(r', g')}{2} \tag{B9.12}$$

BOX 9.2 (*continued*)

(the factor $1/2$ reflects ambiguity in assigning mother and father in any given mating pair). In addition, we assume that mutation is described by two normal mutation kernels $G_{M,r}$ and $G_{M,g}$ with variances $\sigma^2_{M,r}$ and $\sigma^2_{M,g}$. Overall, the rate at which offspring with phenotype (r, g) are produced is then given by

$$\beta \int \int G_{M,r}(r - r')G_{M,g}(g - g')P(r', g')dr'dg', \tag{B9.13}$$

where β is the birth rate, and $P(r', g')$ is given by (B9.12).

- For the pathogen, birth is determined by successful infection. For pathogen type e', the probability that it successfully attacks and subsequently infects a host of type (r, g) is proportional to

$$h(r, g) \exp\left(\frac{g - r}{\sigma_I}\right) a(g - e'), \tag{B9.14}$$

where a is the attack kernel. Therefore, the total probability for pathogen type e' to successfully infect any host plant is

$$\int \int h(r, g) \exp\left(\frac{g - r}{\sigma_I}\right) G_a(g - e') \, dr \, dg. \tag{B9.15}$$

For simplicity, we assume that reproduction is asexual in the pathogen, but as in the plant species we include mutation given by a normal function $G_{M,e}$ with variance $\sigma^2_{M,e}$, so that the total rate of production of offspring of type e is

$$\alpha \delta_I \int G_{M,e}(e - e')p(e') \left[\int \int h(r, g) \exp\left(\frac{g - r}{\sigma_I}\right) G_a(g - e') \, dr \, dg\right] de'. \tag{B9.16}$$

Here the conversion coefficient c indicates how many new parasites are produced from an infected host.

- Finally, the death rate of the pathogen is assumed to be

$$-d_P p(e), \tag{B9.17}$$

for some parameter d_P describing the intrinsic death rate that is independent of the pathogen phenotype e.

Collecting all the birth and death terms into two equations describing the dynamics of the plant and pathogen density distributions yields the following system of coupled partial differential equations:

$$\frac{\partial h(r, g)}{\partial t} = \beta \int \int G_{M,r}(r - r')G_{M,g}(g - g')P(r', g')dr'dg'$$

$$- \delta_I h(r, g) \exp\left(\frac{g - r}{\sigma_I}\right) \int p(e)G_a(e - g)de$$

$$- \delta_I h(r, g) \exp\left(\frac{r - g}{\sigma_{AI}}\right) - \delta_C \frac{h(r, g)H(t)}{K(r, g)}, \quad \text{(B9.18)}$$

$$\frac{\partial p(e)}{\partial t} = \alpha\delta_I \int G_{M,e}(e - e')p(e')$$

$$\times \left[\int \int h(r, g) \exp\left(\frac{g - r}{\sigma_I}\right) G_a(g - e')drdg\right] de'$$

$$- d_P p(e). \quad \text{(B9.19)}$$

Numerical integration of this system of coupled partial differential equations reveals regimes that lead to multimodal equilibrium distributions in the host plant, as illustrated in Figure 9.B1. In such scenarios, which occur for a broad range of parameters (Ispolatov & Doebeli, 2009b), the host-pathogen interaction leads to adaptive speciation in the host. In Figure 9.B1, each of the main peaks in the host plant distribution, with maxima located at (r_1, g_1) and (r_2, g_2), respectively, corresponds to an emerging host species with a distinct genetic makeup of guard and guardee proteins. Note that $r_1 \approx g_1$ and $r_2 \approx g_2$, which means that in each of the strains, the function of the guard and guardee proteins are geared toward both efficient immune response to the pathogen and low likelihood of autoimmune reactions. The equilibrium distribution also shows two secondary peaks located approximately at (r_1, g_2) and (r_2, g_1), corresponding to hybrids between the two emerging species. These hybrids inherit their guard and guardee genes from parents of different strains and thus are either easy victims of pathogen attack (when $g > r$) or exhibit strong hybrid necrosis (when $r > g$). Despite the fact that all individuals have the same birth rates, the hybrid peaks are much smaller because of the higher death rate of hybrids, that is, due to postzygotic isolation between the emerging species. Thus, the two main host peaks along the diagonal are separated by postzygotic isolation, which can be viewed as an example of speciation due to Dobzhansky-Muller incompatibilities (Gavrilets, 2004). Scenarios with more than two host plant peaks along the diagonal in (r, g)-space are also possible (Ispolatov & Doebeli, 2009b).

Challenge: Derive and analyze the adaptive dynamics of the traits (g, r) in the prey and the trait e in the predator based on the ecological interactions defined before.

BOX 9.2 (*continued*)

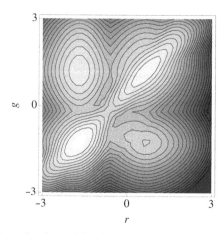

FIGURE 9.B1. Example of a multimodal equilibrium distribution of the host for model (B9.18) and (B9.19). The contour plot shows the host population density on a logarithmic scale as function of the two phenotypes r and g. Clearly visible near the $r = g-$ diagonal are two peaks, centered around two coordinates, $(r_1, g_1 \approx r_1)$ and $(r_2, g_2 \approx r_2)$. Two much weaker peaks with coordinates approximately (r_1, g_2) and (r_2, g_1) correspond to hybrid heterozygotes in the r and g genes. The subpopulation of necrotic hybrids lies below the $r = g$ diagonal, whereas the secondary peak above the diagonal corresponds to susceptible hybrids. The pathogen equilibrium distribution is bimodal along the e-axis (not shown), with each mode being specialized on one of the main host modes along the diagonal. The plot corresponds to the solution of equations (B9.18) and (B9.19) at a steady state for the following parameter values: $\beta = 1$, $\alpha = 1$, $\delta_I = 5$, $\delta_C = 1$, $\delta_P = 1$, $\sigma_I = 0.2$, $\sigma_{AI} = 0.2$, $\sigma_a = 0.8$, $\sigma_{M,r} = \sigma_{M,e} = 0.1$, and $\sigma_r = \sigma_g = 1$. From Ispolatov & Doebeli (2009b). Used by permission.

∎

maintenance of diversity. In particular, frequency-dependent competition models of the type studied in the first part of this chapter have been extended to spatial gradient models by a number of researchers (Case & Taper, 2000; Doebeli & Dieckmann, 2003; Kirkpatrick & Barton, 1997; Leimar et al., 2008; Polechova & Barton, 2005). In these models, the spatial gradient is typically assumed to be reflected in the stabilizing component of selection. Thus, if x is the evolving phenotype as before, and z denotes the spatial direction, then the carrying capacity $K(x, z)$ is a function of both x and z in such a way that the position of the maximum in the x-direction depends on the spatial location z. Here I follow the analysis of Ispolatov & Doebeli (2009a) and assume for simplicity that space is one-dimensional (i.e., that z is a scalar), and that the position of the maximal carrying capacity in the x-direction changes linearly with z. In principle, it is straightforward to extend the models to two-dimensional geographic space, and to more complicated gradient shapes.

Extending from the Gaussian nonspatial case, I assume that the carrying capacity function has the form

$$K(x, z) = K_0(z) \exp\left(-\frac{(x - gz)^2}{2\sigma_K^2}\right), \tag{9.25}$$

so that for any z, the position of the maximal carrying capacity in the x-direction is at $x = gz$, where g is the steepness of the gradient. Moreover, to avoid biologically unrealistic scenarios with potentially infinite species ranges, we also assume that the value of the maximal carrying capacity, given by $K_0(z)$ in eq. (9.25), decreases toward the margins of the spatial domain. Thus, we assume that

$$K_0(z) = K_0 \exp\left(-\frac{z^2}{2\sigma_z^2}\right). \tag{9.26}$$

Biologically, this ensures that populations at the margins of the geographical range (i.e., for $z \to \pm\infty$) are not sustainable, regardless of their phenotypic composition. This is important for the numerical integration of the pde's to be formulated below, for which one has to chose a finite system size, that is, a finite subset of the plane on which the numerical integration takes place. One must then chose boundary conditions to specify the numerical procedure implemented at the boundaries of the chosen finite system. For example, the boundaries could be chosen to be reflecting, which corresponds to assuming that the population densities outside the finite system are the same as the population density at the boundaries of the system, or the boundaries could be chosen to be absorbing, which corresponds to assuming that the population densities outside the finite system are zero. However, the fact that the carrying capacities tend to zero with increasing distance from the origin and in all directions of the two-dimensional (x, z)-plane implies that a finite system size can always be chosen such that the carrying capacities at the boundaries of the system are very small. Therefore, the finite system can always be chosen large enough for the boundary conditions to be irrelevant. Thus, besides the fact that carrying capacity functions of the form (9.25) make sense biologically, they can be used, in conjunction with a large enough system size, to eliminate any numerical artifacts due to boundary conditions.

Just as in the nonspatial model (9.2), we also assume that the strength of competition between individuals declines with phenotypic distance, as described by a function $\alpha_{pheno}(x, x')$, where x and x' are phenotypes of competing individuals, and where the subscript $_{pheno}$ indicates that this function measures competition in phenotype space. This is in contrast to competition in geographical space, which is measured by a different function $\alpha_{geo}(z, z')$, which is also assumed to describe a situation in which the strength of competition between individuals is a decreasing function of their geographical distance. For example, both α_{pheno} and α_{geo} could be Gaussian functions of the distances $|x - x'|$ and $|z - z'|$, respectively, but other functions with a maximum at zero are also feasible. Overall, the strength of competition between individuals (x, z) and (x', z') is assumed to be given by the product $\alpha_{pheno}(x, x')\alpha_{geo}(z, z')$. Finally, we also assume that migration of individuals is described by a dispersal kernel $D(z, z')$, such that $D(z, z')dz$ is the probability that an individual at location z' migrates to a location in the interval $(z, z + dz)$. We assume that there is no death due to migration, and hence that $D(z, z')$ is a normal function with variance σ_D^2, which is a measure of how far individuals migrate.

If $\phi(x, z, t)$ is the density of phenotypes x at spatial location z and at time t, these assumptions result in the following extension of eq. (9.13) for the dynamic behavior of $\phi(x, z, t)$ (where for notational convenience we have dropped the variable t from the arguments of $\phi(x, z)$):

$$
\frac{\partial \phi(x, z)}{\partial t} = r \int B_\phi(x')\phi(x', z)dx'
$$
$$
- \frac{\phi(x, z)}{K(x, z)} \int \int \alpha_{pheno}(x, x')\alpha_{geo}(z, z')\phi(x', z')dx'dz'
$$
$$
+ m \left(\int D(z, z')\phi(x, z')dz' - \phi(x, z) \right) \tag{9.27}
$$

Here the parameter r is the birth rate as before, and the function B_ϕ describes the birth process according to whether reproduction is sexual or asexual, as specified below. Note that we assume no migration at birth, but including such a migration term in the birth term would not have a qualitative effect, since there is migration throughout an individual's life, as described by the last term of the right-hand side of eq. (9.27), in which the parameter m is the per capita rate of migration, and $D(z, z')$ is a migration kernel of normal form with variance σ_D^2. The double integral in the middle term of the right-hand side is the effective density experienced by an individual with phenotype x at location z.

If reproduction is asexual, the function B_ϕ is simply a mutation kernel, which we assume to be normal with variance σ_{mut}. In this case the pde model (9.27) is very similar to the ones used in Doebeli & Dieckmann (2004, Box 7.2), Polechova & Barton (2005), and Leimar et al. (2008) and captures the essential ingredients of those models, with the additional effect of limiting the geographical range of the evolving population (by incorporating the function $K_0(z)$ given by eq. (9.26)).

If reproduction is sexual, the function B_ϕ describes assortative mating in both the phenotypic and the geographical dimension. In the phenotypic dimension, we assume as before that assortment is described by a Gaussian preference function

$$A_{pheno}(x, x') = \frac{1}{\sqrt{2\pi}\,\sigma_{A,pheno}} \exp\left[-\frac{(x-x')^2}{2\sigma_{A,pheno}^2}\right], \qquad (9.28)$$

which describes the relative preference an individual of a given phenotype x has for mates with phenotype x'. Similarly, in the geographic dimension assortment is described by a function

$$A_{geo}(z, z') = \frac{1}{\sqrt{2\pi}\,\sigma_{A,geo}} \exp\left[-\frac{(z-z')^2}{2\sigma_{A,geo}^2}\right]. \qquad (9.29)$$

Here the parameters $\sigma_{A,pheno}$ and $\sigma_{A,geo}$ measure how fast the probability of mating decreases with phenotypic and geographical distance, respectively. For example, large values of $\sigma_{A,pheno}$ correspond to random mating with respect to phenotypes. If populations occupy geographical areas of reasonable sizes, it is unlikely that mating is random in the geographical dimension, for this would imply that individuals have equal access to all spatial locations when searching for mates. Thus, we will assume that $\sigma_{A,geo}$ is fixed at a value that is low enough to describe spatially localized mating, whereas $\sigma_{A,pheno}$ may vary between values coding for strong assortment and values coding for random mating with respect to the phenotype x.

To consider the baseline case of no costs to assortment (whether phenotypic or spatial), we again use the quantity

$$N(x, z) = \int_{x'} \int_{z'} A_{pheno}(x, x') A_{geo}(z, z') \phi(x', z') dx' dz' \qquad (9.30)$$

as a measure of the total amount of mates available to an individual with phenotype x at location z. Then

$$\frac{1}{N(x, z)} A_{pheno}(x, x') A_{geo}(z, z') \phi(x', z') \qquad (9.31)$$

is the probability of an individual with phenotype x at location z to choose a mate with phenotype in the interval $(x', x' + dx')$ and located in the interval $(z', z' + dz')$. Normalizing by the quantity $N(x, z)$ ensures that all individuals have the same per capita birth rate r, that is, that there are no costs to having a rare phenotype or to being spatially isolated. This is reasonable as long as population sizes are large. (It is worth keeping in mind that since $\phi(x, z)$ is a population density, we assume infinite population sizes throughout.) However, it is straightforward to introduce costs to being rare in the form of Allee effects by adding a constant η to the denominator in eq. (9.31), as was done earlier in this chapter for spatially unstructured populations.

The final ingredient for constructing the birth term for sexual populations is the phenotypic segregation kernel, and we again assume that the distribution of offspring from parents with phenotypes x and x' is normal with mean the mid-parent value $(x + x')/2$ and constant variance σ_S^2. Then the function B_ϕ for sexual populations has the form

$$B_\phi(x') = \frac{1}{N(x', z)} \int_{x''} \int_{z'} A_{pheno}(x', x'') A_{geo}(z, z') \phi(x'', z') S_{\frac{x'+x''}{2}}(x) dx'' dz'.$$

(9.32)

With this expression for B_ϕ, the integral

$$r \int_{x'} B_\phi(x') \phi(x', z) dx'$$

(9.33)

comprises all offspring that are born with phenotype x at location z, as stipulated by the first term on the right hand side of the pde (9.27).

9.3.1 ASEXUAL SPATIAL MODELS

We first investigate pattern formation in asexual populations, in which the function B_ϕ is the mutational kernel. We consider the following class of competition kernels:

$$\alpha_{pheno}(x, x') = \exp\left(-\frac{-(x - x')^{2+\epsilon_{pheno}}}{2\sigma_{\alpha,pheno}^{2+\epsilon_{pheno}}}\right)$$

(9.34)

$$\alpha_{geo}(z, z') = \exp\left(-\frac{-(z - z')^{2+\epsilon_{geo}}}{2\sigma_{\alpha,geo}^{2+\epsilon_{geo}}}\right).$$

(9.35)

For $\epsilon_{pheno} = \epsilon_{geo} = 0$, all kernels are Gaussian. In this case, pattern formation in the form of multimodal equilibrium density distributions does not occur as long as the system size is sufficiently larger than the width σ_z of the function

$K_0(z)$, eq. (9.26), that is, as long as the system size is sufficiently larger than the species range, so that boundary effects do not play a role. Instead, in this case the equilibrium distributions always consist of a ridge along the gradient $x = gz$ with decreasing height toward the edges of the species range, as illustrated in Figure 9.11(a) and (b). This result is in agreement with Leimar et al. (2008), who studied models with $\sigma_z = \infty$, that is, with infinite species ranges. Infinite species ranges eliminate boundary effects, and Leimar et al. (2008) did not find multimodal equilibrium distributions for Gaussian competition kernels. This is also in agreement with Polechova & Barton (2005), who studied Gaussian models without boundary effects by considering boundary conditions that wrap onto themselves, that is, by considering pde dynamics unfolding on a torus (with phenotype and geographical space as the two dimensions). Polechova & Barton (2005) concluded that multimodal phenotypic pattern formation in spatial models, for example, as observed in the individual-based models of Doebeli & Dieckmann (2003), is an artifact of boundary conditions. According to these authors, pattern formation should not occur if boundary effects are eliminated.

This contention is wrong in general, for it depends on the assumption of Gaussian kernels in deterministic models. In fact, we have seen repeatedly in this book that the assumption of Gaussian kernels can lead to nongeneric results. Spatial models are no exception, as can be seen, for example, by assuming $\epsilon_{pheno} = \epsilon_{geo} > 0$ in eqs. (9.34) and (9.35) for the competition kernels. Figure 9.11(c), (d) illustrates that in such cases, multimodal phenotypic patterns can easily emerge even if the system size is much larger than σ_z, that is, even if the species range is small enough for boundary conditions to be irrelevant. Similar conclusions have been reached by Leimar et al. (2008) for models with infinite system size and infinite species ranges, in which boundary effects are also absent. Thus, for certain classes of competition kernels, and in particular for competition kernels that have a significantly lower kurtosis than Gaussian kernels (i.e., when either ϵ_{pheno} or ϵ_{geo} is positive enough), diversification in the form of multimodal pattern formation in asexual spatial models is a robust phenomenon that is not generated by effects of the system boundary.

9.3.2 SEXUAL SPATIAL MODELS

Recall from the nonspatial competition models discussed earlier in this chapter that sexual reproduction and assortative mating can also remove the degeneracy of the Gaussian case in the sense that pattern formation is possible in sexual populations even if all relevant kernels are assumed to be Gaussian (cf. Figure 9.4). The same is true in sexual spatial models, in which the function

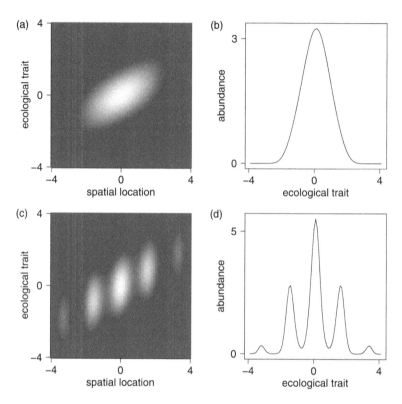

FIGURE 9.11. Equilibrium distributions in asexual, spatially structured models of competition along an environmental gradient. Panel (a) shows an example of the unimodal distributions obtained with Gaussian carrying capacities and Gaussian competition kernels in both the spatial and the phenotypic direction ($\epsilon_{pheno} = \epsilon_{geo} = 0$). The distribution is given as a function of geographical location (x-axis) and of ecological phenotype (y-axis). Panel (b) is the projection of the distribution shown in panel (a) onto the phenotype axis, showing a unimodal phenotype distribution. Panel (c) shows an example of the multimodal distributions obtained with Gaussian carrying capacities, but non-Gaussian competition kernels in the spatial and the phenotypic direction ($\epsilon_{pheno} = \epsilon_{geo} = 1$). Panel (d) is the projection of the distribution shown in panel (c) onto the phenotype axis, showing a multimodal phenotype distribution. Other parameter values were the same in both scenarios and were $r = 1$, $K_0 = 1$, $\sigma_K = 1$, $\sigma_z = 1$, $\sigma_{\alpha,pheno} = 1$, $\sigma_{\alpha,geo} = 1$, $\sigma_{A,pheno} = 1$, $\sigma_{A,geo} = 1$, $m = 5$, $\sigma_D = 0.2$, and $\sigma_{mut} = 0.05$. The long-term dynamics appear to be independent of initial conditions.

B_ϕ determining the birth term is given by expression (9.32). In this case, multimodal equilibrium distributions can emerge even if the competition kernels are Gaussian (i.e., $\epsilon_{pheno} = \epsilon_{geo} = 0$), and even if the system size is significantly larger than the species range, so that boundary effects are not present. This is illustrated in Figure 9.12(a) and (b), which shows

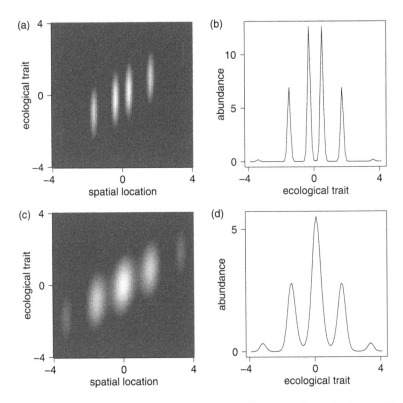

FIGURE 9.12. Equilibrium distributions in sexual, spatially structured models of competition along an environmental gradient. This figure is the same as Figure 9.11, except that reproduction was sexual with assortative mating. Panels (a) and (b) show that in contrast to asexual models, sexual models can produce pattern formation even if all ecological functions are Gaussian. Panels (c) and (d) show multimodal pattern formation with non-Gaussian competition kernels. Ecological parameter values were the same as in Figure 9.11, and assortative mating was given by $\sigma_{A,pheno} = \sigma_{A,geo} = 0.15$ in both scenarios. The long-term dynamics appear to be independent of initial conditions.

that diversification can occur in spatial sexual populations for intermediate strengths of assortment, that is, for intermediate values of the parameter $\sigma_{A,pheno}$. Note again that the different phenotypic modes emerging at equilibrium are largely isolated reproductively due to assortative mating with respect to both phenotype and geographical space. In addition, adaptive diversification and speciation in sexual spatial models can of course also emerge for non-Gaussian competition kernels, as illustrated in Figure 9.12(c) and (d).

The spatial models described here incorporate competition and assortative mating with respect to the geographical dimension in much the same way as they are incorporated with respect to the phenotypic dimension. In fact, the spatial model without phenotypic variability is mathematically equivalent to the nonspatial model with phenotypic variability studied at the beginning of this chapter. Therefore, the spatial model alone is able to produce pattern formation in sexual populations. Such pattern formation corresponds to geographical rather than phenotypic clustering. This spatial source of multimodality often facilitates multimodality in the phenotypic dimension if populations are phenotypically variable and if there is an environmental gradient in stabilizing selection of the form envisaged in the spatial models described above. In that case, clusters that form due to spatial interactions could align themselves in different positions along the environmental gradient and could thus become separated not only in the spatial direction, but also in the phenotypic direction. Therefore, phenotypic diversification can be generated by spatial pattern formation along an environmental gradient even if frequency-dependence in the phenotypic interactions alone would not be strong enough to generate phenotypic pattern formation.

This is illustrated in Tables 9.1 and 9.2, which show results for spatial sexual populations for a set of parameters for which the nonspatial sexual model (9.13) showed multimodality below a threshold value $\sigma_\alpha \approx 1$ of the phenotypic competition width. We determined the threshold value $\sigma^*_{\alpha,pheno}$ of the phenotypic competition width for which the corresponding spatial sexual models (9.27) showed multimodal pattern formation. This was done for different values of the spatial competition width $\sigma_{\alpha,geo}$ for two different slopes g of the spatial gradient ($g = 1$ for Table 9.1, and $g = 2$ for Table 9.2). Note that in both the nonspatial and the spatial models all kernels were Gaussian. In both tables, a value of $\sigma^*_{\alpha,pheno} > 1$ indicates facilitation of multimodal pattern formation in spatial models, and the results indicate that facilitation of diversification indeed occurs. In fact, if spatial competition is localized enough, then for a certain range of slopes of the environmental gradient, phenotypic diversification occurs even in the absence of phenotypic frequency-dependence (i.e., even for very large $\sigma_{\alpha,pheno}$), as is apparent from the first columns of Tables 9.1 and 9.2.

Phenotypic diversification due to purely spatial interactions cannot occur without an environmental gradient, and it is therefore intuitively clear that facilitation of phenotypic diversification due to spatial interactions increases as the slope of the environmental gradient is increased from very shallow gradients to steeper gradients. In fact, this result can be understood in the context of diversification in multidimensional phenotype spaces discussed in Box 9.1. As mentioned, formally the geographical dimension is equivalent to the phenotypic dimension in the models considered here, and the environmental gradient assumed in the carrying capacity function $K(x, z)$ given by

TABLE 9.1. Threshold value of the phenotypic competition width $\sigma^*_{\alpha,pheno}$ below which phenotypic clusters emerges in the spatial pde models for sexual populations, as a function of spatial competition width $\sigma_{\alpha,geo}$ and with slope of the environmental gradient $g = 1$. Other parameter values were $r = 1$ for the intrinsic growth rate, $\sigma_{A,pheno} = 0.15$ and $\sigma_{A,geo} = 0.15$ for assortment, $\sigma_S = 0.05$ for the segregation kernel, $\sigma_D = 0.2$ and $m = 5$ for migration, and $K_0 = 1$, $\sigma_z = 1$ and $\sigma_K = 1$ for the carrying capacity function. In the corresponding nonspatial sexual models, multimodal equilibrium distributions occur for the competition width below the threshold of $\sigma^*_\alpha \approx 1$.

$\sigma_{\alpha,geo}$	0.5	1	2
$\sigma^*_{\alpha,pheno}$	∞	2	1.1

TABLE 9.2. Threshold value of trait competition width $\sigma^*_{\alpha,pheno}$, below which phenotypic clusters emerges in the spatial pde models for sexual populations, as a function of spatial competition width $\sigma_{\alpha,geo}$ and with slope of the environmental gradient $g = 2$. Other parameter values as for Table 9.1. In the corresponding nonspatial sexual models, multimodal equilibrium distributions occur for the competition width below the threshold of $\sigma^*_\alpha \approx 1$.

$\sigma_{\alpha,geo}$	0.5	1	2
$\sigma^*_{\alpha,pheno}$	∞	1.25	0.75

(9.25) can be viewed as incorporating an interaction between the phenotypic and the spatial dimensions, as envisaged in Box 9.1. This interaction relaxes the condition for diversification in the direction of the environmental gradient, that is, along the line $x = gz$ (where g is the steepness of the gradient). In other words, the interaction between the spatial and the phenotypic dimension widens the carrying capacity, and hence relaxes the force of stabilizing selection, along the gradient, which in turn increases the scope for frequency dependent interactions to generate diversification along the gradient.

On the other hand, Doebeli & Dieckmann (2003) reported that facilitation of phenotypic diversification should decrease again when the environmental gradient becomes very steep, that is, that facilitation is greatest for some intermediate slope of the environmental gradient. In fact, this phenomenon is also confirmed by the results from the pde models for sexual populations, as Tables 9.1 and 9.2 indicate that facilitation of diversification is more pronounced for a gradient slope of $g = 1$ (Table 9.1) than for a gradient slope of $g = 2$ (Table 9.2).

The reason for this decline of facilitation with very steep gradient slopes lies in an interaction between the environmental gradient and the spatial cluster formation. As the gradient gets steeper (higher g in (9.25)), the range of geographical locations allowing for high carrying capacities decreases, which eventually prevents the formation of different clusters due to local competition along the spatial dimension, because a single cluster already occupies the whole range of locations with appreciable carrying capacities. If frequency dependence in the phenotypic dimension is weak, this prevention of cluster formation in the spatial dimension due to steep gradients also prevents phenotypic pattern formation.

Overall, the results for spatial models reported here and elsewhere show that multimodal pattern formation in phenotype space, that is, phenotypic diversification and adaptive speciation, is a robust outcome of models incorporating frequency-dependent and spatially localized competition in geographically structured populations. Moreover, spatial gradients in the optimal phenotype lead to a facilitation of such diversification, which seems to be most pronounced for gradients of intermediate slopes. It is also clear that due to the construction of the models, these results are not artifacts of boundary conditions.

9.4 A GENERAL THEORY OF DIVERSIFICATION IN PARTIAL DIFFERENTIAL EQUATION MODELS

The results reported so far in this chapter essentially constitute proof by example: one shows that a phenomenon, in this case multimodal pattern formation in partial differential equation models, occurs in certain examples, from which it follows that the phenomenon can occur in principle. In this approach, the generality of the phenomenon can only be supported by providing an array of different examples, as, for example, provided in this chapter with regard to the phenomenon of pattern formation. It would be desirable, however, to have a proof of the genericity of pattern formation that is independent of the specific choice of the pde model. We have recently made an attempt at providing such a model-independent proof, which I will briefly describe in the conclusion of this chapter.

Our theory, which is described in Elmhirst et al. (2008) and Elmhirst & Doebeli (2010), is based on the observation that any sufficiently well-behaved phenotype distribution $\phi(x)$ can be approximated by a sum of Gaussian distributions

$$\phi(x) \approx \sum_{i=1}^{N} \psi_{x_i,y_i}(x), \tag{9.36}$$

where for simplicity we assume that the trait x is one-dimensional, and where the distributions $\psi_{x_i,y_i}(x)$ are of the form

$$\psi_{x_i,y_i}(x) = \frac{y_i}{\sqrt{2\pi}\sigma} \exp\left(-\frac{(x-x_i)^2}{2\sigma^2}\right). \tag{9.37}$$

Here x_i is the mean of ψ_{x_i,y_i}, and y_i is the total weight of this distribution (i.e., $\int \psi_{x_i,y_i} = y_i$). The component distributions ψ_{x_i,y_i}, $i = 1, \ldots, N$ are called "pods" (Cohen & Stewart, 2000; Stewart et al., 2003), and N is the total number of pods used for approximating $\phi(x)$. The parameter σ determines the width of the pods, which is assumed to be the same for all pods. In general, the approximation (9.36) can be made arbitrarily close by increasing N and decreasing σ. Using this approximation, following the temporal dynamics of $\phi(x)$ becomes equivalent to following the temporal dynamics of the N pods ψ_{x_i,y_i}, which in turn is equivalent to following the dynamics of the means x_i and the weights y_i. Thus, a partial differential equation for $\phi(x)$ can be approximated by a system of $2N$ ordinary differential equations

$$\frac{dx_i}{dt} = F(x_1, \ldots, x_N, y_1, \ldots, y_N) \quad i = 1, \ldots, N \tag{9.38}$$

$$\frac{dy_i}{dt} = G(x_1, \ldots, x_N, y_1, \ldots, y_N) \quad i = 1, \ldots, N. \tag{9.39}$$

Note that the sum of pods in the approximation (9.36) is not unique. For example, a pod ψ_{x_i,y_i} can itself be replaced by a sum of two pods, $\psi_{x_i,y_i} = \psi_{x_i,y_{1,i}} + \psi_{x_i,y_{2,i}}$, where $y_{1,i} + y_{2,i} = y_i$. Nevertheless, different collections of pods that yield the same sum must yield identical dynamics of the form (9.38) and (9.39) for consistency reasons. This implies certain constraints for the functions F and G defining the dynamics (9.38) and (9.39). In other words, only certain types of functions F and G occur on the right-hand side of pod dynamics that approximate the dynamics of pde's. It is possible to derive the general form, called "normal form," that the functions F and G must have in order to define a pod dynamics. Deriving this normal form is mathematically rather challenging and is the subject of Elmhirst et al. (2008).

If any pde can be approximated by ordinary differential equations defined by normal forms F and G, studying the general properties of differential equations given by functions of normal form yields a model independent approach to understanding qualitative features of pde's. In the context of pattern formation, it is of particular interest to understand what happens when unimodal equilibrium distributions $\phi(x)$ become unstable. Unimodal distributions centered at, say, x_0 can be approximated by sums of pods ψ_{x_i,y_i} in which all means x_i are the same, $x_i = z_0$ for $i = 1, \ldots, N$. (Note that this is also true for equilibrium distributions that are delta peaks, as would, e.g., be expected in asexual populations under stabilizing selection and without mutations; such distributions have to be approximated by pods for which the width $\sigma \to 0$.) Thus, a unimodal equilibrium distribution corresponds to an equilibrium of the system of differential equations (eqs. (9.38) and (9.39)) at which $x_i = z_0$ for all i.

The question is, what happens to this equilibrium when the corresponding unimodal equilibrium distribution becomes unstable?

The basis for answering this question is the observation that the pod dynamics (9.38) and (9.39) must satisfy certain symmetry conditions: because only the sum of the pods matters, the dynamics of a pod system cannot depend on the particular numbering of the pods. Thus, the dynamics must be invariant for any renumbering of the pods. Mathematically speaking, this means that pod dynamics of the form (9.38) and (9.39) must be invariant under the action of the symmetric group S_N, which consists of all permutations of the pod indexes. This puts severe constraints on the dynamics of pod systems, which can be understood in the framework of "equivariant bifurcation theory" (Golubitsky & Stewart, 2002). In general, the goal of bifurcation theory is to classify all possible regimes of dynamical systems, as well as the transition between different regimes. If a dynamical system satisfies certain symmetry conditions, that is, is invariant under the action of a symmetry group such as S_N, the set of possible dynamic behaviors becomes more limited, and studying the resulting restricted dynamic regimes and the transitions between them is the subject of equivariant bifurcation theory.

Applying this theory to pod systems of the form (9.38) and (9.39) is highly nontrivial, and is the subject of Elmhirst & Doebeli (2010). Nevertheless, the most important result of this analysis can be easily summarized by saying that if an equilibrium of a pod system with $x_i = z_0$ for all i loses stability through a bifurcation, then in a vicinity of the bifurcation point the pod system has an equilibrium at which $x_i = z_1$ for $i = 1, \ldots, P$ and $x_i = z_2$ for $i = P + 1, \ldots, N$, where $1 < P < N$, and where $z_1 < z_0 < z_2$ (Elmhirst & Doebeli, 2010). Thus, in the vicinity of a bifurcation point, there is an equilibrium consisting of two modes, one made of pods with mean z_1, and the other made up of pods with mean z_2. This result is true generically, that is, for an open subset of the space of all pod systems. For pde's whose dynamics are approximated by pod systems, this implies that when a unimodal equilibrium becomes unstable as a system parameter is moved through a bifurcation point, there exists a bimodal equilibrium solution of the pde in the vicinity of the bifurcation point. Again, this is true generically, that is, for an open subset of the space of all pde's that have a unimodal equilibrium and can be approximated by a pod system.

While equivariant bifurcation theory reveals the type of equilibria occurring near a bifurcation point, it does not necessarily say anything about the stability of these equilibria. Nevertheless, the theory uncovers the generic existence of bimodal equilibrium solutions of pod systems, and hence of pde's, in the vicinity of bifurcation points at which unimodal equilibrium distributions lose stability. This corresponds to finding bimodal equilibrium distributions in the asexual and sexual pde models described earlier in this chapter. For example, in models for competition in asexual populations, bimodality can occur as a parameter determining the strength of frequency dependence moves through a bifurcation point (i.e., as the strength of frequency-dependence is increased above a certain threshold). That this does not occur for asexual models in which the carrying capacity and the competition kernel are Gaussian illustrates the meaning of "generic": bimodal equilibrium distributions arising through bifurcations do not occur in all pde models, but they do occur in a large class of pde models. Similarly, in pde models for sexual populations, bimodality can occur as a parameter determining the strength of assortative mating moves through a bifurcation point (even if the underlying ecological functions are of Gaussian form, see Fig. 9.4).

It is important to realize that the bimodal equilibria predicted by equivariant bifurcation theory for pod systems are model-independent, because they occur for generic pod systems of the type (9.38) and (9.39), where F and G have the normal form needed for the pod system to approximate the dynamics of a pde. Therefore, the diversifications observed in the specific pde models investigated in the previous sections are instantiations of a general, model-independent principle. In general, much theoretical insight in ecology and evolution comes from the use of specific models for particular questions. This is problematic to some extent, because it often leaves one wondering whether the results obtained are qualitatively independent of the specific model used, that is, whether the results are generic from a modeling point of view. It is therefore encouraging that there are indeed circumstances where the use of sophisticated mathematical theories allows us to uncover general, model-independent principles underlying biological diversification. Of course, we have used such a mathematical theory all along in this book: adaptive dynamics. While adaptive dynamics can of course be applied to specific models, this theory also makes general and model-independent predictions. In particular, adaptive dynamics theory shows that in principle, adaptive diversification through evolutionary branching is a generic outcome of frequency-dependent selection.

CHAPTER TEN

Experimental Evolution of Adaptive Diversification in Microbes

This is a book about the theory of adaptive diversification, but I want to close with a chapter describing some laboratory experiments with which we attempted to test parts of this theory. In general, conclusively describing the process of adaptive diversification in empirical studies is difficult, because this process typically unfolds over at least hundreds of generations, which for organisms of human scale implies nonhuman time scales. However, experimental evolution with microbes has emerged as a very attractive alternative to overcome the problem of long time scales in empirical studies of evolution. This is exemplified by the famous long-term evolution experiments of Richard Lenski, whose experimental *Escherichia coli* lines have evolved for more than 40,000 generations to date. Lenski and his many collaborators have used these lines to address a great variety of fundamental questions about evolution and adaptation (Elena & Lenski, 2003). Sympatric diversification in well-mixed bacterial populations is one of the phenomena they observed (Rozen & Lenski, 2000), albeit in only one of their experimental lines. They convincingly argued that the diversified strains have coexisted over long time periods, and hence that this diversification represents a case of asexual speciation (Cohan, 2002; Rozen et al., 2005). The ecological mechanism for diversification in this case appears to be related to crossfeeding, a scenario in which one strain or species persists by scavenging on nutrients that accumulate in the environment as metabolic by-products of the coexisting strain. In fact, crossfeeding is one of the classic examples for the maintenance of polymorphism in microbial populations (Helling et al., 1987; Rosenzweig et al., 1994; Treves et al., 1998), which has been used to refute Gause's exclusion principle. With crossfeeding, polymorphisms can be maintained even in simple environments with a single limiting resource, such as glucose, because the consumption of glucose results in secondary compounds that become alternative resources. This is an excellent example of frequency-dependent selection, as the fitness of the cross-feeder depends on the presence or absence of the glucose specialist. In the absence of the glucose specialist, evolution tends to maximize performance on the single resource glucose, and the fitness of the crossfeeder, which is specializing on by-products, is low. However, once the glucose specialist is present, crossfeeders can invade. The advantage of crossfeeding in the presence of a glucose specialist is probably influenced by rate-yield trade-offs (MacLean & Gudelj, 2006; Pfeiffer et al., 2001), because optimization of competitive performance on glucose may often involve the maximization of the growth rate at the expense of efficient use of the resource. Thus, specialization on glucose may often lead to fast, but wasteful consumption of glucose, which in turn creates a niche for types that can thrive on the metabolic waste products (such as acetate).

Another by now classic example of diversification in bacterial populations was first described by Rainey & Travisano (1998) in the bacterium *Pseudomonas fluorescens*. Essentially, if these bacteria are left undisturbed in a jar containing the right mix of nutrients, they diversify extremely rapidly, with different types coming to occupy different spatial locations in the jar, such as the air-broth interface, or the oxygen-depleted bottom of the jar. Much of the work studying this diversification has concentrated on the diversification that is due to the colonization of the air-broth interface by a new type, the "wrinkly spreader" (WS) (Rainey & Travisano, 1998). In particular, the genetic and metabolic changes underlying this diversification have been studied in detail (Bantinaki et al., 2007; Goymer et al., 2006; Spiers et al., 2002), and it has become clear that a single mutation suffices to produce WS type (even though subsequent evolution at the air-broth interface may lead to the fixation of further mutations). This is also the reason why this diversification happens extremely fast. Essentially, one only needs to await the occurrence of a single mutation in the ancestral population living inside the broth. Once the right mutation has occurred, the mutant can colonize the air-broth interface, hence the population diversifies. On first thought, it appears that there may not be much conceptual interest in this diversification itself, as it simply seems to involve colonization of an empty spatial niche. However, the story may be more complicated, because it appears that the empty niche only becomes an attractive place to live once growth of the ancestral strain in the broth interior has led to oxygen depletion in the broth, and hence to a gradient of sharply declining oxygen concentration as one moves from the air-broth interface to the interior of the broth (P. Rainey, pers. comm.). Thus, frequency-dependent interactions may play an important role for this diversification. For example, models of the type described in Chapter 2 may be useful for theoretical descriptions of the *Pseudomonas* system.

Nevertheless, in the experiments described below we chose to pursue a different approach to investigate adaptive diversification in bacteria. Similar to Lenski's evolution experiments, we used serial transfers for long-term propagation of well-mixed bacterial populations, but rather than being limited by a single resource, the evolving populations experienced an environment consisting of a mixture of resources. In our experiments, the ancestral strain was the same asexual *E. coli* B clone that was used as ancestor in Lenski's long-term evolution experiments, and a number of replicate lines founded by this ancestral strain evolved during approximately 1,000 generations of serial batch culture in a well-mixed environment containing the two carbon sources glucose and acetate. (Note to the reader: unless otherwise stated, I refer to our papers Friesen et al. (2004), Spencer et al. (2007a,b), Le Gac et al. (2008), and Spencer et al. (2008) for details about the methods and protocols used for obtaining the experimental results described below.)

For the daily serial transfer between batches, a 1/100 of the population in the previous batch is used to inoculate a new batch containing a virgin environment containing the two resources. In each daily batch, the population grows from this inoculate to reach stationary phase, in which no further population growth occurs. In an environment containing glucose and acetate, metabolic trade-offs prevent the bacteria from concomitantly using the two carbon sources. Roughly speaking, when the bacterial cells consume glucose, through the metabolic pathway of glycolysis, some metabolites accumulate within the cell. To consume more glucose, a cell has to get rid of these intermediate metabolites, either by secreting them into the environment, or by consuming them in secondary pathways. The latter is typically much more time consuming than the former, and hence as long as glucose is abundant, cells secrete the intermediate metabolites into the environment and shut down the secondary pathways through the mechanism of catabolite repression (White, 2000).

It so happens that the secondary pathways used to metabolize the by-products of glycolysis largely overlap with the pathways used to metabolize acetate, the second resource in the evolution environment. Thus, shutting down these pathways means that no acetate is consumed during glycolysis. In fact, acetate is one of the metabolites that is secreted into the environment during glycolysis. However, once glucose has been depleted, the only carbon source left in the environment is acetate, and hence the bacterial cells now try to switch to consuming acetate by turning on the secondary pathways that have previously been repressed. As a result, the bacteria use the two carbon resources sequentially, which is a common occurrence when microorganisms grow on mixed substrates (Harder & Dijkhuizen, 1982). Sequential resource

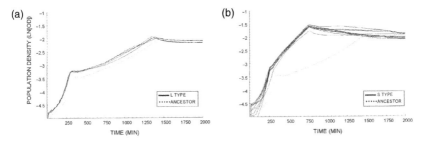

FIGURE 10.1. Growth curves of different ecotypes of *E. coli*. The two panels show the dynamics of the optical density of a population of L types (panel (a)) and a population of S types (panel (b)) during a single batch of growth in liquid medium containing a mixture of glucose and acetate in 9:1 proportions. The L and S types diversified in a 1,000-generation evolution experiment from a common ancestor (whose growth rate is indicated by the dotted line; see text for an explanation of the L and S types). All strains exhibit diauxic growth patterns, but the pattern differs between L, S, and ancestor. From Friesen et al. (2004), to which I refer for further details. Used by permission.

use in turn results in a diauxic growth pattern of the bacterial population, that is, in a growth profile showing two distinct growth phases. As illustrated in Figure 10.1, the bacteria first grow fast by consuming glucose and then more slowly after switching to acetate consumption when glucose is depleted. It turns out that this diauxic switch is the basis for the diversification that we observed in our experiments.

Determining relevant phenotypes is one of the difficulties of working with microbes. One straightforward phenotypic assay consists of plating bacteria after appropriate dilution onto agar plates, on which isolated bacterial colonies can be observed and their morphology determined. If this is done with samples from our experimental populations after 1,000 generations of evolution (corresponding to approximately 6 months of daily transfers), a striking pattern emerges: there is a clear dimorphism in colony size, with some strains (called L-strains) forming large colonies, and some strains (called S-strains) forming small colonies (Figure 10.2). Thus, on the basis of colony morphology the population has diversified. This phenotypic diversification is very similar to the one reported in Rozen & Lenski (2000) for one of their replicate lines, one difference being that in our experiments all replicate lines diversified. The problem with using colony size as phenotype is that on the agar plates on which colony size is assayed, both the biotic (i.e., the nutritional) and the abiotic (i.e., geographical) environments are very different from the liquid batch culture environment experienced by the evolving populations. Thus, the phenotypic diversification observed on the plates may be a fluke,

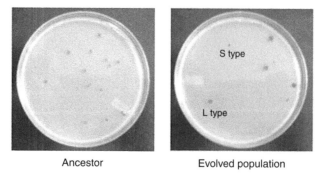

| Ancestor | Evolved population |

FIGURE 10.2. Examples of colony morphology of *E. coli* on agar plates. The plates illustrate the difference in the colony morphologies between the ancestor (left plate) and a population after 1,000 generations of evolution in glucose/acetate medium. The evolved population shows a clear dimorphism of large and small colonies, corresponding to the two ecotypes L and S. After Friesen et al. (2004), to which I refer for further details.

and it is important to have a phenotypic assay that can be directly related to the evolution environment. Such assays can be obtained by measuring diauxic growth curves.

Growth profiles of single strains, such as the ones shown in Figure 10.1, can be used to extract a number of different parameters, such as maximal growth rate on glucose, maximal growth rate on acetate, and switching lag between the two growth phases. Note that such parameters can be measured for growth in various types of media, for example, only glucose, only acetate, or mixtures of glucose and acetate in various proportions. Each such assay may serve to highlight relevant phenotypic properties. For a detailed description of how one can extract strain-specific ecological parameters from growth profiles see Friesen et al. (2004).

Analysis of growth profiles of the L- and S-strains, which exhibit different colony morphologies on agar plates, shows that these strains also have distinctly different growth profiles. Most importantly, the two types differ in the switching lag, that is, in the amount of time it takes them to make the diauxic switch from growth on glucose to growth on acetate in environments containing both carbon sources. Specifically, the L-type tends to have a very long switching lag, whereas the S-type tends to have a short switching lag (Friesen et al., 2004; Tyerman et al., 2008). For this reason, the L-type is also called the Slow Switcher (SS), while the S-type of also called the Fast Switcher (FS). The two types also differ in other growth parameters. For example, the SS type tends to have higher maximal growth rates on glucose than the FS type, but the

FS type tends to have higher maximal growth rates on acetate. Overall, analysis of many different ecological parameters reveals a clear clustering of the two types in ecological phenotype space (Friesen et al., 2004; Tyerman et al., 2008). (Curiously, we have not yet been able to determine why the ecological properties of the SS type in the evolution environment would cause this type to form large colonies on agar plates, or why the properties of the FS type would cause it to form small colonies.)

It is natural to ask whether the different ecological properties of the SS and the FS strains reflect a trade-off that underlies their coexistence. First of all, competition experiments readily reveal frequency-dependent growth rates: when SS is rare it can invade an FS resident, and vice versa (Friesen et al., 2004; Tyerman et al., 2005). Thus, each type has an advantage when rare, a hallmark of frequency-dependence. Moreover, Friesen et al. (2004) have shown that when the two types grow together in a single daily batch, the SS type has an advantage (i.e., a higher growth rate than FS) in the first phase of diauxie, that is, in the glucose phase, whereas the FS type has an advantage in the second phase, that is, after glucose has been exhausted from the medium. Thus, coexistence of SS and FS seems to be mediated by a trade-off between the performances in the two phases of diauxic growth. This is supported by the results of Spencer et al. (2007b), who performed evolution experiments in which the two strains were propagated in batch cultures containing only one of the two diauxic phases. A single batch in the original evolution experiment can be viewed as a seasonal environment. At the beginning of the batch, the environment contains both glucose and acetate in abundance, and the bacteria start out by consuming mainly glucose. This is the glucose season. But since no nutrients are added, glucose is exhausted after some time, which marks the beginning of the acetate season. In the experiments of Spencer et al. (2007b), we propagated the two strains either only in the glucose season by always transferring them to a new batch at the end of the glucose season, or only in the acetate season by always transferring them into medium obtained after the end of the glucose season (i.e., into medium obtained by growing separate batches containing the competing strains and filtering the batch after the glucose season). Over many batches, that is, over many generations, the SS type outcompeted the FS type when the environment consisted only of the glucose season, and the FS type outcompeted the SS type when the environment consisted only of the acetate season. This suggests that the two seasons are necessary for the maintenance of coexistence, and that coexistence is based on a trade-off across both seasons. Doebeli (2002) has shown that in theory, such trade-offs can generate evolutionary branching in adaptive dynamics models for the evolution of uptake efficiencies in mixed-resource batch cultures. These models for adaptive diversification are explained in Box 10.1 (see Friesen et al. (2004) for a related, but different formulation of such models; adapted from Chapter 14 in Dieckmann et al. (2004)).

Kunihiko Kaneko and Tetsuya Yomo (Kaneko & Yomo (2000, 2002)) have developed an alternative theory of adaptive diversification in microbial populations, according to which diversification first occurs only at the pheotypic level, so that genetically homogenous populations first become phenotypically dimorphic. Such phenotypic diversification is possible in systems exhibiting phenotypic bistability, so that the same genome can produce different phenotypes depending on initial concentration of metabolites and gene products. Moreover, phenotypic diversity can be stabilized by ecological interactions between the different phenotypes. In Kaneko's theory, the initial phenotypic diversity is transformed over time into genetic diversity in a process of genetic assimilation, so that eventually, the co-existing phenotypes are produced by different genotypes. It is in principle possible that such genetic assimilation of diversity that is initially only present at the phenotypic level is operating in our evolution experiments, because it is known that *E. coli* can exhibit phenotypic bistability with respect to pathways in the carbohydrate metabolism (M. Heinemann, pers. comm.).

But in any case, coexistence between different ecotypes should be mediated by trade-off. The question then becomes: what are the possible metabolic and genetic causes of such trade-offs? Quite clearly, the answer should involve the metabolic pathways involved in glucose and acetate consumption, that is, glycolysis, as well as the secretion of intermediate metabolites and the secondary pathways to further metabolize the intermediate metabolites. These central pathways for carbohydrate consumption are illustrated in Figure 10.3.

Recall that in the ancestral state, *E. coli* use catabolite repression when glucose is abundant, that is, they repress the secondary pathways and instead secrete the metabolic byproducts of glycolysis. Our basic hypothesis is that such strong catabolite repression trades off with weak catabolite repression. Specifically, we think it likely that the SS type, whose ecological properties resemble those of the ancestral strain, uses strong catabolite repression when glucose is abundant, and consequently has a high throughput in the glycolytic pathway because it secretes intermediate metabolites rather than decomposing them in slow, secondary pathways. On the other hand, the FS type may use weaker catabolite repression, so that the slower secondary pathways are active at least to some extent even when glucose is abundant. In the glucose phase, this may constitute a cost, because it slows down glycolysis and hence implies a lower growth rate. However, at the end of the glucose season, when it is time to switch to consumption of acetate using the secondary pathways, weak

BOX 10.1

A MODEL FOR EVOLUTIONARY BRANCHING DUE TO RESOURCE COMPETITION
IN BACTERIAL BATCH CULTURES

This box describes how a trade-off in the efficiencies of using a primary and
a secondary resource can generate adaptive diversification in an environment
that consists of two seasons, a first one in which both resources are abundant,
and a second one in which the primary resources has been exhausted, leaving
only the secondary resource. The modeling follows Doebeli (2002) (see also
Friesen et al., 2004), and the adaptive dynamics is based on ecological models
for Michaelis-Menten kinetics for growth in batch cultures. If n is the density of a
bacterial population living on a single resource, whose concentration is denoted
by g (e.g., glucose), Michaelis-Menten kinetics in a single batch is given by

$$\frac{dn}{dt} = \frac{r_g g}{k_g + g} n \qquad (B10.1)$$

$$\frac{dg}{dt} = -\frac{1}{Y_g} \frac{r_g g}{k_g + g} n. \qquad (B10.2)$$

Here the parameters r_g, k_g, and Y_g characterize the particular strain of bacteria whose ecological dynamics is given by eqs. (B10.1) and (B10.2): r_g is
the maximal growth rate, k_g is the concentration of glucose needed to attain a
growth rate of $r_g/2$, and Y_g is the yield (see, e.g., Edelstein-Keshet, 1988, for a
detailed account of Michaelis-Menten dynamics). To use the dynamics (B10.1)
and (B10.2) for batch cultures, one simply assumes a high initial concentration g_0 of the resource, and a low initial population density n_0, representing the
newly inoculated batch after transfer. The dynamics (B10.1) and (B10.2) will
reach its equilibrium when the resource concentration approaches zero due to
consumption.

Including a secondary resource, whose concentration is denoted by a (e.g.,
acetate) is straightforward and results in the following dynamics for a single
batch:

$$\frac{dn}{dt} = \frac{r_g g}{k_g + g} n + \frac{r_a a}{k_a + a} n \qquad (B10.3)$$

$$\frac{dg}{dt} = -\frac{1}{Y_g} \frac{r_g g}{k_g + g} n \qquad (B10.4)$$

$$\frac{da}{dt} = -\frac{1}{Y_a} \frac{r_a a}{k_a + a} a. \qquad (B10.5)$$

BOX 10.1 (*continued*)

As above, the parameters r_a, k_a, and Y_a characterize consumption of the secondary resource, and to describe the dynamics of a single batch, one assumes
some initial concentrations g_0 and a_0 for the two resources, and a low initial
population density n_0, representing the newly inoculated batch after transfer. In
Doebeli (2002) it was also assumed that consumption of the primary resource
g results in production of the secondary resource a as a metabolic by-product.
This assumption is necessary if one wants to study the phenomenon of crossfeeding, but for the purposes of studying the effect of trade-offs in seasonal
environments, the simplified system (B10.3)–(B10.5) is sufficient. To incorporate trade-offs, we assume that all parameters are externally fixed, except for the
maximal growth rates r_g and r_a, which we assume to be evolving under a trade-
off described by some function $r_a = \phi(r_g)$, for which we assume $\phi'(r_g) < 0$.
This means that a high growth rate on g comes at a cost of low growth rates
on a, and vice versa. Such a trade-off generates frequency dependence, because
specialization on glucose (high r_g) leads to higher rates of glucose consumption and hence to quicker exhaustion of glucose, that is, a quicker onset of the
acetate season, on which strains with lower r_g and correspondingly higher r_a
do better.

The adaptive dynamics of the trait r_g can be studied as follows. First, the
ecological dynamics of a resident strain are calculated over a number of serial
batch cultures by assuming that each batch has the same initial resource concentrations g_0 and a_0, and by inoculating each new batch with a fraction $pn(t_{end})$
of the population density $n(t_{end})$ that the strain reached at the end of the previous patch. (In evolution experiments, the proportion p is typically 1/100.) This
serial iteration is run until $n(t_{end})$ reaches an equilibrium, that is, does not change
anymore from one batch to the next. The success of a mutant strain is then determined by assuming that the mutant is initially very rare, so that the ecological
dynamics in a single batch of the quantities $n(t)$, $g(t)$, and $a(t)$, where $n(t)$ is
the population density of the resident strain, are the same as just described,
and the dynamics of the mutant density $n_{mut}(t)$ in a single batch are simply
determined by

$$\frac{dn_{mut}}{dt} = \frac{r_{g,mut}g}{k_g + g}n_{mut} + \frac{r_{a,mut}a}{k_a + a}n_{mut}, \tag{B10.6}$$

where $r_{g,mut}$ is the mutant trait value, and $r_{a,mut} = \phi(r_{g,mut})$. Starting from a small
initial condition $n_{mut}(0)$, one calculates $n_{mut}(t_{end})$ in the environment determined
by the dynamics of the resident during a single batch. The inoculate size of the
mutant in the next batch is then $pn_{mut}(t_{end})$, and the mutant can invade, that is,

the invasion fitness $f(r_g, r_{g,mut}) > 0$, if and only if

$$\frac{pn_{mut}(t_{end})}{n_{mut}(0)} > 1. \tag{B10.7}$$

By plotting the sign of the quantity $(pn_{mut}(t_{end})/n_{mut}(0)) - 1$ one can then construct pairwise invasibility plots (see, e.g., Chap. 2), from which one can determine the adaptive dynamics of the trait r_g, and in particular the existence of evolutionary branching points r_g^*. It turns out that evolutionary branching can only occur if the trade-off function $\phi(r_g)$ is convex at the singular point, that is, if it satisfies $\phi''(r_g^*) > 0$ at a singular point r_g^* (cf. Doebeli (2002)). This is reminiscent of the results for the evolution of habitat specialists discussed in Chapter 2 (see Fig. 2.2), and of the results of Claessen et al. (2007) for a consumer-resource model with two different types of resources, in which convex trade-offs between resource use generate evolutionary branching. In the model presented here, evolutionary branching occurs for substantial regions in parameter space if the trade-off function is convex at the singular point. In particular, adaptive diversification can be found for parameters describing primary and secondary resources as envisaged for our experiments with *E. coli*, such a glucose and acetate, which differ greatly in their utility, and hence in the parameters (k_g, Y_g) and (k_a, Y_a).

Challenge: Prove these statements.

In an environment with a primary and a secondary resource, evolutionary branching is expected to generate coexistence between a "glucose specialist," characterized by a high r_g and a correspondingly low r_a, and a more generalist strain with lower r_g and correspondingly higher r_a. This second strain is outcompeted on the primary resource, but outcompetes the specialist in the second season, that is, once the primary resource has been exhausted, which corresponds to the experimental findings of Friesen et al. (2004) and Spencer et al. (2007b). The model thus shows how a trade-off in maximal growth rates on primary and secondary resources can lead to the gradual evolution of adaptive diversification in resource use from a single ancestral bacterial strain (see Friesen et al. (2004) for alternative models of this process).

■

catabolite repression may constitute an advantage, because the cells are already primed for acetate consumption. In contrast, the SS type may have to undergo major regulatory changes to be able to consume acetate, which may lead to a long switching lag, and hence to a delay of growth in the second phase of diauxie.

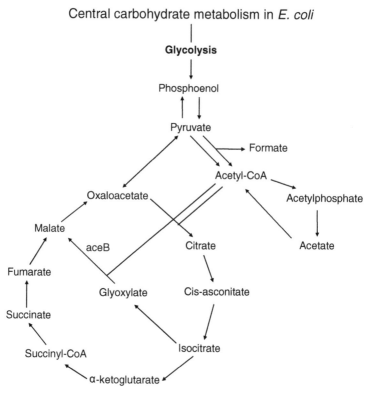

FIGURE 10.3. Schematic representation of some of the main pathways and enzymes of the carbohydrate metabolism in *E. coli*, including the circular TCA cycle, the glyoxylate shunt, as well as acetate consumption/excretion pathways. After Le Gac et al. (2008).

We first investigated this hypothesis by measuring expression of candidate genes in the secondary pathways. We expected that during growth on glucose, such genes are over expressed in the FS strains compared to both the ancestor and the SS strain. Indeed, this is what we found when measuring expression of *aceB*, a gene involved in the glyoxylate shunt (Figure 10.3), in one of the diversified populations (Spencer et al., 2007a). When growing on acetate, this gene is expressed in all three strains, but when growing on glucose, this gene is only active in the FS strain. This indicates that the secondary pathways are active in the FS strain even when growing on glucose, which does not seem to be the case for the SS strain and the ancestral strain. We were also able to identify a genetic change in the FS strain that likely leads to over expression of *aceB*. The *iclR* gene is a regulatory gene that represses expression of *aceB*, and we found that the FS strain that over expresses *aceB* had a large insertion in the *iclR* gene, essentially knocking out *iclR*, and hence knocking out at least part of the downregulation of *aceB* (Spencer et al., 2007a). Insertion elements can play important roles in microbial evolution (e.g., Schneider & Lenski, 2004), and this may be one example where insertion elements are important for evolutionary diversification. However, we did not find this insertion in FS strains from other replicate lines, which indicates that other genetic changes are also needed to create an FS type. This is corroborated by the fact that when we transformed the FS strain with the wild type *iclR* gene from the ancestral strain (i.e., when we inserted the wild type *iclR* gene into FS), its growth profile became more similar to that of the ancestral strain, yet still had some of the FS characteristics.

We thus turned to a more global analysis of gene expression using microarray technology (Le Gac et al., 2008). This analysis confirmed that during growth on glucose, the secondary pathways are generally overexpressed in the FS strains (Fig. 10.4). It is known that a knockout of the global regulator gene *arcA* has similar effects on these metabolic pathways, which led us to test for mutations in this gene. Indeed, we found that one of our FS strains has a point mutation in this gene, which is also important for growth under anaerobic conditions, that is, when there is little oxygen in the medium. In our batch cultures, low oxygen concentration is a consequence of growth on glucose, because the glucose season already results in substantial population densities with a concomitant scarcity of oxygen. Thus, switching from the first to the second season in our batch cultures may not only be associated with switching from glucose to acetate consumption, but also with switching from growth in aerobic to growth in anaerobic conditions. This may generate further metabolic trade-offs across the two seasons. In fact, it is known that oxygen limitation represses secondary carbohydrate pathways that are required for growth on acetate (White, 2000), and hence oxygen limitation may exacerbate the negative effects of strong catabolite repression on the ability to switch from glucose to acetate consumption.

Our microarray analysis (Le Gac et al., 2008) also revealed other interesting patterns of evolutionary change. The majority of the differences in gene expression occur between the SS and FS strain on the one hand, and the ancestral strain on the other hand. This indicates that the SS and FS strains have many changes in common, and that these changes likely occurred before the diversification into the SS and FS types. Before diversification, the ancestral lineage probably simply adapted to the glucose/acetate environment by improving in various ways, for example, in translation efficiency, glucose uptake capacity and survival rate during to stationary phase, which are adaptations that have been found in previous evolution experiments with *E. coli*. A

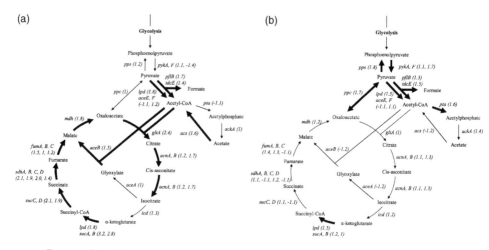

FIGURE 10.4. TCA, glyoxylate shunt, and acetate consumption/excretion metabolic pathways showing different trends in FS vs. Ancestor (a) and SS vs. Ancestor (b) during exponential growth on glucose. Enzymes are indicated by italics and fold changes are in parentheses. A thick arrow indicates a fold change 1.5. From Le Gac et al. (2008), to which I refer for more details. Used by permission.

minority of differences in gene expression observed in our microarray analysis are differences between the two evolved types, SS and FS. In accordance with our trade-off hypothesis, these differences indicate an upregulation of the secondary carbohydrate pathways, of acetate consumption and of anaerobic respiration genes in FS, and of acetate secretion in SS (Fig. 10.4).

Even though these patterns of changes in gene expression were obtained from only three strains (the ancestor, as well as the derived strains SS and FS), they allow us to speculate about the temporal unfolding of the dynamics of diversification in our evolution experiments. Since a majority of the differences in gene expression occur between the pooled evolved strains (SS and FS) and the ancestral strain, it seems likely that there was a prolonged period of local adaptation to the mixed resource environment before diversification occurred. Thus, it seems that the evolving population first needed to be adaptively "primed" before diversification could occur. This would correspond to the pattern of evolutionary branching seen in many adaptive dynamics models, in which adaptive diversification typically only occurs after a prolonged period of directional adaptation, that is, of evolution to the branching point. Indeed, if we use the "fossil" record of our experimental lines (i.e., samples frozen during different time points of the evolutionary trajectory) to document the evolutionary change, we find a pattern that resembles evolutionary branching. This is

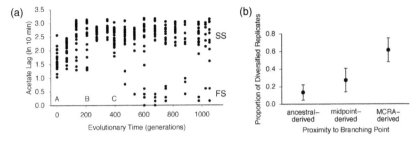

FIGURE 10.5. Evolutionary branching in acetate lag. Panel (a) shows the acetate lag (defined as the time from the switching point to the maximum growth rate on acetate during diauxic growth; cf. Fig. 10.1) as a function of evolutionary time in an evolving *E. coli* lineage. Acetate lag increases from the ancestral state (generation 0) to the branching point (generation 395), where the population splits into the two ecotypes SS (long lag) and FS (short lag). Data points represent randomly selected isolated genotypes from the frozen record of the evolutionary lineage. Letters on the *x*-axis represent the ancestral (A), midpoint (B), and Most Recent Common Ancestor (MRCA; C) populations used in the rediversification experiment shown in panel (b). In this experiment, replicate populations started with single SS types from the different time points were reevolved for approximately 140 generations, after which the fraction of replicates containing both SS and FS types was determined. Populations started with "older" genotypes, that is, farther from the branching point, are less likely to diversify. Modified from Spencer et al. (2008), to which I refer for further details. Copyright 2008 National Academy of Sciences, USA.

illustrated in Figure 10.5, which shows the distribution of switching lags in an evolving population at different time points during the evolution experiment. It is apparent that the switching lag first increases directionally and then branches into two clusters representing the SS type (long switching lag) and the FS type (short switching lag).

The pattern that the phenotypic split only occurs after prior directional adaptation is a hallmark of diversification due to frequency-dependent interactions: during directional adaptation, the fitness landscape, which depends on the current position of the population in phenotype space, changes such that diversification eventually becomes adaptive. This is in contrast to classic theory for static, frequency-independent fitness landscapes, in which diversification can occur when an ancestral population is near a trough in the landscape, and different subpopulations evolve toward different fitness peaks. In this case, it becomes increasingly difficult to cross the fitness valleys as the ancestral population evolves closer to a fitness peak. As a consequence, diversification becomes less likely with increasing adaptation, as has been argued experimentally by Buckling et al. (2003).

To test whether the propensity to diversify increases over time in our evolutionary lines, as predicted by the theory of evolutionary branching, or decreases, as predicted by classic theory, we used single genotypes from three different time points along the evolutionary trajectory to found new experimental lines (Spencer et al., 2008). Thus, we replayed the evolutionary tape from different starting points, which were chosen prior to the diversification event in the original evolution experiment. All founding genotypes from all three time points showed a growth profile that was similar to that of the ancestor and of the SS strain. After propagation serial batch culture for 200 generations, replicates lines started with genotypes extracted at generation 0 (i.e., ancestral genotype) and at generation 208 of the original experiment showed a significantly lower probability of diversification than lines started with genotypes extracted from generation 395, just prior to the diversification event in the original experimental line (Fig. 10.5). Thus, as the directional phase progresses in this line, the evolving population becomes more likely to diversify. These results correspond to the temporal patterns of the propensity to diversify predicted by theory, and hence support the notion that this bacterial lineage has undergone evolutionary branching.

There are (of course) possible alternative explanations for the observed increase in the likelihood of diversification over time. For example, purely genetic constraints could imply that a mutation causing FS-like growth patterns can only arise after a number of other mutations have occurred. Or clonal interference (Rozen et al., 2002) could prevent invasion of FS-like types until the population is well-adapted to the mixed resource environment. However, we argued that neither of these alternatives is likely to fully explain the observed patterns (Spencer et al., 2008). Instead, we think it likely that the change in the competitive environment brought about by directional adaptive change in the evolving population prior to diversification was a precondition for diversification, that is, for the invasion of more FS-like types.

It would be desirable if these arguments could be supported by identifying actual mutations responsible for generating the different ecotypes, and then following the invasion history of such mutations in the frozen "fossil" record of the evolution experiments. I have already mentioned two mutations that we identified in some of our FS strains, one in the local regulator iclR, and one in the global regulator arcA. Because arcA is known to affect central pathways of the carbohydrate metabolism, and in particular acetate consumption, we used this mutation to investigate how selective pressures might have changed over the time course of the evolutionary process taking place in our experiments.

The *arcA* mutation was identified in an FS-strain at generation 1,000 in an evolution experiment in which diversification into FS and SS occurred before generation 200 and persisted until the end of the evolution experiment. Our transcription analysis (Le Gac et al., 2008) revealed that numerous genes controlled by *arcA* are differentially expressed in the two ecotypes after 1,000 generations. Sequencing *arcA* in FS clones isolated at 1,000 and 1,200 generations, but not before that, revealed a nonsynonymous point mutation responsible for substitution of threonine by alanine at the protein level, thus replacing a polar amino-acid by a nonpolar one. This mutation occurs at a site forming a side chain involved in a hydrogen bond with the phosphate responsible for the activation of *arcA* (Toro-Roman et al., 2005). Therefore, it is likely that the mutation affects the activity of *arcA*. To understand the significance of this mutation for the evolutionary dynamics, we characterized the effects of this mutation on phenotype and fitness at different points in time of the diversifying evolutionary lineage by replacing the ancestral allele by the mutant allele, or vice versa, in a number of different FS and SS clones frozen during the evolution experiments (Le Gac & Doebeli, 2010). That is, we tested the effects of the mutation in specimens from the "fossil" record.

If we denote the ecotypes isolated at different time points without the *arcA* mutation by FS and SS, and the corresponding ecotypes with the mutation by FS* and SS*, then comparing growth profiles of SS and SS*, and of FS and FS*, shows the phenotypic effects of the mutation. Moreover, competing SS against SS* (in the presence or absence of FS), or FS against FS* (in the presence or absence of SS), reveals the fitness effects of the mutation in various ecological contexts.

For example, the comparison of growth profiles and competition assays between SS and SS* indicated that the *arcA* mutation is highly adaptive when occurring in the SS genomic background (Le Gac & Doebeli, 2010). Essentially, in SS the mutation broadens the ecological niche and leads to a more generalist strategy by improving the consumption of the nonpreferred resource acetate. However, in the FS genomic background the mutation does not generate significantly different growth profiles and seems to be neutral, except at generations 200 and 1,000, in which it is slightly adaptive (the mutation actually invaded the evolutionary FS-lineage around generation 1,000). This indicates that in the FS genomic background, the mutation does not affect the width of the ecological niche, but may affect the competitiveness within the niche by generating some fine-tuning adaptation. These results represent an interesting case of strong epistasis, showing that the effect of the mutation strongly depends on the genomic background in which it occurs.

Interestingly, three-way competition experiments between SS, FS, and SS* on the one hand, and between SS, FS, and FS* on the other hand show that the *arcA* mutation not only exhibits intragenomic epistasis, but also intergenomic epistasis, that is, that the fitness effects of the mutation depend on the other genomes present in the population. Specifically, while SS* readily outcompetes SS in the absence of FS, the mutation becomes neutral, and in particular SS* does not outcompete SS anymore, in the presence of FS (Le Gac & Doebeli, 2010). Such intergenomic epistasis is of course just another name of frequency dependence: the fitness of the *arcA* mutation depends on which other phenotypes are present in the population. When SS* is competing in the absence of FS, the mutation allows exploitation of the acetate niche, thus generating a selective advantage. However, in the presence of FS, the acetate niche is already occupied by an efficient generalist, which eliminates the selective advantage of the *arcA* mutation in SS. Not surprisingly, such frequency-dependence is much less pronounced when the mutation is introduced in FS, as the fitness of FS* compared to FS is essentially independent of whether SS is present or absent. Because SS probably consumes virtually no acetate in the presence of FS, the fitness of a mutation affecting competitiveness during acetate consumption (i.e., the fitness of FS*) is not affected by the presence of SS.

Although the arcA mutation is not the mutation responsible for the ecological diversification observed in the evolutionary lineage in which it was found (SS* does not correspond to FS, and the mutation arose after the initial split into FS and SS), the results of Le Gac & Doebeli (2010) shed light on the fate of adaptive mutations during adaptive diversification: the fitness of mutations is strongly affected by the genomic background in which they occur, as well as by the genomes of the other individuals present in the population. Overall, and even though many open questions remain (e.g., understanding the complete sequence of genetic changes associated with the diversification), I think that the diversification observed in our experimental lines is a clear example of adaptive diversification due to frequency-dependent competition for resources, and that these experiments therefore support the theory of adaptive diversification described and developed in this book.

Basic Concepts in Adaptive Dynamics

Adaptive dynamics was invented by Hans Metz (Metz et al., 1996) to describe the long-term evolution of quantitative traits using systems of differential equations. In essence, adaptive dynamics is a gradient dynamics on temporally varying fitness landscapes. Fitness landscapes change over time because they are determined by the evolving trait values, that is, because of frequency dependence. Gradient dynamics has of course a long tradition in evolutionary theory, starting with optimization models, that is, gradients on static fitness landscapes. Russ Lande's framework for quantitative genetics (Lande, 1979) is a well-known and general gradient dynamics that allows for incorporating frequency-dependence. In the context of the problem of diversification, this framework has been most prominently used by Peter Abrams and colleagues (Abrams, 2001). Another type of adaptive gradient dynamics has been introduced by Karl Sigmund and Martin Nowak on the basis of evolutionary game theory (Nowak & Sigmund, 1990). The innovation advocated by Hans Metz consisted in defining fitness explicitly as an ecological quantity at the individual level (Metz et al., 1992). Specifically, fitness is the long-term per capita growth rate of individuals carrying a rare mutant trait. This ecological definition of fitness is in rather stark contrast to the traditional population genetic definition that is based on the change in allele frequency over one generation. An ecological definition of fitness has the great advantage of closing the feedback loop between ecology and evolution: ecological dynamics determines the selective pressures on phenotypes, which in turn determine the ecological dynamics.

One very nice feature of adaptive dynamics is that using certain approximations, its equations can be derived from a stochastic master equation. Roughly speaking, adaptive dynamics can be obtained from an underlying individual-based stochastic birth-death process in the limit of large populations sizes and small, rare mutations. I refer to the seminal paper by Ulf Dieckmann and Richard Law (Dieckmann & Law, 1996) and to Champagnat et al. (2006) for the mathematical details of these derivations. Perhaps somewhat surprisingly, no textbooks on adaptive dynamics are available to date, but there is a nice primer by Odo Diekmann (Diekmann, 2004). Nevertheless, for the sake of completeness I will briefly recapitulate the most basic concepts and analytical features of adaptive dynamics in this Appendix. Many notions and concepts from dynamical systems theory, such as fixed points and asymptotic stability, translate into useful concepts in adaptive dynamics. However, these traditional notions are insufficient to capture the behavior of an adaptive dynamical system that undergoes evolutionary branching and gradually splits into two distinct lineages. To investigate this phenomenon analytically one needs a different stability concept, called evolutionary stability, which is familiar from evolutionary game theory. Evolutionary instabilities may lead to an increase in the dimensionality of the deterministic system needed to describe the adaptive dynamics.

A.1 SINGULAR POINTS AND CONVERGENCE STABILITY

In general, an adaptive dynamical system describes the evolution of real-valued traits x_i, $i = 1, \ldots, m$ in m different species. Here the trait $x_i = (x_{i1}, \ldots, x_{in_i}) \in \mathbb{R}^{n_i}$ is a n_i-dimensional vector. The differential equations for the adaptive dynamics then have the form

$$\frac{dx_i}{dt} = g_i(x, t), \quad i = 1, \ldots, m, \tag{A.1}$$

where $x = (x_1, \ldots, x_m) \in \mathbb{R}^{\sum_{i=1}^m n_i}$ is the vector of all trait values, and the $g_i(x, t) = (g_{i1}(x, t), \ldots, g_{in_i}(x, t))$ are vector-valued functions $\mathbb{R}^{\sum_{i=1}^m n_i} \times \mathbb{R} \to \mathbb{R}^{n_i}$ with the current trait values in the m monomorphic resident populations as well as the time t as arguments. This latter argument is only explicitly needed if there is some external parameter in the system that is time-dependent, for example, describing a seasonal temperature regime, or long-term fluctuations in rain fall, etc. In the following we restrict our attention to autonomous differential equations, in which the functions g_i do not explicitly depend on t. Moreover, in most examples described in this book the dimension of the trait space n_i in each species is small, as is the number of species m.

In general, the functions g_i are determined by the gradients of the invasion fitness functions, as well as as by the mutational process. In each of the m species, the invasion fitness function $f_i(x, y_i)$ is the long-term per capita growth rate of a rare mutant type $y_i = (y_{i1}, \ldots, y_{in_i})$ of species i, appearing in the community in which each species is monomorphic for the resident type given by the vector $x = (x_1, \ldots, x_m)$. This resident community determines the ecological environment in which the long-term growth rate of the mutant is assessed. Note that the invasion fitness f_i can also be interpreted as the probability of the mutant type y_i not going extinct (Dieckmann & Law, 1996). Also note that we must have $f_i(x, x_i) = 0$ for all i, because the long-term growth rate of a mutant that has the same trait values as the resident must have the same long-term growth rate as the resident. But the resident is assumed to persist ecologically (i.e., to neither go extinct nor to increase indefinitely), and hence its long-term growth rate must be zero. (Strictly speaking, this is only true if the ecological dynamics unfolds in continuous time, for in discrete time long-term persistence is equivalent to a long-term growth rate of one. The corresponding consistency condition for the invasion fitness function is then $f_i(x, x_i) = 1$ for all i, which can be translated into $f_i(x, x_i) = 0$ for all i by taking logarithms.)

The functions g_i are given as follows:

$$
\begin{pmatrix} g_{i1}(x) \\ \dots \\ g_{in_i}(x) \end{pmatrix} = B_i(x) \cdot \begin{pmatrix} \left. \dfrac{\partial f_i(x, y_i)}{\partial y_{i1}} \right|_{y_i=x_i} \\ \dots \\ \left. \dfrac{\partial f_i(x, y_i)}{\partial y_{in_i}} \right|_{y_i=x_i} \end{pmatrix}. \tag{A.2}
$$

Here the selection gradient $\left. \dfrac{\partial f_i(x, y_i)}{\partial y_{ij}} \right|_{y_i=x_i}$ is the partial derivative of the invasion fitness function with respect to mutant trait value y_{ij} evaluated at the resident trait x, and $B_i(x)$ is the mutational variance-covariance matrix, describing the frequency, size and effect of mutations occurring in species i. $B_i(x)$ is symmetric, and nonzero off-diagonal entries in $B_i(x)$ imply pleiotropic mutational effects, that is, nonzero covariances in mutational effects on different phenotypic components in species i. In general, the total mutation rate in a population depends on populations size, which is why $B_i(x)$ depends in general on the current resident trait values, as these determine the ecology, and hence the population sizes of the various species. The dynamics given by eqs (A.1) and (A.2) is formally similar to Lande's equations for quantitative genetics (Lande, 1979). However, the genetic variance-covariance matrices appearing in Lande's equations (and corresponding to the matrices $B_i(x)$ above) describe standing genetic variation rather than the effects of mutations, and the fitness functions whose gradients are used in Lande's equations usually lack an ecological underpinning based on per capita growth rates.

A solution to a differential equation of the form (A.1) is a vector-valued function $t \mapsto x(t) = (x_1(t), \dots, x_m(t))$ satisfying $\frac{dx_i}{dt} = g_i(x(t))$ at all times t. In general, given any differential equation, the goal is to determine all its solutions and their quantitative and qualitative behaviors. For example, if the functions $g_i(x)$ in (A.1) are linear in each argument x_i, it is easy to determine all solutions of this so-called linear differential equation explicitly (e.g., Edelstein-Keshet, 1988). Most differential equations arising in adaptive dynamics are nonlinear, and finding explicit solutions is usually impossible. In these cases the emphasis is on trying to obtain qualitative information about the long-term behavior of solutions. For example, one might ask whether solutions are periodic, that is, whether there is a number T such that the solution satisfies $x(t) = x(t + T)$ for all t. The simplest case of a periodic solution occurs when $T = 0$, that is, when the periodic solution is stationary and consists of a single point $x^* = (x_1^*, \ldots, x_m^*)$ in trait space. In dynamical systems theory such a point is called a fixed point or a stationary point. Note that a stationary point of (A.1) is a solution of the following system of equations:

$$g_1(x_1^*, \ldots, x_m^*) = \cdots = g_m(x_1^*, \ldots, x_m^*) = 0. \tag{A.3}$$

Definition: A point $x^* = (x_1^*, \ldots, x_m^*)$ in trait space satisfying $g_1(x_1^*, \ldots, x_m^*) = \cdots = g_m(x_1^*, \ldots, x_m^*) = 0$ is called *singular point* of the adaptive dynamics (A.1), or an *evolutionary singularity*.

Note that as long as all the mutational matrices $B_i(x)$ have nonzero determinants, the functions $g_i(x)$, $i = 1, \ldots, m$ can only become zero if all the selection gradients $\left. \frac{\partial f_i(x, y_i)}{\partial y_{ij}} \right|_{y_i = x_i}$ are zero. When a mutational matrix has a zero determinant, then its kernel (i.e., the set of all vectors that are mapped to zero by the matrix) contains nonzero vectors, and it is therefore possible that a function g_i is zero even if the corresponding selection gradients in the ith species are nonzero. However, such cases will not play a role in the sequel, so that for all practical purposes we can assume that singular points are given as points in trait space at which all selection gradients vanish. Note also that as we will see below, singular points of an adaptive dynamical system (A.1) need not be the endpoint of the evolutionary process.

Evolutionary singularities are a central focus of attention in adaptive dynamics, and the most basic question is whether a singular point is an attractor for the adaptive dynamics for which it is a stationary point. There are many notions of stability in the theory of ordinary differential equations that could, in principle, be used to define the dynamical stability of a singular point. The most straightforward approach (Christiansen, 1991; Dieckmann & Law, 1996) is to use the concept of asymptotic stability from dynamical systems theory. Thus, given an adaptive dynamical system of the form (A.1) and a singular point $x^* = (x_1^*, \ldots, x_m^*)$, one considers the linearization of the dynamical system, which is obtained using a Taylor expansion of the functions g_i around the singular point. For x_i-values that are sufficiently close to x_i^*, the

Taylor expansion yields

$$\frac{dx_i}{dt} = g_i(x_1, \ldots, x_m)$$

$$= g_i(x_1^*, \ldots, x_m^*) + \sum_{j=1}^{m} \frac{\partial g_i}{\partial x_j}(x_1^*, \ldots, x_m^*) \cdot (x_j - x_j^*) + o(|x|) \qquad \text{(A.4)}$$

where $o(|x|)$ are second and higher-order terms (i.e., terms which, when divided by the distance $|x_j - x_j^*|$, tend to zero as x tends to x^*). Note that in expression (A.4), the quantities $\partial g_i/\partial x_j$ are $n_i \times n_j$-matrices:

$$\frac{\partial g_i}{\partial x_j}(x_1^*, \ldots, x_m^*) = \begin{pmatrix} \frac{\partial g_{i1}}{\partial x_{j1}}(x_1^*, \ldots, x_m^*) & \cdots & \frac{\partial g_{i1}}{\partial x_{jn_j}}(x_1^*, \ldots, x_m^*) \\ \cdots & \cdots & \cdots \\ \frac{\partial g_{in_i}}{\partial x_{j1}}(x_1^*, \ldots, x_m^*) & \cdots & \frac{\partial g_{in_i}}{\partial x_{jn_j}}(x_1^*, \ldots, x_m^*) \end{pmatrix}, \qquad \text{(A.5)}$$

and that $\partial g_i/\partial x_j(x_1^*, \ldots, x_m^*) \cdot (x_j - x_j^*)$ is the vector obtained by applying the matrix $\partial g_i/\partial x_j(x_1^*, \ldots, x_m^*)$ to the vector $(x_j - x_j^*)$. Since x^* is singular, we have $g_i(x_1^*, \ldots, x_m^*) = 0$ for $i = 1, \ldots, m$, and disregarding higher-order terms we get the system of linear differential equations

$$\frac{dx_i}{dt} = \sum_{j=1}^{m} \frac{\partial g_i}{\partial x_j}(x_1^*, \ldots, x_m^*) \cdot (x_j - x_j^*). \qquad \text{(A.6)}$$

This system is called linear because the functions on the right-hand side, describing the rate of change in the traits x_i, are linear functions of the variables $x_i, i = 1, \ldots, m$. In matrix notation, this linear system can be written as

$$\frac{dx}{dt} = J(x^*) \cdot (x - x^*), \qquad \text{(A.7)}$$

where

$$J(x^*) = \begin{pmatrix} \frac{\partial g_1}{\partial x_1}(x_1^*, \ldots, x_m^*) & \cdots & \frac{\partial g_1}{\partial x_m}(x_1^*, \ldots, x_m^*) \\ \cdots & \cdots & \cdots \\ \frac{\partial g_m}{\partial x_1}(x_1^*, \ldots, x_m^*) & \cdots & \frac{\partial g_m}{\partial x_m}(x_1^*, \ldots, x_m^*) \end{pmatrix} \qquad \text{(A.8)}$$

is the Jacobian matrix of the adaptive dynamical system at the singular point x^*. There are two reasons why considering the linear system is very useful. First, systems of linear differential equations are very well understood analytically. Second, the dynamic behavior of the linear system yields a good description of the behavior of the full system (A.1) if the trait values $x = (x_1, \ldots, x_m)$ are close to the singular point $x^* = (x_1^*, \ldots, x_m^*)$. In other words, the linear system is an approximation to the full system around the singular point x^* that is much easier to handle analytically.

For example, consider one-dimensional systems of adaptive dynamics, that is, systems of the form

$$\frac{dx}{dt} = g(x) = B(x) \left. \frac{\partial f(x, y)}{\partial y} \right|_{y=x}, \tag{A.9}$$

where x is a scalar trait in a single species (e.g., body size), $f(x, y)$ is the invasion fitness function for a mutant type y appearing in a resident population that is monomorphic for trait value x, and the scalar function $B(x)$ describes the mutational process in the trait x. A trait value x^* is an evolutionary singularity if $g(x^*) = 0$. The Jacobian at a singular point is simply the derivative of g at this point: $J(x^*) = \frac{dg}{dx}(x^*)$. If $J(x^*) < 0$, and if the adaptive dynamics is started sufficiently close to x^*, then any solution will approach the singular trait value x^* over time. This is because $\frac{dg}{dx}(x^*) < 0$ together with $g(x^*) = 0$ implies that the rate of change in the trait value x is positive, $g(x) > 0$, if $x(t)$ is sufficiently close to, but smaller than x^*, and the rate of change in x is negative, $g(x) < 0$, if $x(t)$ is sufficiently close to, but bigger than x^*. Conversely, if $J(x^*) > 0$, then any sufficiently small initial distance to x^* will monotonously increase over time. In sum, if $J(x^*) < 0$ then x^* is an attractor for the adaptive dynamics, and if $J(x^*) > 0$ then x^* is a repellor for the adaptive dynamics. In fact, it is well known that any solution of a one-dimensional differential equation of the form (A.9) in the long term either approaches ∞ or $-\infty$, or it approaches an evolutionary singular point with a negative Jacobian. In particular, no oscillatory behavior can occur in one-dimensional adaptive dynamical systems.

For general systems of differential equations of the form (A.7) it can be shown that if all the eigenvalues of the Jacobian matrix $J(x^*)$ at a singular point x^* have negative real parts, then all solutions of the linear system converge to the singular point x^*. For the full system (A.1) this implies that if the traits have starting values sufficiently close to x^*, then they will remain close to x^* at all times and will eventually converge to the singular point. Thus x^* is an attractor for the adaptive dynamics in this case.

Definition: A singular point x^* for an adaptive dynamical system of the form (A.1) is called convergent stable if all the eigenvalues of the Jacobian $J(x^*)$ at the singular point have negative real parts.

We have seen many examples of convergent stable singular points throughout this book. In dynamical systems theory convergence stability is usually called linear stability or asymptotic stability. It is important to note that in general, convergence stability is not determined solely by the invasion fitness functions, because the Jacobian matrix (eq. (A.8)) also depends on the mutational matrices B_i. Therefore, convergence stability is in general not a property of the selection gradients alone. For example, it can happen that for a given set of selection gradients, all eigenvalues of the Jacobian matrix at a singular point have negative real parts for some choices of the mutational quantities $B(x_i^*)$, but not for other choices. However, there are certain circumstances in which this difficulty does not arise, and hence in which the selection gradients contain all the information about the convergence stability of a singular point. This situation has been captured in the definition of strong convergence stability by Leimar (2001, 2005, 2009) as follows.

Recall that if the mutational matrix B has a nonzero determinant, all selection gradients must vanish at a singular point x^*. Then the Jacobian at x^* has the form

$$J(x^*) = B(x^*) \cdot S(x^*), \tag{A.10}$$

where the $S(x^*)$ is the Jacobian of the fitness gradients, that is,

$$S(x^*) = \begin{pmatrix} \dfrac{\partial F_1}{\partial x_1}(x_1^*, \ldots, x_m^*) & \cdots & \dfrac{\partial F_1}{\partial x_m}(x_1^*, \ldots, x_m^*) \\ \cdots & \cdots & \cdots \\ \dfrac{\partial F_m}{\partial x_1}(x_1^*, \ldots, x_m^*) & \cdots & \dfrac{\partial F_m}{\partial x_m}(x_1^*, \ldots, x_m^*), \end{pmatrix} \tag{A.11}$$

where

$$F_i(x_1, \ldots, x_m) = \left. \frac{\partial f_i(x_1, \ldots, x_m, y_i)}{\partial y_i} \right|_{y_i = x_i} \tag{A.12}$$

is the selection gradient in the ith species (note that in general, the F_i are vector-valued functions, and the $\partial F_i / \partial x_j$ are matrices).

We recall that a matrix A is called negative definite if all the eigenvalues of the symmetric matrix $(A + A^t)$ are < 0, where A^t denotes the transposed matrix.

Definition: A singular point x^* is called strongly convergent stable if the Jacobian of the selection gradients $S(x^*)$ is negative definite.

The significance of this definition lies in the following result of Leimar (2001): a singular point x^* is convergent stable for all mutational matrices $B(x^*)$ if x^* is strongly convergent stable (see also Leimar (2009)). Thus, a strongly convergent stable singular point is convergent stable independent of the mutational process that drives the evolutionary dynamics. It should be noted, however, that strong convergence stability is a rather restrictive condition that cannot be expected to be encountered very often in general, multidimensional adaptive dynamical systems. Thus, in most situations convergence stability of a singular point will depend explicitly on the mutational process driving the system, as is for example the case even in relatively simply systems such as the two-dimensional adaptive dynamics of predator-prey systems discussed in Chapter 5.

An alternative approach to reducing the problem of convergence stability to properties of selection gradients alone consists of trying to reduce it to properties along single trait axes. For example, in the two-dimensional predator-prey system of Chapter 5, in which a scalar trait is evolving in each species, one can investigate whether the prey trait converges to the singular value when the predator trait is fixed at the singular value, and vice versa. One can then ask whether convergence to the singular point along the two trait axes tells us something about convergence stability of the singular point in the full adaptive dynamical system.

In general, this approach also fails. For example, in the predator-prey system it can happen that the singular point is attracting in both trait directions, yet is not convergent stable: despite converging to the singular point on both trait axes, the adaptive dynamics escapes from the singular point in other directions. In terms of the Jacobian matrix $J(x^*)$, in this two-dimensional system convergence stability is equivalent to the determinant of $J(x^*)$ being positive and the trace of $J(x^*)$ (i.e., the sum of its diagonal elements) being negative, while convergence to x^* along the trait axes is described by the two diagonal entries in $J(x^*)$ alone. If these are negative, that is, if x^* is attracting in the directions of the two trait axis, then trace of $J(x^*)$ is negative as required, but the off-diagonal elements of $J(x^*)$ can still cause its determinant to be negative as well, so that x^* would not convergent stable. Thus, convergence of traits considered in isolation while the other traits are fixed at their singular values is not a sufficient condition for convergence stability. On the other hand, it is possible that some diagonal elements of the Jacobian are positive, yet the Jacobian is negative definite. For example, in the predator-prey system above it can happen that the singular prey trait is convergent unstable for the adaptive dynamics in the prey when the predator trait is fixed at the singular value, yet the combined system converges to the singular point (see Chap. 5). Thus, convergence stability along single trait axes is neither a necessary nor a sufficient condition for convergence stability of a singular point in general, multidimensional adaptive dynamical system.

Olof Leimar has pointed out that even if a singular point is strongly convergent stable, there are in general at least some adaptive escape routes from the singular point, in the sense that there are particular sequences of phenotype substitutions through which the population can evolve away from a strongly convergent stable singular point (the Darwinian Demon, see Leimar (2001, 2009)). To exclude this possibility, one has to require absolute convergence stability for the singular point (Leimar, 2009), which is essentially the requirement that the selection gradients are proportional to the gradient of a potential function with a unique minimum at the singular point. The condition of absolute convergence stability is stronger than the condition of strong convergence stability, and hence even less likely to be satisfied in any given system.

In the context of adaptive dynamics, the term convergence stability is used to distinguish linear asymptotic stability from *evolutionary stability*, which will be defined below and which refers to whether populations at a singular point can be invaded by nearby mutants. Convergent stable singular points need not be the endpoint of the evolutionary process, because a convergent stable singular point may be evolutionarily unstable, and hence may be the starting point for the evolution of a polymorphism. If a polymorphism evolves from a convergent stable singular point, then the deterministic dynamical system in which convergence stability is observed ceases to be a valid description of the adaptive dynamics, and instead a higher dimensional adaptive dynamical system must be used in which each evolving morph of the polymorphism is represented by a separate differential equation. This "growth of the adaptive tree" has been beautifully described in Geritz et al. (1998) and will be the subject of the next section.

A.2   EVOLUTIONARY STABILITY

Mutations are the raw material for adaptive dynamics. In particular, once an adaptive dynamical system has reached a singular point in phenotype space, it will still be constantly challenged by new mutations. The fate of these mutants are determined by the invasion fitness function at the singular point. For example, consider a one-dimensional adaptive dynamical system (A.9). The function $g(x)$ is proportional to the selection gradient $\frac{df}{dy}(x, y)|_{y=x}$. If x^* is an a evolutionary singular point, then $\frac{df}{dy}(x^*, y)|_{y=x^*} = 0$, and the fate of rare mutants occurring in a resident population that is monomorphic for x^* is therefore determined by the second derivative $H(x^*) = \frac{d^2f}{dy^2}(x^*, y)|_{y=x^*}$. If $H(x^*) < 0$, then, as a function of mutant trait values x', the invasion fitness function has a maximum at x^*, which means that x^* is uninvadable by nearby mutants, that is, all nearby mutants have a negative invasion fitness. In this case x^* is called *evolutionarily stable*. On the other hand, if $H(x^*) > 0$, then the invasion fitness function has a minimum at x^*, which means that x^* is invasible by all nearby mutants, that is, all nearby mutants have a positive invasion fitness. In this case x^* is called *evolutionarily unstable*.

The extension of the notion of evolutionary stability of a singular point from one to higher dimensions is intuitively straightforward: a singular point is evolutionarily stable if, in each of the interacting species, no nearby mutant can invade the resident populations that are monomorphic for the trait combinations specified by the singular point. If one is considering the coevolution of scalar traits in different species, then the condition for evolutionary stability of a singular point $x^* = (x_1^*, \ldots, x_m^*)$ is simply that each value x_i^* represent a fitness maximum for the invasion fitness function $f_i(x^*, y)$, that is,

$$\frac{\partial^2 f_i}{\partial y^2}(x^*, y)|_{y=x_i^*} < 0 \tag{A.13}$$

for all $i = 1, \ldots, m$. On the other hand, if there is only one species involved, and if one is considering the evolution of a multidimensional trait $x = (x_1, \ldots, x_n)$ in a single species determined by the invasion fitness function $f(x, y)$, the condition for evolutionary stability is that the Hessian matrix of second derivatives of the invasion fitness function with respect to mutant trait values at the singular point is negative definite. In other words, all eigenvalues of the matrix

$$H(x^*) = \begin{pmatrix} \dfrac{\partial^2 f}{\partial y_1 \partial y_1}(x^*, y)\bigg|_{y=x^*} & \cdots & \dfrac{\partial^2 f}{\partial y_1 \partial y_n}(x^*, y)\bigg|_{y=x^*} \\ \dots & \dots & \dots \\ \dfrac{\partial^2 f}{\partial y_n \partial y_1}(x^*, y)\bigg|_{y=x^*} & \cdots & \dfrac{\partial^2 f}{\partial y_n \partial y_n}(x^*, y)\bigg|_{y=x^*} \end{pmatrix} \tag{A.14}$$

must be negative. The extension of these two observations to the general case of evolution of multidimensional traits in interacting species leads to the general definition of evolutionary stability:

Definition: A singular point x^* of an adaptive dynamical system is called *evolutionarily stable* if in each of the evolving species the Hessian matrix of second partial derivatives of the invasion fitness function with respect to mutants and evaluated at the singular point is negative definite.

Note that in contrast to the definition of convergence stability given above, the definition of evolutionary stability only involves the invasion fitness functions, but not the mutational matrix B. This notion of evolutionary stability conforms with the classical definition of an *Evolutionary Stable Strategy* (ESS) (Maynard Smith & Price, 1973): if almost all individuals in coevolving populations have phenotypes specified by an evolutionary stable singular point, then no (nearby) mutant has a positive invasion fitness, and hence no nearby mutant can invade this population. Thus, once such a strategy becomes established in a population, no further evolutionary change through small mutations occurs deterministically.

It is important to make a clear distinction between evolutionary stability and convergence stability of a singular point. For example, it can happen that a point in trait space at which all selection gradients vanish is evolutionarily stable, but not convergent stable. Such singular points are called Garden-of-Eden configurations (Nowak & Sigmund, 1990): if a population is at such a singularity, it will stay there forever, because no mutant can invade; however, if the population instead happens to be monomorphic for a phenotype that is only slightly different from the trait value given by the singular point, then the population will never reach the uninvasible singular point and instead will evolve away from it, because the singular point is not convergent stable.

Since Garden-of-Eden configurations are not convergent stable they are of little practical interest. Of considerably more interest are cases in which a singular point is both convergent stable and evolutionarily stable. Such points represent endpoints of the evolutionary process and are classically called Continuously Stable Strategies (CSS) (Christiansen, 1991; Eshel, 1983). Note that even though the condition for evolutionary stability does not depend on the mutational variance-covariance matrix, in multidimensional trait spaces the problem again arises that a singular point maybe a CSS for some, but not all mutational matrices, because in general convergence stability is affected by the mutational matrix. For two-dimensional trait spaces, Cressman (2010) has derived conditions for a singular point to be a CSS independently of the mutational matrix, but more general results appear to be hard to obtain. Of course, one can always require strong convergence stability to ensure independence of the the mutational matrix (see earlier), and in fact Lessard (1990) and Leimar (2009) have suggested to restrict the notion of CSS to strongly convergent stable ESSs. I tend to disagree with such a strict definition, for it leaves no terminology for ESSs that are convergent stable for some, but not all mutational matrices, even though such singular points might be important in many adaptive dynamical systems (e.g., in the predator-prey systems of Chap. 5). Thus, I think it might be preferable to define CSSs as convergent stable ESSs (so that a singular point might be a CSS for some mutational schemes, but not for others), and then extend this term to strongly convergent stable CSS (or simply strong CSS) if the singular point is indeed strongly convergent stable.

As a counterpoint to CSSs, singular points that are convergent stable, but evolutionarily unstable are of central importance in adaptive dynamics. For example, consider the one-dimensional adaptive dynamics of a scalar trait x in a single species (i.e., $m = 1$ and $n_1 = 1$). Then the mutational matrix $B(x)$ is a real number, which for simplicity we assume to be 1. The adaptive dynamics is then given by

$$\frac{dx}{dt} = g(x) = \left.\frac{\partial f(x, y)}{\partial y}\right|_{y=x}, \tag{A.15}$$

where $f(x, y)$ is the invasion fitness function. If x^* is a singular point, the condition for convergence stability is

$$\left.\frac{dg}{dx}\right|_{x=x^*} = \left.\frac{\partial^2 f(x, y)}{\partial x \partial y}\right|_{y=x=x^*} + \left.\frac{\partial^2 f(x, y)}{\partial y^2}\right|_{y=x=x^*} < 0, \tag{A.16}$$

whereas the condition for evolutionary stability is

$$\left.\frac{\partial^2 f(x^*, y)}{\partial y^2}\right|_{y=x^*} < 0. \tag{A.17}$$

Clearly, if $\left.\frac{\partial^2 f(x,y)}{\partial x \partial y}\right|_{y=x=x^*} < 0$, it is possible that the first condition is satisfied, but the second is not, and hence that the singular point is convergent stable, but evolutionarily unstable. Once the trait value in such an adaptive system reaches the basin of attraction of the singular point, the trait will converge to x^*. Since x^* is also evolutionarily unstable, this system will therefore exhibit convergence toward a fitness minimum, that is, convergence to a point in trait space at which the resident population can be invaded by all nearby mutants. This seemingly paradoxical evolutionary process has been found in many adaptive dynamics models, as is exemplified by the contents of this book.

A.3   EVOLUTIONARY BRANCHING

It is natural to investigate what happens after a system has converged to a fitness minimum, and since all nearby mutants can now invade the resident, the most important question is whether such nearby mutants are able to coexist, thus forming a resident dimorphism. Moreover, it is then important to know whether in this dimorphic resident, evolution leads to gradual divergence of the two resident strains, eventually resulting in two very distinct and coexisting phenotypic clusters. This phenomenon is called *evolutionary branching*.

For scalar traits in a single species, that is, for adaptive dynamics of the form (A.9), it is fairly easy to show that convergence stability and evolutionary instability imply evolutionary branching. First, the fact that the singular point is both convergent stable and evolutionary unstable implies that phenotypes on either side of x^* can mutually invade each other. The argument is as follows. Let $x_1 < x^* < x_2$ be two trait values that are close to a singular point that is both convergent stable and evolutionarily unstable, and consider the invasion fitness function $f(x_2, x_1)$ describing the fate of x_1-mutants in a x_2-resident. A Taylor expansion with respect to x_1 gives

$$f(x_2, x_1) = f(x_2, x_2) + (x_1 - x_2) \cdot \frac{\partial f}{\partial x_1}(x_2, x_1)\Big|_{x_1=x_2}$$

$$+ \frac{(x_1 - x_2)^2}{2} \cdot \frac{\partial^2 f}{\partial x_1^2}(x_2, x_1)\Big|_{x_1=x_2} + o(|x_1|) \qquad (A.18)$$

The first summand on the left-hand side is zero, because $f(x, x) = 0$ for all x. To check the sign of the second summand, note that $\frac{\partial f}{\partial x_1}(x_2, x_1)\Big|_{x_1=x_2} = g(x_2)$. Since x^* is a singular point, we have $g(x^*) = 0$, and because x^* is convergent stable, we have $\frac{dg}{dx}(x^*) < 0$, and hence $g(x_2) < 0$ for x_2 close to and bigger than x^*. Since $x_1 < x_2$, that is, $x_1 - x_2 < 0$, it follows that $(x_1 - x_2) \cdot \frac{\partial f}{\partial x_1}(x_1, x_2)|_{x_1=x_2} > 0$, hence the second summand on the left-hand side of eq. (A.18) is positive. Finally, since we assumed that the singular point is evolutionarily unstable, $\frac{\partial^2 f}{\partial x_1^2}(x^*, x_1)\Big|_{x_1=x^*} > 0$, and because x_2 is close to x^*, it follows that $\frac{\partial^2 f}{\partial x_1^2}(x_2, x_1)|_{x_1=x_2} > 0$ as well, and hence the third summand on the left-hand side of eq. (A.18) is also greater than zero. Therefore, $f(x_2, x_1) > 0$, and an analogous argument with the roles of x_1 and x_2 reversed shows that $f(x_1, x_2) > 0$ is also true. Thus, x_1 can invade x_2, and x_2 can invade x_1.

Mutual invasibility leads to a protected polymorphism in the vicinity of the singular point, but to complete the process of evolutionary branching, the two coexisting lineages must diverge evolutionarily. That is, in each of the coexisting strains x_1 and x_2, selection should drive the trait value further away from the singular point x^*. If $f_1(x_1, x_2, y)$ and $f_2(x_1, x_2, y)$ are the invasion fitness functions in the two resident strains x_1 and x_2 with $x_1 < x^* < x_2$, the two-dimensional adaptive dynamics describing evolution in the dimorphic resident is given by

$$\frac{dx_1}{dt} = g_1(x_1, x_2) = \frac{\partial f_1}{\partial y}(x_1, x_2, y)\Big|_{y=x_1} \qquad (A.19)$$

$$\frac{dx_2}{dt} = g_2(x_1, x_2) = \frac{\partial f_2}{\partial y}(x_1, x_2, y)\Big|_{y=x_2} \qquad (A.20)$$

where for simplicity I have assumed that the mutational matrix B is the identity matrix (in particular, the following analysis assumes that mutations in the two resident strains occur independently). First note that the invasion fitness functions f_1 and f_2 both reduce to the invasion fitness function $f(x, y)$ of the one-dimensional system if $x_1 = x_2$, because in that case we are simply considering invasion into a single monomorphic population. Thus, we have

$$f_1(x, x, y) = f(x, y) \qquad\qquad (A.21)$$

$$f_2(x, x, y) = f(x, y) \qquad\qquad (A.22)$$

for all x. In particular, it follows from this that $g_1(x^*, x^*) = g_2(x^*, x^*) = g(x^*) = 0$, hence that the point (x^*, x^*) is a singular point of the adaptive dynamics eqs. (A.19) and (A.20). What we want to show is that in the first branch, lying to the left of the singular point x^* in phenotype space, smaller mutants are favored, that is, mutants that are farther away from x^* than the current resident in that branch, and similarly in the second branch, where mutants that are larger than the current resident should be favored if dimorphic divergence is to hold. This is equivalent to saying that the singular point (x^*, x^*) in system (A.19) and (A.20) is convergent unstable. Using the consistency relations (A.21) and (A.22) it is easy to see that the Jacobian $J(x^*, x^*)$ of system (A.20) and (A.21) at the singular point is of the form

$$J(x^*, x^*) = \begin{pmatrix} a+b & b \\ b & a+b \end{pmatrix}, \qquad\qquad (A.23)$$

where $a = \left.\dfrac{\partial^2 f(x^*, y)}{\partial y^2}\right|_{y=x^*}$ determines evolutionary stability of the singular point x^* in the one-dimensional adaptive dynamics, and where $a + b = \left.\dfrac{\partial^2 f(x^*, y)}{\partial y^2}\right|_{y=x^*} + \left.\dfrac{\partial^2 f(x, y)}{\partial x \partial y}\right|_{y=x=x^*}$ determines convergence stability of the singular point x^* in the one-dimensional adaptive dynamics. This Jacobian has eigenvalues a and $a + 2b$, corresponding to eigenvectors in the antidiagonal and diagonal direction, respectively. If the singular point x^* is convergent stable and evolutionarily unstable, then $a + b < 0$, $a > 0$ and hence $b < 0$. Thus, the eigenvalue in the direction of the antidiagonal $(x, -x)$ is positive, while the eigenvalue along the diagonal (x, x) is negative. This means that the two-dimensional adaptive dynamics (eqs. (A.19) and (A.20)) converges to (x^*, x^*) along the diagonal, but diverges from (x^*, x^*) along the antidiagonal, which in turn implies the evolutionary divergence of two mutually invasible branches emerging on either side of the singular point x^*. It is worthwhile noting that when the singular point is not only convergent stable, but also evolutionarily stable, then it is still possible that mutual invasibility holds for neighboring strategies on either side of the singular point. However, in this case evolution in two coexisting branches is always convergent, so that both branches evolve back to the singular value (see, e.g., Diekmann (2004) for details).

The previous arguments are valid for generic adaptive dynamics generated by invasion fitness functions whose second derivative at singular points is nonzero. It is interesting to note that in cases where this second derivative is zero, it is possible that a singular point is convergent stable and evolutionarily stable, yet mutual invasibility around the singular point leads to divergent evolution of coexisting branches. An example of this is described in Box A.1.

BOX A.1

AN EXAMPLE FOR EVOLUTIONARY BRANCHING AFTER CONVERGENCE TO AN
EVOLUTIONARILY STABLE STRATEGY

The example given here follows Doebeli & Ispolatov (2010b). We consider models for frequency-dependent competition in which the invasion fitness $f(x, y)$ of a rare mutant y in a resident x is given by eq. (3.5) in Chapter 3:

$$f(x, y) = 1 - \frac{\alpha(x, y)K(x)}{K(y)}, \tag{BA.1}$$

where $\alpha(x, y)$ is the competition kernel, $K(x)$ is the carrying capacity function, and where we have assumed that the growth rate b appearing in eq. (3.5) is independent of the phenotype and equal to one. We further assume that competition is symmetric, so that the adaptive dynamics of the trait x is given by eq. (3.9). If the carrying capacity function has a unique maximum at x^*, then x^* is, in any case, a convergent stable singular point, as explained in Chapter 3.

Instead of the functions given by expressions (3.13) and (3.14) in Chapter 3, we now choose the following functions for the competition kernel and the carrying capacity:

$$\alpha(x, y) = \exp\left[\frac{-(x - y)^4}{4\sigma_\alpha^4} \right]$$

$$K(x) = \exp\left[-\frac{x^4}{4\sigma_K^4} \right]. \tag{BA.2}$$

Without loss of generality we have assumed that the maximum of the carrying capacity, and hence the singular point, is at $x^* = 0$. Just as for the Gaussian functions used in Chapter 3, it is then easy to see that for $\sigma_\alpha > \sigma_K$, the invasion fitness function $f(0, y)$ has a maximum at $y = 0$, and hence that the convergent stable singular point is an ESS for $\sigma_\alpha > \sigma_K$. In contrast to the Gaussian case,

this does not preclude evolutionary diversification when the ecological functions
have the forms (BA.2), as we will now see.

We first consider mutual invasibility of phenotypes $-\epsilon$ and ϵ on either
side of the singular point zero. Taylor expansion of the invasion fitness function
around zero reveals that all terms of order 3 or less are zero, and that

$$f(\epsilon, -\epsilon) = \frac{4}{\sigma_\alpha^4} \epsilon^4 + h.o.t \qquad (BA.3)$$

(where h.o.t are higher order terms in ϵ). It follows that $f(\epsilon, -\epsilon) > 0$ for all
ϵ with $|\epsilon|$ sufficiently small, and hence that mutual invasibility holds for trait
values on either side of the singular point at least in some neighborhood of the
singular point. As mentioned in the main text, such mutual invasibility can occur
near convergent stable ESSs even if the second derivative of the invasion fitness
function at the singular point is nonzero and negative. What cannot happen in
that case is that the adaptive dynamics diverges along the antidiagonal $(\epsilon, -\epsilon)$.
However, this can happen if the invasion fitness function (BA.1) is based on the
ecological functions (BA.2), in which case the second derivative of the invasion
fitness function at the singular point is zero.

To see this, we assume that two resident traits ϵ and $-\epsilon$ coexist at equilibrium and calculate the invasion fitness function $F(\epsilon, -\epsilon, z)$ of a mutant z
appearing in either of the residents. To obtain F one first needs to calculate
the equilibrium densities N_ϵ and $N_{-\epsilon}$ of the resident strains. Straightforward
calculations of the ecological (logistic) dynamics of the two residents yield

$$N_\epsilon = N_{-\epsilon} = \frac{K(\epsilon) - \alpha(\epsilon, -\epsilon)K(\epsilon)}{1 - \alpha(\epsilon, -\epsilon)^2}, \qquad (BA.4)$$

and the invasion fitness for mutants z becomes

$$F(\epsilon, -\epsilon, z) = 1 - \frac{(\alpha(\epsilon, z) + \alpha(-\epsilon, z))N_\epsilon}{K(z)}$$

$$= 1 - \frac{\exp\left[\frac{4\epsilon^4}{\sigma_\alpha^4} + \frac{z^4 - \epsilon^4}{4\sigma_K^4}\right]\left(\exp\left[-\frac{(\epsilon - z)^4}{4\sigma_\alpha^4}\right] + \exp\left[-\frac{(\epsilon + z)^4}{4\sigma_K^4}\right]\right)}{1 + \exp\left[\frac{4\epsilon^4}{\sigma_\alpha^4}\right]}.$$

$$(BA.5)$$

Note that $F(0, 0, z) = f(0, z)$, which has a fourth-order maximum at $z = 0$ for
$\sigma_\alpha > \sigma_K$ (see before). In particular, for $\epsilon = 0$, $F(\epsilon, -\epsilon, z) = 0$ has a solution at
$z = 0$ of multiplicity 4. For $\epsilon \neq 0$, and depending on the values of the parameters
σ_α and σ_K, this solution can lose its degeneracy (i.e., its multiplicity) and the
single solution of multiplicity 4 is replaced by four different solutions, each with

BOX A.1 (*continued*)

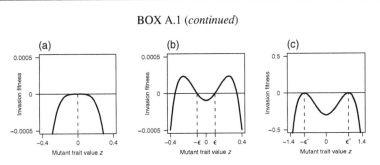

FIGURE A.B1. Invasion fitness function $F(\epsilon, -\epsilon, z)$ as a function of the mutant trait value z for different values of the two resident traits ϵ and $-\epsilon$. In panel (a), $\epsilon = 0$, so that $F(0, 0, z) = f(0, z)$ corresponds to the invasion fitness function with a single resident at the singular trait value 0. For $\sigma_\alpha > \sigma_K$ the invasion fitness has a maximum at zero, that is, the singular trait is an ESS. Panel (b) shows $F(\epsilon, -\epsilon, z)$ for $\epsilon = 0.1$, for which there are four distinct solutions of $F(\epsilon, -\epsilon, z) = 0$. The two solutions lying closest to the singular point correspond to the two resident strains. Note that the selection gradients, defined as the slope of the invasion fitness function at the resident values, point away from the singular point and generate divergent evolution of the resident traits. Panel (c) shows the singular coalition of trait values $(\epsilon^*, -\epsilon^*)$ resulting from the two-dimensional adaptive dynamics (the value of ϵ^* is given by eq. BA.8). The singular coalition is reached once the two maxima of the invasion fitness function shown in panel (b) come to lie on the z-axis as a result of adaptive diversification. Parameter values for the ecological functions (BA.2) were $\sigma_\alpha = 1.1$ and $\sigma_K = 1$.

multiplicity 1. This is illustrated in Figure A.B1(a) and (b). Note that two of these solutions are given by ϵ and $-\epsilon$, that is, by the two resident strains. These are the solutions that lie closest to $z = 0$ (Figure A.B1). Also note that it is evident from Figure A.B1(b) that the selection gradient at these resident values, given as the slopes of $F(\epsilon, -\epsilon, z)$ with respect to z, is negative at the resident trait that is smaller than the singular value zero, and positive at the resident trait that is larger than the singular value. It immediately follows that the two coexisting branches will diverge evolutionarily.

To determine the range of the parameters σ_α and σ_K for which this phenomenon occurs, we calculate the selection gradients at the resident trait values explicitly:

$$\frac{\partial F}{\partial z}(\epsilon, -\epsilon, z)\bigg|_{z=\epsilon} = -\frac{\epsilon^3\left(\left(1 + \exp\left[\frac{4\epsilon^4}{\sigma_\alpha^4}\right]\right)\sigma_\alpha^4 - 8\sigma_K^4\right)}{\left(1 + \exp\left[\frac{4\epsilon^4}{\sigma_\alpha^4}\right]\right)\sigma_\alpha^4\sigma_K^4}$$

(BA.6)

$$\frac{\partial F}{\partial z}(\epsilon, -\epsilon, z)\bigg|_{z=-\epsilon} = -\frac{\partial F}{\partial z}(\epsilon, -\epsilon, z)\bigg|_{z=\epsilon}.$$

Assuming $\epsilon > 0$ without loss of generality, it immediately follows that the selection gradient is positive at ϵ and negative at $-\epsilon$ for all ϵ small enough if and only if $\sigma_\alpha^4 < 8\sigma_K^4$, that is, if and only if

$$\sigma_\alpha < \sqrt{2}\sigma_K. \tag{BA.7}$$

We note that condition BA.7 is also the condition for $F(\epsilon, -\epsilon, z) = 0$ to have four different solutions in the first place (proof left as an exercise). It now follows that for $\sigma_K < \sigma_\alpha < \sqrt{2}\sigma_K$ we have the following situation. On the one hand, the singular point $x^* = 0$ is a convergent stable ESS. On the other hand, phenotypes on either side of the singular point can coexist, and two such coexisting phenotypic branches diverge evolutionarily. In fact, one can calculate the endpoint of the two-dimensional adaptive dynamics ensuing after evolutionary branching by determining the solution of $\partial F/\partial z(\epsilon, -\epsilon, z)|_{z=\epsilon} = 0$, that is, by determining the resident values at which the selection gradients vanish. This yields

$$\pm\epsilon^* = \pm\frac{1}{\sqrt{2}}\sigma_\alpha \ln\left[\frac{8\sigma_K^4 - \sigma_\alpha^4}{\sigma_\alpha^4}\right]^{1/4} \tag{BA.8}$$

This solution (which again only exists when B.A7 is satisfied) is indicated in Figure A.B1(c), which shows the invasion fitness when the resident is given by the singular coalition $(-\epsilon^*, \epsilon^*)$.

   Challenge: Prove that the singular coalition given by BA.8 is convergent stable. Investigate its evolutionary stability.

   It is important to note that in the competition models considered here, evolutionary branching from a CSS occurs even in the limit of infinitely small mutation size. More precisely, as $\epsilon \to 0$, and considering residents $x^* - \epsilon$, mutual invasibility and evolutionary divergence holds for mutants $x^* + \epsilon(1 + \delta)$, where δ can vary in an interval that is independent of ϵ (see Doebeli & Ispolatov (2010b) for a proof of this statement). This implies that as the resident trait approaches the singular value, the proportion of mutations that lead to mutual coexistence and divergent evolution is positive and independent of the resident value. Moreover, as the resident approaches the singular point, the size of these mutations tends to 0.

   The example discussed here illustrates the complications that can arise when the invasion fitness function vanishes to higher than first order at a singular point. In particular, the example shows that the notion of evolutionary stability has to be treated with caution in such cases, and convergence to the

---

BOX A.1 (*continued*)

vicinity of the ESS can result in coexistence of diverging phenotypic clusters. While coexistence of different phenotypes is already possible if the invasion fitness only vanishes to first order at an ESS, in that case evolution in coexisting phenotypic clusters would be convergent and would take the population back to the singular point. However, with more degenerate invasion fitness functions, such as the one considered here, evolutionary branching can lead to adaptive diversification despite the fact that the convergent stable singular point of the one-dimensional adaptive dynamics is also evolutionarily stable.

Challenge: Show that evolutionary branching can occur for $\sigma_\alpha < \sqrt{2}\sigma_K$ in individual-based models of frequency-dependent competition based on the ecological functions B.A2 (see Chaps 3 and 4 for the construction of individual-based models).

Challenge: Show that evolutionary diversification can occur for $\sigma_\alpha < \sqrt{2}\sigma_K$ in the partial differential equation models (9.2) and (9.8) for frequency-dependent competition based on the ecological functions B.A2.

∎

In general multidimensional adaptive dynamics the conditions for evolutionary branching are a straightforward generalization from the one-dimensional case just described. Thus, a singular point x^* of an adaptive dynamical system of the form (A.1) is called an *evolutionary branching point* if the following four conditions are satisfied:

1. x^* is convergent stable.
2. x^* is evolutionarily unstable in at least one direction of trait space.
3. mutual invasibility holds around x^* in the directions of evolutionary instability given by 2.
4. the different phenotypic branches constituting the protected polymorphism given by 3. diverge evolutionarily from x^*.

Condition 1 is necessary for the system to be attracted to the singularity at which branching is to take place (note again that, in general, convergence stability may depend on the mutational variance-covariance matrix). Condition 2 says that nearby mutants are able to invade populations that are monomorphic for the singular trait values specified by x^*. Condition 3 is necessary for nearby mutants to be able to coexist, and condition 4 guarantees that the different branches constituting a polymorphism around the singular point according to 3 diverge from each other over evolutionary time.

As we have seen above, for one-dimensional adaptive dynamics conditions 1 and 2 imply conditions 3 and 4. The situation is slightly different in the case of m coevolving scalar traits in m different species. In such systems evolutionary instability is measured along the various trait axes: a singular point $x^* = (x_1^*, \ldots, x_m^*)$ is evolutionarily unstable in the i-th direction if x_i^* represents fitness minimum for the invasion fitness function in species i. As we have seen in Chapter 5, convergence stability of x^* does not necessarily imply convergence stability of x^* along every single trait axis. Thus, it can happen that $x^* = (x_1^*, \ldots, x_m^*)$ is convergent stable, but that for one or more (but not all!) i, x_i^* is not convergent stable for trait x_i when all other traits are fixed at their singular values. As a consequence, convergence stability of x^* together with evolutionary instability of x_i^* do not automatically guarantee mutual invasibility for nearby mutants in the i-th species (i.e., mutual invasibility of mutants x_{i1} and x_{i2} with $x_{i1} < x^* < x_{i2}$). Therefore, condition 3 in the definition of an evolutionary branching point becomes an independent condition in higher-dimensional systems. Note, however, that convergence stability of a singularity x^* and evolutionary instability along a trait axis x_i is enough for mutual invasibility in species i if x^* is also convergent stable along the trait axis x_i, just as in the one-dimensional case.

Once a singular point x^* in a system of n coevolving scalar traits satisfies the global condition 1 (i.e., convergence stability), as well as conditions 2 and 3 along one or more trait axes x_i, it can be shown that condition 4, that is, dimorphic divergence along trait axis x_i, is automatically satisfied. Thus, in a system of coevolving scalar traits in different species, a singular point $x^* = (x_1^*, \ldots, x_m^*)$ is a branching point if it is convergent stable, and if at least one of the species is sitting at a fitness minimum such that mutants on either side of the singular value of that species can invade and coexist.

Challenge: Prove this statement.

In general, the more complicated case of evolutionary branching in the adaptive dynamics of multidimensional traits is not yet well understood. As we have seen, in the adaptive dynamics of one-dimensional traits in two interacting species, convergence stability and evolutionary instability of a singular point need not necessarily imply mutual invasibility around the singularity. This is also true for the evolution of multidimensional traits in a single or multiple species. Thus, conditions 1–3 in the definition for evolutionary branching are in general independent requirements in such systems. Moreover, in contrast to systems of coevolving scalar traits, for which we know that dimorphic divergence, condition 4 automatically follows from conditions 1–3 (see earlier), it is not known whether conditions 1–3 imply condition 4 in the adaptive dynamics of multidimensional traits. Thus, it is an open question whether the conditions 1–4 are independent requirements for evolutionary branching of multidimensional traits. More generally, classifying all possible adaptive dynamical behaviors resulting from a convergent stable, evolutionarily unstable singular points is an open problem (see Box 9.1 for some pertinent results).

If a scalar trait undergoes evolutionary branching, different types of long-term evolutionary dynamics are possible, as the examples discussed in this book illustrate. However, the short-term result is always a diverging polymorphism in the trait that is branching. It is interesting to note that when evolutionary branching occurs in multidimensional traits, then the emergent polymorphism is not necessarily a dimorphism, that is, does not necessarily consist of only two phenotypic branches. In principle, if m is the dimension of the trait that undergoes evolutionary branching, then up to $m + 1$ new branches can from simultaneously (Meszéna & Metz, 1999). For instance, Ruxton & Doebeli (1996) have studied the evolution of dispersal in a metapopulation consisting of two habitats with dispersal rates d_1 from habitat 1 to habitat 2 and dispersal rate d_2 from habitat 2 to habitat 1. In this case the two-dimensional dispersal trait (d_1, d_2) can branch simultaneously into three different branches at an evolutionary branching point (Ruxton & Doebeli, 1996). Thus, the simultaneous emergence of many different lineages during evolutionary branching is possible in principle.

As a brief aside, I mention that function-valued traits as introduced in Dieckmann et al. (2006) and Parvinen et al. (2006) are a very interesting example of essentially infinite-dimensional traits that describe the dependence of a phenotype on the environment in which it is expressed. Dieckmann et al. (2006) and Parvinen et al. (2006) have shown how to extend the framework of adaptive dynamics to function-valued traits. Box A.2 briefly describes their approach by means of an example of evolutionary branching in flowering schedules.

The stability concepts introduced above allow for a complete classification of adaptive dynamics in one-dimensional trait spaces. Apart from scenarios in which traits evolve to ever bigger or ever smaller values, the only long term dynamical behavior possible for a scalar trait in a single species is convergence to a singular point in trait space, because these are the only types of dynamic behavior possible for autonomous one-dimensional differential equations. As noted, even though convergent unstable points may be evolutionarily stable, they are not relevant because unless started exactly at such a point, evolution will drive the system away from them. Thus, the only singular points of practical interest are the convergent stable ones. As noted above,

BOX A.2

AN EXAMPLE FOR THE ADAPTIVE DYNAMICS OF FUNCTION-VALUED TRAITS

Here I briefly describe an extension of adaptive dynamics and evolutionary branching to function-valued traits. Classically, function-valued traits are often called reaction norms (Stearns, 1992) and occur whenever the expression of a phenotype depends on a continuous variable such as environmental quality, time of season, age, etc. In this case, the phenotype is given as a function of the continuous variable, and the question is how this function evolves. Thus, instead of studying adaptive dynamics in finite-dimensional phenotype spaces, one needs a formalism for adaptive dynamics in infinite-dimensional function spaces. Such a formalism has been developed in Dieckmann et al. (2006) and Parvinen et al. (2006), and I will explain the basic ingredients of this formalism by means of a specific example given in Dieckmann et al. (2006).

In this example, the function-valued trait is a flowering schedule $x(a)$, where a is the time of the seasons (rescaled so that a ranges from zero to one, and $x(a)$ is a measure for the flowering intensity at a given time x. It is assumed that plants can produce a fixed amount of flowers per seasons, so that $\int_0^1 x(a)\,da$ is a constant, which is set to one. It is further assumed that flowers produced at different times of the season face different ecological conditions, and hence have different probabilities of setting seed, and given the ecological constraints, the problem is to understand the evolution of the flowering schedule $x(a)$.

The ecological conditions are best understood by considering a population that is monomorphic for a flowering schedule $x(a)$. Then, if n_x is the density of that population, it is assumed that the probability of seed set of flowers produced at time a is given by

$$p_x(a) = \exp\left(-\frac{n_x x(a)}{K(a)}\right). \qquad (BA.9)$$

Here the function $K(a)$ can be thought of as a time-dependent carrying capacity: if $K(a)$ is large, the probability of seed set will be high, and vice versa for small $K(a)$. For example, $K(a)$ could describe time-dependent pollinator abundance. The quantity $n_x x(a)$ appearing in eq. (BA.9) is the amount of flowers produced at time a in the population, and (BA.9) implies that the probability of seed set decreases with increasing number of flowers, for example, as a consequence of competition for pollination.

Mutant individuals having a flowering schedule $x'(a)$ should have an advantage over the resident $x(a)$ at those times a at which the mutant invests more into producing flowers than the resident, that is, at which $x'(a) > x(a)$. This advantage is described by a function $c(e(a)) = 2/(1 + \exp[-\alpha e(a)])$, where

---

BOX A.2 (*continued*)

$e(a) = [x'(a) - x(a)]/x(a)^\beta$ is a measure of the difference between the mutant and resident investments into flowering at time a. α and β are system parameters. Note that $c(e(a)) \to 0$ for $e(a) \to -\infty$ and $c(e(a)) \to 2$ for $e(a) \to +\infty$. Thus, the function $c(e(a))$ reflects asymmetric competition, with larger flowering investments having an advantage. Also note that even though it would be advantageous to increase flowering intensity at all times a, this is not possible because of the constraint that the total seasonal investment into flowers is constant, so that an increase at one time necessarily comes with a decrease at other times.

The invasion fitness function $f(x, x')$ is now obtained by summing the probability of seed set for the mutant $x'(a)$ over all possible times, assuming that the mutant is rare and the resident is at its population dynamical equilibrium. Taking asymmetric competition into account, the probability of seed set of the rare mutant at any given time a is $x'(a)c(e(a))\bar{p}_x(a)$, where the expression $\bar{p}_x(x) = \exp\left(-\frac{\bar{n}_x x(a)}{K(a)}\right)$ is determined by the equilibrium population density $\bar{n}_x$ of the resident:

$$f(x, x') = \int_0^1 x'(a) \left(c \left[x'(a) - x(a) \right] / x(a)^\beta \right) \bar{p}_x(a) da . \qquad \text{(BA.10)}$$

The equilibrium population size $\bar{n}_x$ of the resident is defined by $f(x, x) = 0$. Expression (BA.10) extends the notion of invasion fitness to function-valued traits.

For invasion fitness functions that are given as integrals as (BA.10), the selection gradient $D_x(a)$ can be calculated "locally" at each time point a by taking the derivative of the integrand in (BA.10) with respect to the mutant strategy (Dieckmann et al., 2006):

$$D_x(a) = \frac{\partial}{\partial y} \left[c \left(y - x(a)/x(a)^\beta \right) \bar{p}_x(a) \right] \Bigg|_{y=x(a)}$$

$$= \bar{p}_x(a) \left(1 + \frac{1}{2}\alpha x(a)^{1-\beta} \right). \qquad \text{(BA.11)}$$

To obtain the evolutionary dynamics of the function-valued trait $x(a)$, one now has to integrate the selection gradients (BA.11) over the season, taking into account the variance-covariance structure generated by the mutational process in the trait $x(a)$. This is explained in detail in Dieckmann et al. (2006), who showed that the canonical equation of function-valued adaptive dynamics is given by

$$\frac{d}{dt}x(a) = \frac{1}{2}\mu_x \bar{n}_x \int \sigma_x^2(a', a)D_x(a') da'. \qquad \text{(BA.12)}$$

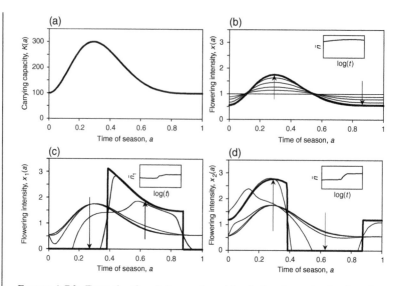

FIGURE A.B2. Example of evolutionary branching in function-valued traits: adaptive dynamics of seasonal flowering schedules. (a) Carrying capacity $K(a)$, used in expression (BA.9), as a function of time of season. (b) Monomorphic adaptive dynamics of the flowering schedule $x(a)$, with the thick line representing the equilibrium solution $x^*(a)$ obtained by setting the right-hand side of eq. (BA.14) to zero. This solution represents a singular point (in function space) of the adaptive dynamics, and for the parameters chosen, this singular point is an evolutionary branching point, as illustrated in panels (c) and (d). These panels show the dynamics of dimorphic evolution of flowering schedules $x_1(a)$ and $x_2(a)$ initialized close to the singular schedule $x^*(a)$. Initial and intermediate flowering schedules are shown as thin curves, and the equilibrium schedules as thick curves. Insets show changes in equilibrium population sizes resulting from the evolutionary change. Reprinted from the Journal of Theoretical Biology, Volume 241, Issue 2, by Ulf Dieckmann, Mikko Heino, and Kalle Parvinen, *The adaptive dynamics of function-valued traits*, Pages 370–389, Copyright (2006), with permission from Elsevier. See Dieckmann et al. (2006) for further details.

As in the adaptive dynamics of scalar traits, the multiplier in front of the integral in (BA.12) reflects the frequency with which new mutations occur and generally only affects the speed of evolution. The variance-covariance function σ_x^2 essentially describes the correlation between mutational changes in the flowering schedule at different times of the season. In the present case, this function must reflect the assumption of constant total investment into flowering across the whole season, that is, the assumption $\int x(a)\, da = 1$ for all schedules x. The

---

BOX A.2 (*continued*)

simplest function respecting this constraint is

$$\sigma_x(a', a) = \sigma_2\left(\delta(a' - a) - 1\right), \tag{BA.13}$$

where $\delta(a' - a)$ is the Dirac delta function, and σ_2 is the (unconstrained) mutational variance that is assumed to be independent of a. (BA.13) implies that there is no cohesion of trait values $x(a)$ between adjacent times a, and that any mutational increase of $x(a)$ at a given time a is compensated by a uniform decrease at all other times a so as to keep the integral $\int x(a)\, da$ invariant (Dieckmann et al., 2006).

Finally, to derive the adaptive dynamics of the function-valued trait $x(a)$, one has to take into account the constraint that $x(a) \geq 0$ for all times a. This results in

$$\frac{d}{dt}x(a) = \frac{1}{2}\mu_x \bar{n}_x \int \sigma_x^2(a', a)C(x, a, a')D_x(a')\, da' \tag{BA.14}$$

where $C(x, a, a')$ is a function that is zero whenever $D_x(a) < 0$ and $x(a) \leq 0$ or $D_x(a') < 0$ and $x(a') \leq 0$, and one otherwise (Dieckmann et al., 2006).

The adaptive dynamics (BA.14) has to be solved numerically, and Figure A.B2 describes some salient results from Dieckmann et al. (2006). In particular, the adaptive dynamics of the function-valued trait $x(a)$ can exhibit evolutionary branching, which results in coexistence of two flowering schedules that are segregated in time, so that at any given time during the flowering season, only one of the two schedules is positive, that is, only one of the phenotypes produces flowers. This adaptive diversification in function-valued traits is ultimately brought about by asymmetric competition generating frequency-dependent selection for flower production.

∎

---

such singular points may or may not be evolutionarily stable. If x^* is convergent stable, and if the invasion fitness function $f(x, y)$ satisfies $\frac{\partial^2 f}{\partial y^2}(x^*, y) < 0$, then the singular point is also evolutionarily stable. In this case evolution comes to a halt at x^*, that is, x^* is the final evolutionary outcome. Singular points that are convergent stable and evolutionarily stable were called continuously stable strategies (CSS) by Eshel (1983) to distinguish them from evolutionary stable strategies that are convergent unstable. Joel Brown and his collaborators have suggested to redefine the ESS as singular points that are both convergent stable and evolutionarily stable (see, e.g., McGill & Brown, 2007; Vincent & Brown, 2005). That is, they suggested to call the CSS an ESS.

This is unfortunate, both because CSSs have already been defined, and because the notion of ESSs as evolutionarily stable (and not necessarily convergent stable) points in trait space has a long tradition and should not be redefined. Moreover, it is conceptually important to distinguish between the two forms of stability.

Besides convergent unstable singular points, which act as evolutionary repellors, and convergent stable singular points that are also evolutionarily stable (i.e., CSSs), the only other possibility in one-dimensional adaptive dynamics is convergent stable singular points that are evolutionarily unstable, that is, evolutionarily attracting fitness minima. As shown above, for scalar traits in a single species such points are automatically evolutionary branching points, because evolutionary instability of a convergent stable singular point implies mutual invasibility of phenotypes that are close to the singular points.

In sum, the evolutionary attractors in one-dimensional adaptive dynamics come in two flavors: either a convergent stable singular point is also evolutionarily stable, that is, it is a CSS, in which case it is a final stop for the adaptive dynamics; or it is evolutionarily unstable, in which case it is also an evolutionary branching point, that is, the starting point for adaptive diversification. These two scenarios are illustrated summarized in Figure 2.2 in Chapter 2. Geritz et al. (1998) have given a more fine-grained classification of one-dimensional adaptive dynamics, but for many purposes the classification given here suffices. In particular, this classification shows that evolutionary branching is a generic feature of one-dimensional adaptive dynamical systems, essentially because convergence stability and evolutionary stability are generally independent conditions. Both of these conditions are given by inequalities and therefore generally occur for open subsets of parameter space (at least when selection is frequency-dependent). The intersection of these open sets, which is again an open (i.e., a large) set, is the region in parameter space where evolutionary branching occurs. Therefore, evolutionary branching is a robust feature of one-dimensional adaptive dynamics, and the significance of this evolutionary phenomenon in both asexual and sexual populations has been one of the main topics in this book.

The classification given before illustrates that the starting point for the analysis of adaptive dynamics is the identification of singular points and the investigation of their convergence stability. This is in contrast to traditional ESS analysis, in which the starting point is to investigate the evolutionary stability of equilibrium points. While traditional ESSs always represent singular points of a suitable adaptive dynamics, an ESS may not be convergence stable. In this case, the singular point is of little interest: it has the same status as an unstable equilibrium in population dynamics. In the past, the predominant approach to study frequency-dependent selection was based on game theory, and in particular on the notion of an ESS, but it now seems clear that the framework of adaptive dynamics provides a much more general theory of evolutionary dynamics. This framework encompasses both convergence stability and evolutionary stability and therefore allows to study two fundamentally different ways in which an evolutionary equilibrium, that is, a singular point, can become unstable and give rise to an evolutionary bifurcation. The first type of bifurcation concerns the adaptive dynamics in a given dimension, that is, the evolutionary dynamics of a given number of traits in a given number of species. Such bifurcations occur when singular points lose their convergence stability due to changes in system parameters that in turn change the real parts of one or more eigenvalues of the Jacobian matrix at the singular point from negative to positive. As a consequence, the adaptive dynamics changes qualitatively. In contrast to one-dimensional adaptive dynamics, where no oscillatory behavior is possible, bifurcations in higher dimensional systems can lead to complicated dynamics. For example, in a Hopf bifurcation a singular point loses its convergence stability because the real parts of a pair of complex conjugate eigenvalues of the Jacobian matrix at a singular point become positive. At the parameter value at which the bifurcation occurs, a periodic orbit, that is, a periodic solution of the adaptive dynamical system, is born. Thus, during the bifurcation the adaptive dynamics can change from converging toward the singular point to tracing out a periodic trajectory in trait space. Such bifurcations can for example be observed in the adaptive dynamics of predator-prey systems (see Chap. 5, or, e.g., Dieckmann et al., 1995). Note that during such a bifurcation, the dimensionality of the adaptive dynamical system does not change.

The second type of evolutionary bifurcation occurs when the selection gradients cease to satisfy the ESS condition at a singular point, that is, when a singular point becomes evolutionarily unstable as system parameters are changed. As in the one-dimensional case discussed before, this can happen even if the singular point remains convergent stable (and indeed evolutionary instability is only interesting in that case). If a convergent stable singular point becomes evolutionarily unstable, evolutionary branching is possible. If it occurs, the dynamical system used to identify the evolutionary branching point ceases to be an adequate description of the adaptive dynamics after the branching event, and instead a higher dimensional system of differential equations must be used to model the ensuing adaptive dynamics of the diversified system. As we have seen in the examples throughout this book, evolutionary bifurcations generating evolutionary branching points are at the core of understanding adaptive diversification.

Bibliography

Abrams, P. A. 1987a. Alternative models of character displacement and niche shift. 1. Adaptive shifts in resource use when there is competition for nutritionally nonsubstitutable resources. *Evolution* **41(3)**, 651–661.

Abrams, P. A. 1987b. Alternative models of character displacement and niche shift. 2. Displacement when there is competition for a single resource. *American Naturalist* **130(2)**, 271–282.

Abrams, P. A. 2001. Modelling the adaptive dynamics of traits involved in inter- and intraspecific interactions: An assessment of three methods. *Ecology Letters* **4(2)**, 166–175.

Abrams, P. A., Matsuda, H., & Harada, Y. 1993. Evolutionarily unstable fitness maxima and stable fitness minima of continuous traits. *Evolutionary Ecology* **7(5)**, 465–487.

Ackermann, M., & Doebeli, M. 2004. Evolution of niche width and adaptive diversification. *Evolution* **58(12)**, 2599–2612.

Almeida, C. R., & de Abreu, F. V. 2003. Dynamical instabilities lead to sympatric speciation. *Evolutionary Ecology Research* **5(5)**, 739–757.

Anderson, R. M., & May, R. M. 1992. *Infectious Diseases of Humans: Dynamics and Control.* Oxford: Oxford University Press.

Arnegard, M. E., & Kondrashov, A. S. 2004. Sympatric speciation by sexual selection alone is unlikely. *Evolution* **58(2)**, 222–237.

Bantinaki, E., Kassen, R., Knight, C. G., Robinson, Z., Spiers, A. J., & Rainey, P. B. 2007. Adaptive divergence in experimental populations of pseudomonas fluorescens. iii. Mutational origins of wrinkly spreader diversity. *Genetics* **176(1)**, 441–453.

Baptestini, E. M., de Aguiar, M. A. M., Bolnick, D. I., & Araujo, M. S. 2009. The shape of the competition and carrying capacity kernels affects the likelihood of disruptive selection. *Journal of Theoretical Biology* **259(1)**, 5–11.

Barabas, G., & Meszéna, G. 2009. When the exception becomes the rule: The disappearance of limiting similarity in the lotka-volterra model. *Journal of Theoretical Biology* **258(1)**, 89–94.

Barton, N. H. 1992. On the spread of new gene combinations in the 3rd phase of wright shifting-balance. *Evolution* **46(2)**, 551–557.

Bell, G. 1978. Evolution of anisogamy. *Journal of Theoretical Biology* **73(2)**, 247–270.

Bellows, T. S. 1981. The descriptive properties of some models for density dependence. *Journal of Animal Ecology* **50(1)**, 139–156.

Berlocher, S. H., & Feder, J. L. 2002. Sympatric speciation in phytophagous insects: Moving beyond controversy? *Annual Review of Entomology* **47**, 773–815.

Bever, J. D. 1999. Dynamics within mutualism and the maintenance of diversity: Inference from a model of interguild frequency dependence. *Ecology Letters* **2(1)**, 52–62.

Boldin, B., & Diekmann, O. 2008. Superinfections can induce evolutionarily stable coexistence of pathogens. *Journal of Mathematical Biology* **56(5)**, 635–672.

Boldin, B., Geritz, S. A. H., & Kisdi, E. 2009. Superinfections and adaptive dynamics of pathogen virulence revisited: A critical function analysis. *Evolutionary Ecology Research* **11(2)**, 153–175.

Bolnick, D. I. 2004. Waiting for sympatric speciation. *Evolution* **58(4)**, 895–899.

Bolnick, D. I. 2006. Multi-species outcomes in a common model of sympatric speciation. *Journal of Theoretical Biology* **241(4)**, 734–744.

Bolnick, D. I., & Doebeli, M. 2003. Sexual dimorphism and adaptive speciation: Two sides of the same ecological coin. *Evolution* **57(11)**, 2433–2449.

Bolnick, D. I., & Fitzpatrick, B. M. 2007. Sympatric speciation: Models and empirical evidence. *Annual Review of Ecology Evolution and Systematics* **38**, 459–487.

Bomblies, K., Lempe, J., Epple, P., Warthmann, N., Lanz, C., Dangl, J. L., & Weigel, D. 2007. Autoimmune response as a mechanism for a dobzhansky-muller-type incompatibility syndrome in plants. *PLoS Biology* **5**, 1962–1972.

Bonsall, M. B. 2006. The evolution of anisogamy: The adaptive significance of damage, repair and mortality. *Journal of Theoretical Biology* **238(1)**, 198–210.

Boucher, D. H. 1985. Lotka-volterra models of mutualism and positive density-dependence. *Ecological Modelling* **27(3–4)**, 251–270.

Boyd, R., & Richerson, P. J. 1985. *Culture and the evolutionary process*. Chicago: University of Chicago Press.

Brauer, F., & Castillo-Chavez, C. 2000. *Mathematical Models in Population Biology and Epidemiology*. New York: Springer.

Britton, N. F. 2003. *Essential Mathematical Biology*. London: Springer.

Bronstein, J. L., Wilson, W. G., & Morris, W. E. 2003. Ecological dynamics of mutualist/antagonist communities. *American Naturalist* **162(4)**, S24–S39.

Brown, J. S., & Pavlovic, N. B. 1992. Evolution in heterogeneous environments: effects of migration on habitat specilaization. *Evolutionary Ecology* **6**, 360–382.

Buckling, A., Wills, M. A., & Colegrave, N. 2003. Adaptation limits diversification of experimental bacterial populations. *Science* **302(5653)**, 2107–2109.

Bulmer, M. G., & Parker, G. A. 2002. The evolution of anisogamy: A game-theoretic approach. *Proceedings of the Royal Society of London Series B-Biological Sciences* **269(1507)**, 2381–2388.

Bürger, R., & Schneider, K. A. 2006. Intraspecific competitive divergence and convergence under assortative mating. *American Naturalist* **167(2)**, 190–205.

Bürger, R., Schneider, K. A., & Willensdorfer, M. 2006. The conditions for speciation through intraspecific competition. *Evolution* **60(11)**, 2185–2206.

Bush, G. L. 1969. Sympatric host race formation and speciation in frugivorous flies of genus *rhagoletis* (Diptera, Tephritidae). *Evolution* **23**, 237–251.

Cadet, C., Ferriére, R., Metz, J. A. J., & Van Baalen, M. 2003. The evolution of dispersal under demographic stochasticity. *American Naturalist* **162**, 427–441.

Case, T. J., & Taper, M. L. 2000. Interspecific competition, environmental gradients, gene flow, and the coevolution of species' borders. *American Naturalist* **155(5)**, 583–605.

Cavalli-Sforza, L. L., & Feldman, M. 1981. *Cultural Evolution*. Princeton: Princeton University Press.

Champagnat, N., Ferrière, R., & Meleard, S. 2006. Unifying evolutionary dynamics: From individual stochastic processes to macroscopic models. *Theoretical Population Biology* **69(3)**, 297–321.

Charlesworth, B. 1978. Population-genetics of anisogamy. *Journal of Theoretical Biology* **73(2)**, 347–357.

Chomsky, N. 1972. *Language and Mind*. New York: Harcourt Brace Jovanovich.

Chow, S. N., & Mallet-Paret, J. 1995. Pattern-formation and spatial chaos in lattice dynamical-systems. 1. *IEEE Transactions on Circuits and Systems I-Fundamental Theory and Applications* **42(10)**, 746–751.

Christiansen, F. B. 1991. On conditions for evolutionary stability for a continuously varying character. *American Naturalist* **138(1)**, 37–50.

Claessen, D., Andersson, J., Persson, L., & de Roos, A. M. 2007. Delayed evolutionary branching in small populations. *Evolutionary Ecology Research* **9**, 51–69.

Claessen, D., Andersson, J., Persson, L., & de Roos, A. M. 2008. The effect of population size and recombination on delayed evolution of polymorphism and speciation in sexual populations. *American Naturalist* **172**, E18–E34.

Cohan, F. M. 2002. Sexual isolation and speciation in bacteria. *Genetica* **116(2–3)**, 359–370.

Cohen, J., & Stewart, I. 2000. In *Nonlinear Phenomena in Physical and Biological Sciences* (S. K. Malik, ed.), 1–67. New Delhi: Indian National Science Academy.

Collet, P., & Eckmann, J. P. 1980. *Iterated Maps on the Interval as Dynamical System*. Boston: Birkhuser Boston.

Comins, H. N., Hassell, M. P., & May, R. M. 1992. The spatial dynamics of host parasitoid systems. *Journal of Animal Ecology* **61(3)**, 735–748.

Coyne, J. A., & Orr, H. A. 2004. *Speciation*. Sunderland, MA: Sinauer Associates.

Cressman, R. 2010. Css, nis and dynamic stability for two-species behavioral models with continuous trait spaces. *Journal of Theoretical Biology* **262**, 80–89.

Darwin, C. 1859. *On the Origin of Species by Means of Natural Selection: Or the Preservation of Favored Races in the Struggle for Life*. Cambridge, MA: Harvard University Press. Reprinted 1964.

Dawkins, R. 1976. *The Selfish Gene*. Oxford: Oxford University Press.

De Cara, M. A. R., Barton, N. H., & Kirkpatrick, M. 2008. A model for the evolution of assortative mating. *American Naturalist* **171(5)**, 580–596.

de Mazancourt, C., & Dieckmann, U. 2004. Trade-off geometries and frequency-dependent selection. *American Naturalist* **164(6)**, 765–778.

Dennett, D. 2009. Darwin's "strange inversion of reasoning." *Proceedings of the National Academy of Sciences of the United States of America* **106**, 10061–10065.

Dercole, F., Dieckmann, U., Obersteiner, M., & Rinaldi, S. 2008. Adaptive dynamics and technological change. *Technovation* **28(6)**, 335–348.

Dercole, F., Ferrière, R., Gragnani, A., & Rinaldi, S. 2006. Coevolution of slow-fast populations: evolutionary sliding, evolutionary pseudo-equilibria and complex red queen dynamics. *Proceedings of the Royal Society B-Biological Sciences* **273(1589)**, 983–990.

Dercole, F., Irisson, J. O., & Rinaldi, S. 2003. Bifurcation analysis of a prey-predator coevolution model. *Siam Journal on Applied Mathematics* **63(4)**, 1378–1391.

Dercole, F., & Rinaldi, S. 2008. *Analysis of Evolutionary Processes: The Adaptive Dynamics Approach and Its Applications*. Princeton: Princeton University Press.

Diamond, J. 1997. *Guns, Germs, and Steel: The Fates of Human Societies*. New York: W.W. Norton & Company.

Dickinson, H., & Antonovics, J. 1973. Theoretical considerations of sympatric divergence. *American Naturalist* **107(954)**, 256–274.

Dieckmann, U., & Doebeli, M. 1999. On the origin of species by sympatric speciation. *Nature* **400(6742)**, 354–357.

Dieckmann, U., & Doebeli, M. 2004. In *Adaptive Speciation* (U. Dieckmann, M. Doebeli, J. A. J. Metz, & D. Tautz, eds.), 76–111. Cambridge Studies in Adaptive Dynamics. Cambridge: Cambridge University Press.

Dieckmann, U., Doebeli, M., Metz, J. A. J., & Tautz, D. eds. 2004. *Adaptive Speciation*. Cambridge Studies in Adaptive Dynamics. Cambridge: Cambridge University Press.

Dieckmann, U., Heino, M., & Parvinen, K. 2006. The adaptive dynamics of function-valued traits. *Journal of Theoretical Biology* **241(2)**, 370–389.

Dieckmann, U., & Law, R. 1996. The dynamical theory of coevolution: A derivation from stochastic ecological processes. *Journal of Mathematical Biology* **34(5–6)**, 579–612.

Dieckmann, U., Law, R., & Metz, J. A. J. eds. 2004a. *The Geometry of Ecological Interactions: Simplifying Spatial Complexity*. Cambridge: Cambridge University Press.

Dieckmann, U., Marrow, P., & Law, R. 1995. Evolutionary cycling in predator-prey interactions- population-dynamics and the red queen. *Journal of Theoretical Biology* **176(1)**, 91–102.

Diehl, S. R., & Bush, G. L. 1989. In *Speciation and Its Consequences*. (D. Otte, & J. A. Endler, eds.), 345–365. Sunderland, MA: Sinauer Associates, Inc.

Diekmann, O. 2004. A beginner's guide to adaptive dynamics. *Banach Center Publ.* **63**, 47–86.

Dockery, J., Hutson, V., Mischaikow, K., & Pernarowski, M. 1998. The evolution of slow dispersal rates: a reaction diffusion model. *Journal of Mathematical Biology* **37**, 61–83.

Doebeli, M. 1995. Dispersal and dynamics. *Theoretical Population Biology* **47(1)**, 82–106.

Doebeli, M. 1996a. An explicit genetic model for ecological character displacement. *Ecology* **77(2)**, 510–520.

Doebeli, M. 1996b. A quantitative genetic competition model for sympatric speciation. *Journal of Evolutionary Biology* **9(6)**, 893–909.

Doebeli, M. 2002. A model for the evolutionary dynamics of cross-feeding polymorphisms in microorganisms. *Population Ecology* **44(2)**, 59–70.

Doebeli, M. 2005. Adaptive speciation when assortative mating is based on female preference for male marker traits. *Journal of Evolutionary Biology* **18(6)**, 1587–1600.

Doebeli, M., Blok, H. J., Leimar, O., & Dieckmann, U. 2007. Multimodal pattern formation in phenotype distributions of sexual populations. *Proceedings of the Royal Society B-Biological Sciences* **274(1608)**, 347–357.

Doebeli, M., & Dieckmann, U. 2000. Evolutionary branching and sympatric speciation caused by different types of ecological interactions. *American Naturalist* **156**, S77–S101.

Doebeli, M., & Dieckmann, U. 2003. Speciation along environmental gradients. *Nature* **421(6920)**, 259–264.

Doebeli, M., & Dieckmann, U. 2004. In *Adaptive Speciation* (U. Dieckmann, M. Doebeli, J. A. J. Metz, & D. Tautz, eds.), 140–165. Cambridge Studies in Adaptive Dynamics. Cambridge: Cambridge University Press.

Doebeli, M., & Hauert, C. 2005. Models of cooperation based on the prisoner's dilemma and the snowdrift game. *Ecology Letters* **8(7)**, 748–766.

Doebeli, M., Hauert, C., & Killingback, T. 2004. The evolutionary origin of cooperators and defectors. *Science* **306(5697)**, 859–862.

Doebeli, M., & Ispolatov, Y. 2010a. Complexity and diversity. *Science* **328**, 493–497.

Doebeli, M., & Ispolatov, Y. 2010b. Continuously stable strategies as evolutionary branching points. *Journal of Theoretical Biology* **266(4)**, 529–537.

Doebeli, M., & Ispolatov, Y. 2010c. A model for the evolutionary diversification of religions. *Journal of Theoretical Biology*, **267(4)**, 676–684.

Doebeli, M., & Knowlton, N. 1998. The evolution of interspecific mutualisms. *Proceedings of the National Academy of Sciences of the United States of America* **95(15)**, 8676–8680.

Doebeli, M., & Ruxton, G. D. 1997. Evolution of dispersal rates in metapopulation models: Branching and cyclic dynamics in phenotype space. *Evolution* **51(6)**, 1730–1741.

Doebeli, M., & Ruxton, G. D. 1998. Stabilization through spatial pattern formation in metapopulations with long-range dispersal. *Proceedings of the Royal Society of London Series B-Biological Sciences* **265(1403)**, 1325–1332.

Drossel, B., & McKane, A. 2000. Competitive speciation in quantitative genetic models. *Journal of Theoretical Biology* **204(3)**, 467–478.

Durinx, M., Metz, J., & Meszéna, G. 2008. Adaptive dynamics for physiologically structured models. *Journal of Mathematical Biology* **56**, 673–742.

Durinx, M., & Van Dooren, T. J. M. 2009. Assortative mate choice and dominance modification: Alternative ways of removing heterozygote disadvantage. *Evolution* **63(2)**, 334–352.

Edelstein-Keshet, L. 1988. *Mathematical Models in Biology*. New York: Birkhaeuser.

Eizaguirre, C., Lenz, T. L., Traulsen, A., & Milinski, M. 2009. Speciation accelerated and stabilized by pleiotropic major histocompatibility complex immunogenes. *Ecology Letters* **12(1)**, 5–12.

Elena, S. F., & Lenski, R. E. 2003. Evolution experiments with microorganisms: The dynamics and genetic bases of adaptation. *Nature Reviews Genetics* **4(6)**, 457–469.

Elmhirst, T., & Doebeli, M. 2010. Generic bifurcations from unimodal steady states in spatial dynamical systems. *In Review*.

Elmhirst, T., Stewart, I., & Doebeli, M. 2008. Pod systems: An equivariant ordinary differential equation approach to dynamical systems on a spatial domain. *Nonlinearity* **21(7)**, 1507–1531.

Endler, J. A. 1977. *Geographic Variation, Speciation, and Clines*. Princeton, NJ: Princeton University Press.

Erban, R., Chapman, J., & Maini, P. 2007. *A practical guide to stochastic simulations of reaction-diffusion processes.* http://www.citebase.org/abstract?id=oai:arXiv.org: 0704.1908.

Eshel, I. 1983. Evolutionary and continuous stability. *Journal of Theoretical Biology* **103**, 99–111.

Estes, S., & Arnold, S. J. 2007. Resolving the paradox of stasis: Models with stabilizing selection explain evolutionary divergence on all timescales. *American Naturalist* **169(2)**, 227–244.

Ewald, P. W. 1996. *Evolution of Infectious Disease.* Oxford: Oxford University Press.

Felsenstein, J. 1981. Evolutionary trees from gene-frequencies and quantitative characters—finding maximum-likelihood estimates. *Evolution* **35(6)**, 1229–1242.

Ferrière, R., Bronstein, J. L., Rinaldi, S., Law, R., & Gauduchon, M. 2002. Cheating and the evolutionary stability of mutualisms. *Proceedings of the Royal Society of London Series B-Biological Sciences* **269(1493)**, 773–780.

Fisher, R. A. 1930. *The Genetical Theory of Natural Selection.* Oxford: Clarendon Press.

Fort, H., Scheffer, M., & van Nes, E. H. 2009. The paradox of the clumps mathematically explained. *Theoretical Ecology* **2(3)**, 171–176.

Friesen, M. L., Saxer, G., Travisano, M., & Doebeli, M. 2004. Experimental evidence for sympatric ecological diversification due to frequency-dependent competition in *Escherichia coli*. *Evolution* **58(2)**, 245–260.

Fry, J. D. 2003. Multilocus models of sympatric speciation: Bush versus rice versus felsenstein. *Evolution* **57(8)**, 1735–1746.

Gavrilets, S. 1997. Coevolutionary chase in exploiter-victim systems with polygenic characters. *Journal of Theoretical Biology* **186(4)**, 527–534.

Gavrilets, S. 2004. *Fitness Landscapes and the Origin of Species.* Princeton, NJ: Princeton University Press.

Gavrilets, S., & Waxmann, D. 2002. Sympatric speciation by sexual conflict. *Proceedings of the National Academy of Science of the USA* **99**, 10533–10538.

Geritz, S. A. H. 2005. Resident-invader dynamics and the coexistence of similar strategies. *Journal of Mathematical Biology* **50(1)**, 67–82.

Geritz, S. A. H., & Kisdi, E. 2000. Adaptive dynamics in diploid, sexual populations and the evolution of reproductive isolation. *Proceedings of the Royal Society of London Series B-Biological Sciences* **267(1453)**, 1671–1678.

Geritz, S. A. H., Kisdi, E., Meszéna, G., & Metz, J. A. J. 1998. Evolutionarily singular strategies and the adaptive growth and branching of the evolutionary tree. *Evolutionary Ecology* **12(1)**, 35–57.

Geritz, S. A. H., Kisdi, E., & Yan, P. 2007. Evolutionary branching and long-term coexistence of cycling predators: Critical function analysis. *Theoretical Population Biology* **71(4)**, 424–435.

Geritz, S. A. H., Metz, J. A. J., Kisdi, E., & Meszéna, G. 1997. Dynamics of adaptation and evolutionary branching. *Physical Review Letters* **78(10)**, 2024–2027.

Gilchrist, M. A., & Coombs, D. 2006. Evolution of virulence: Interdependence, constraints, and selection using nested models. *Theoretical Population Biology* **69(2)**, 145–153.

Gillespie, D. T. 1976. General method for numerically simulating stochastic time evolution of coupled chemical-reactions. *Journal of Computational Physics* **22(4)**, 403–434.

Gillespie, D. T. 1977. Exact stochastic simulation of coupled chemical-reactions. *Journal of Physical Chemistry* **81(25)**, 2340–2361.

Golubitsky, M., & Stewart, I. N. 2002. *The Symmetry Perspective* Vol. Progress in Mathematics 200. Verlag: Birkhauser.

Gourbiere, S. 2004. How do natural and sexual selection contribute to sympatric speciation? *Journal of Evolutionary Biology* **17(6)**, 1297–1309.

Goymer, P., Kahn, S. G., Malone, J. G., Gehrig, S. M., Spiers, A. J., & Rainey, P. B. 2006. Adaptive divergence in experimental populations of pseudomonas fluorescens. ii. Role of the GGDEF regulator WspR in evolution and development of the wrinkly spreader phenotype. *Genetics* **173(2)**, 515–526.

Grant, P. R., & Grant, B. R. 2006. Evolution of character displacement in darwin's finches. *Science* **313(5784)**, 224–226.

Gyllenberg, M., & Meszéna, G. 2005. On the impossibility of coexistence of infinitely many strategies. *Journal of Mathematical Biology* **50(2)**, 133–160.

Gyllenberg, M., & Parvinen, K. 2001. Necessary and sufficient conditions for evolutionary suicide. *Bulletin of Mathematical Biology* **63(5)**, 981–993.

Gyllenberg, M., Parvinen, K., & Dieckmann, U. 2002. Evolutionary suicide and evolution of dispersal in structured metapopulations. *Journal of Mathematical Biology* **45(2)**, 79–105.

Gyllenberg, M., Soderbacka, G., & Ericsson, S. 1993. Does migration stabilize local-population dynamics—analysis of a discrete metapopulation model. *Mathematical Biosciences* **118(1)**, 25–49.

Harder, W., & Dijkhuizen, L. 1982. Strategies of mixed substrate utilization in microorganisms. *Philosophical Transactions of the Royal Society of London Series B-Biological Sciences* **297(1088)**, 459–480.

Hassell, M. P., Lawton, J. H., & May, R. M. 1976. Patterns of dynamical behavior in single-species populations. *Journal of Animal Ecology* **45(2)**, 471–486.

Hastings, A. 1993. Complex interactions between dispersal and dynamics—Lessons from coupled logistic equations. *Ecology* **74(5)**, 1362–1372.

Helling, R. B., Vargas, C. N., & Adams, J. 1987. Evolution of *Escherichia coli* during growth in a constant environment. *Genetics* **116(3)**, 349–358.

Hendry, A. 2007. Evolutionary biology—The elvis paradox. *Nature* **446**, 147–149.

Hendry, A. P. 2009. Evolutionary biology speciation. *Nature* **458(7235)**, 162–164.

Henrich, J., Boyd, R., & Richerson, P. J. 2008. Five misunderstandings about cultural evolution. *Human Nature—An Interdisciplinary Biosocial Perspective* **19(2)**, 119–137.

Hernandez-García, E., Lopez, C., Pigolotti, S., & Andersen, K. H. 2009 Species competition: Coexistence, exclusion and clustering. *Philosophical Transactions of the Royal Society A—Mathematical, Physical and Engineering Sciences* **367(1901)**, 3183–3195.

Higashi, M., Takimoto, G., & Yamamura, N. 1999. Sympatric speciation by sexual selection. *Nature* **402(6761)**, 523–526.

Hoekstra, R. F. 1980. Why do organisms produce gametes of only 2 different sizes—Some theoretical aspects of the evolution of anisogamy. *Journal of Theoretical Biology* **87(4)**, 785–793.

Hoekstra, R. F. 1987. In *The Evolution of Sex and Its Consequence* (S. C. Stearns, ed.), 59–91. Basel, Switzerland: Birkhäuser Verlag.

Hofbauer, J., & Sigmund, K. 1998. *Evolutionary Games and Population Dynamics.* Cambridge: Cambridge University Press.

Holt, R. D., & McPeek, M. A. 1996. Chaotic population dynamics favors the evolution of dispersal. *American Naturalist* **148(4)**, 709–718.

Hurst, L. D. 1990. Parasite diversity and the evolution of diploidy, multicellularity and anisogamy. *Journal of Theoretical Biology* **144(4)**, 429–443.

Hurst, L. D., & Hamilton, W. D. 1992. Cytoplasmic fusion and the nature of sexes. *Proceedings of the Royal Society of London Series B-Biological Sciences* **247(1320)**, 189–194.

Ispolatov, Y., & Doebeli, M. 2009a. Diversification along environmental gradients in spatially structured populations. *Evolutionary Ecology Research* **11(2)**, 295–304.

Ispolatov, Y., & Doebeli, M. 2009b. Speciation due to hybrid necrosis in plant-pathogen models. *Evolution* **63(12)**, 3076–3084.

Ito, H. C., & Dieckmann, U. 2007. A new mechanism for recurrent adaptive radiations. *American Naturalist* **170(4)**, E96–E111.

Johnson, P. A., Hoppensteadt, F. C., Smith, J. J., & Bush, G. L. 1996. Conditions for sympatric speciation: A diploid model incorporating habitat fidelity and non-habitat assortative mating. *Evolutionary Ecology* **10(2)**, 187–205.

Jones, E. I., Ferrière, R., & Bronstein, J. L. 2009. Eco-evolutionary dynamics of mutualists and exploiters. *American Naturalist* **174(6)**, 780–794.

Jones, J. D. G., & Dangl, J. L. 2006. The plant immune system. *Nature* **444**, 323–329.

Kaneko, K. 1992. Overview of coupled map lattices. *Chaos* **2(3)**, 279–282.

Kaneko, K., & Yomo, T. 2000. Sympatric speciation: compliance with phenotype diversification from a single genotype. *Proceedings of the Royal Society of London Series B-Biological Sciences* **267**, 2367–2373.

Kaneko, K., & Yomo, T. 2002. Symbiotic sympatric speciation through interaction-driven phenotype differentiation. *Evolutionary Ecology Research* **4**, 317–350.

Kawecki, T. J. 1996. Sympatric speciation driven by beneficial mutations. *Proceedings of the Royal Society of London Series B-Biological Sciences* **263(1376)**, 1515–1520.

Kawecki, T. J. 1997. Sympatric speciation via habitat specialization driven by deleterious mutations. *Evolution* **51(6)**, 1751–1763.

Kawecki, T. J. 2004. In *Adaptive Speciation* (U. Dieckmann, M. Doebeli, D. Tautz, & J. A. J. Metz, eds.), 36–53. Cambridge: Cambridge University Press.

Kiester, A. R., Lande, R., & Schemske, D. W. 1984. Models of coevolution and speciation in plants and their pollinators. *American Naturalist* **124(2)**, 220–243.

Killingback, T., Doebeli, M., & Hauert, C. 2010. Diversity of cooperation in the tragedy of the commons. *Biological Theory*, **5**, 3–6.

Killingback, T., Doebeli, M., & Knowlton, N. 1999. Variable investment, the continuous prisoner's dilemma, and the origin of cooperation. *Proceedings of the Royal Society of London Series B-Biological Sciences* **266(1430)**, 1723–1728.

Kirkpatrick, M., & Barton, N. H. 1997. Evolution of a species' range. *American Naturalist* **150(1)**, 1–23.

Kirkpatrick, M., & Barton, N. H. 2006. Chromosome inversions, local adaptation and speciation. *Genetics* **173**, 419–434.

Kirkpatrick, M., & Nuismer, S. L. 2004. Sexual selection can constrain sympatric speciation. *Proceedings of the Royal Society of London Series B-Biological Sciences* **271(1540)**, 687–693.

Kirkpatrick, M., & Ravigne, V. 2002. Speciation by natural and sexual selection: Models and experiments. *American Naturalist* **159**, S22–S35.

Kisdi, E. 1999. Evolutionary branching under asymmetric competition. *Journal of Theoretical Biology* **197(2)**, 149–162.

Kisdi, E. 2001. Long-term adaptive diversity in levene-type models. *Evolutionary Ecology Research* **3(6)**, 721–727.

Kisdi, E., & Geritz, S. A. H. 1999. Adaptive dynamics in allele space: Evolution of genetic polymorphism by small mutations in a heterogeneous environment. *Evolution* **53(4)**, 993–1008.

Koella, J. C. 2000. The spatial spread of altruism versus the evolutionary response of egoists. *Proceedings of the Royal Society of London Series B-Biological Sciences* **267(1456)**, 1979–1985.

Koella, J. C., & Doebeli, M. 1999. Population dynamics and the evolution of virulence in epidemiological models with discrete host generations. *Journal of Theoretical Biology* **198(3)**, 461–475.

Kondrashov, A. S. 1983a. Multilocus model of sympatric speciation. 1. One character. *Theoretical Population Biology* **24(2)**, 121–135.

Kondrashov, A. S. 1983b. Multilocus model of sympatric speciation. 2. 2 characters. *Theoretical Population Biology* **24(2)**, 136–144.

Kondrashov, A. S. 1984. The intensity of selection for reproductive isolation at the beginning of sympatric speciation. *Genetika* **20(3)**, 408–415.

Kondrashov, A. S. 1986. Multilocus model of sympatric speciation. 3. Computer-simulations. *Theoretical Population Biology* **29(1)**, 1–15.

Kondrashov, A. S., & Kondrashov, F. A. 1999. Interactions among quantitative traits in the course of sympatric speciation. *Nature* **400(6742)**, 351–354.

Kondrashov, A. S., & Mina, M. V. 1986. Sympatric speciation—When is it possible. *Biological Journal of the Linnean Society* **27(3)**, 201–223.

Kondrashov, A. S., & Shpak, M. 1998. On the origin of species by means of assortative mating. *Proceedings of the Royal Society of London Series B-Biological Sciences* **265(1412)**, 2273–2278.

Kooijman, S. A. L. M. 2000. *Dynamic Energy and Mass Budgets in Biological Systems.* Cambridge: Cambridge University Press.

Kopp, M., & Hermisson, J. 2006. The evolution of genetic architecture under frequency-dependent disruptive selection. *Evolution* **60(8)**, 1537–1550.

Kot, M. 2001. *Elements of Mathematical Ecology.* Cambridge: Cambridge University Press.

Lande, R. 1979. Quantitative genetic-analysis of mulitvariate evolution, applied to brain-body size allometry. *Evolution* **33(1)**, 402–416.

Lande, R. 1981. Models of speciation by sexual selection on polygenic traits. *Proceedings of the National Academy of Sciences of the United States of America-Biological Sciences* **78(6)**, 3721–3725.

Law, R., Marrow, P., & Dieckmann, U. 1997. On evolution under asymmetric competition. *Evolutionary Ecology* **11(4)**, 485–501.

Le Gac, M., Brazas, M. D., Bertrand, M., Tyerman, J. G., Spencer, C. C., Hancock, R. E. W., & Doebeli, M. 2008. Metabolic changes associated with adaptive diversification in *Escherichia coli*. *Genetics* **178(2)**, 1049–1060.

Le Gac, M., & Doebeli, M. 2010. Epistasis and frequency dependence influence the fitness of an adaptive mutation in a diversifying lineage. *Molecular Ecology* **19**, 2430–2438.

Leimar, O. 2001. Evolutionary change and darwinian demons. *Selection* **2**, 65–72.

Leimar, O. 2005. The evolution of phenotypic polymorphism: Randomized strategies versus evolutionary branching. *American Naturalist* **165(6)**, 669–681.

Leimar, O. 2009. Multidimensional convergence stability. *Evolutionary Ecology Research* **11**, 191–208.

Leimar, O., Doebeli, M., & Dieckmann, U. 2008. Evolution of phenotypic clusters through competition and local adaptation along an environmental gradient. *Evolution* **62(4)**, 807–822.

Lessard, S. 1990. Evolutionary stability: One concept, several meanings. *Theoretical Population Biology* **37**, 159–170.

Levene, H. 1953. Genetic equilibrium when more than one ecological niche is available. *American Naturalist* **87(836)**, 331–333.

Levins, R. 1968. *Evolution in Changing Environments*. Princeton: Princeton University Press.

Li, T. Y., & Yorke, J. A. 1975. Period three implies chaos. *The American Mathematical Monthly* **82(10)**, 985–992.

Lim, M., Metzler, R., & Bar-Yam, Y. 2007. Global pattern formation and ethnic/cultural violence. *Science* **317(5844)**, 1540–1544.

Ludwig, D., & Levin, S. A. 1991. Evolutionary stability of plant-communities and the maintenance of multiple dispersal types. *Theoretical Population Biology* **40(3)**, 285–307.

Lynch, M., & Walsh, B. 1998. *Genetics and Analysis of Quantitative Traits*. Sunderland, MA: Sinauer Associates, Inc.

MacArthur, R. H., & Levins, R. 1964. Competition, habitat selection and character displacement in patchy environments. *Proceedings of the National Academy of Sciences of the United States of America* **51(6)**, 1207–1210.

MacArthur, R. H., & Levins, R. 1967. Limiting similarity, convergence and divergence of coexisting species. *American Naturalist* **101(921)**, 377–385.

MacLean, R. C., & Gudelj, I. 2006. Resource competition and social conflict in experimental populations of yeast. *Nature* **441(7092)**, 498–501.

Maire, N., Ackermann, M., & Doebeli, M. 2001. On the evolution of anisogamy through evolutionary branching. *Selection* **2(1–2)**, 119–131.

Marrow, P., Dieckmann, U., & Law, R. 1996. Evolutionary dynamics of predator-prey systems: An ecological perspective. *Journal of Mathematical Biology* **34(5–6)**, 556–578.

Mathias, A., Kisdi, E., & Olivieri, I. 2001. Divergent evolution of dispersal in a heterogeneous landscape. *Evolution* **55(2)**, 246–259.

Matsuda, H., & Abrams, P. A. 1999. Why are equally sized gametes so rare? The instability of isogamy and the cost of anisogamy. *Evolutionary Ecology Research* **1**(7), 769–784.

May, R. M. 1974. Biological populations with nonoverlapping generations—Stable points, stable cycles, and chaos. *Science* **186**(4164), 645–647.

May, R. M. 1976. Simple mathematical-models with very complicated dynamics. *Nature* **261**(5560), 459–467.

May, R. M., & Oster, G. F. 1976. Bifurcations and dynamic complexity in simple ecological models. *American Naturalist* **110**(974), 573–599.

Maynard Smith, J. 1962. Disruptive selection, polymorphism and sympatric speciation. *Nature* **195**(4836), 60–62.

Maynard Smith, J. 1966. Sympatric speciation. *American Naturalist* **100**(916), 637–650.

Maynard Smith, J. 1978. *The Evolution of Sex*. Cambridge: Cambridge University Press.

Maynard Smith, J. 1982. *Evolution and the Theory of Games*. Cambridge: Cambridge University Press.

Maynard Smith, J., & Brown, R. L. W. 1986. Competition and body size. *Theoretical Population Biology* **30**(2), 166–179.

Maynard Smith, J., & Price, G. R. 1973. Logic of animal conflict. *Nature* **246**(5427), 15–18.

Maynard Smith, J., & Szathmáry, E. 1997. *The Major Transitions in Evolution*. Oxford: Oxford University Press.

McGill, B. J., & Brown, J. S. 2007. Evolutionary game theory and adaptive dynamics of continuous traits. *Annual Review of Ecology Evolution and Systematics* **38**, 403–435.

McKane, A. J., & Newman, T. J. 2005. Predator-prey cycles from resonant amplification of demographic stochasticity. *Physical Review Letters* 94(21).

McPeek, M. A., & Holt, R. D. 1992. The evolution of dispersal in spatially and temporally varying environments. *American Naturalist* **140**(6), 1010–1027.

Meszéna, G., Czibula, I., & Geritz, S. A. H. 1997. Adaptive dynamics in a 2-patch environment: A toy model for allopatric and parapatric speciation. *Journal of Biological Systems* **5**, 265–284.

Meszéna, G., & Metz, J. A. J. 1999. *Species Diversity and Population Regulation: The Importance of Environmental Feedback Dimensionality*. Working Papers ir99045 http://ideas.repec.org/p/wop/iasawp/ir99045.html: International Institute for Applied Systems Analysis.

Metz, J. A. J., Geritz, S. A. H., Meszéna, G., Jacobs, F. J. A., & van Heerwaarden, J. S. 1996. In *Stochastic and Spatial Structures of Dynamical Systems, Proceedings of the Royal Dutch Academy of Science (KNAW Verhandlingen)* (S. van Strien, & S. Verduyn Lunel, eds.), 183–231. Dordrecht, Netherlands: North Holland.

Metz, J. A. J., Nisbet, R. M., & Geritz, S. A. H. 1992. How should we define fitness for general ecological scenarios. *Trends in Ecology & Evolution* **7**(6), 198–202.

Mitchener, W. G., & Nowak, M. A. 2003. Competitive exclusion and coexistence of universal grammars. *Bulletin of Mathematical Biology* **65**(1), 67–93.

Mylius, S. D., & Diekmann, O. 1995. On evolutionarily stable life histories, optimization and the need to be specific about density dependence. *Oikos* **74**(2), 218–224.

Noest, A. J. 1997. Instability of the sexual continuum. *Proceedings of the Royal Society of London Series B-Biological Sciences* **264(1386)**, 1389–1393.

Nowak, M., & Sigmund, K. 1990. The evolution of stochastic strategies in the prisoners-dilemma. *Acta Applicandae Mathematicae* **20(3)**, 247–265.

Nowak, M. A., & Krakauer, D. C. 1999. The evolution of language. *Proceedings of the National Academy of Sciences of the United States of America* **96(14)**, 8028–8033.

Nowak, M. A., & Sigmund, K. 2004. Evolutionary dynamics of biological games. *Science* **303(5659)**, 793–799.

Nuismer, S. L., Doebeli, M., & Browning, D. 2005. The coevolutionary dynamics of antagonistic interactions mediated by quantitative traits with evolving variances. *Evolution* **59(10)**, 2073–2082.

Olivieri, I., Michalakis, Y., & Gouyon, P. H. 1995. Metapopulation genetics and the evolution of dispersal. *American Naturalist* **146(2)**, 202–228.

Otto, S. P., & Day, T. 2007. *A Biologist's Guide to Mathematical Modeling in Ecology and Evolution.* Princeton: Princeton University Press.

Otto, S. P., Servedio, M. R., & Nuismer, S. L. 2008. Frequency-dependent selection and the evolution of assortative mating. *Genetics* **179(4)**, 2091–2112.

Parker, G. A. 1978. Selection on nonrandom fusion of gametes during evolution of anisogamy. *Journal of Theoretical Biology* **73(1)**, 1–28.

Parker, G. A., Smith, V. G. F., & Baker, R. R. 1972. Origin and evolution of gamete dimorphism and male-female phenomenon. *Journal of Theoretical Biology* **36(3)**, 529–553.

Parvinen, K. 1999. Evolution of migration in a metapopulation. *Bulletin of Mathematical Biology* **61(3)**, 531–550.

Parvinen, K. 2002. Evolutionary branching of dispersal strategies in structured metapopulations. *Journal of Mathematical Biology* **45(2)**, 106–124.

Parvinen, K., Dieckmann, U., & Heino, M. 2006. Function-valued adaptive dynamics and the calculus of variations. *Journal of Mathematical Biology* **52(1)**, 1–26.

Peischl, S., & Bürger, R. 2008. Evolution of dominance under frequency-dependent intraspecific competition. *Journal of Theoretical Biology* **251(2)**, 210–226.

Pennings, P. S., Kopp, M., Meszéna, G., Dieckmann, U., & Hermisson, J. 2008. An analytically tractable model for competitive speciation. *American Naturalist* **171(1)**, E44–E71.

Pfeiffer, T., Schuster, S., & Bonhoeffer, S. 2001. Cooperation and competition in the evolution of atp-producing pathways. *Science* **292(5516)**, 504–507.

Pigolotti, S., Lopez, C., & Hernandez-García, E. 2007. Species clustering in competitive lotka-volterra models. *Physical Review Letters* 98(25).

Pigolotti, S., Lopez, C., Hernandez-García, E., & Andersen, K. H. 2010. How gaussian competition leads to lumpy or uniform species disributions. *Theoretical Ecology* in press.

Pineda-Krch, M. 2010. Gillespiessa: Implementing the gillespie stochastic simulation algorithm in R. *Journal of Statistical Software* p. In Press.

Pineda-Krch, M., Blok, H. J., Dieckmann, U., & Doebeli, M. 2007. A tale of two cycles—Distinguishing quasi-cycles and limit cycles in finite predator-prey populations. *Oikos* **116(1)**, 53–64.

Pinker, S. 2000. *The Language Instinct: How the Mind Creates Language*. New York: Harper Perennial Modern Classics.

Polechova, J., & Barton, N. H. 2005. Speciation through competition: A critical review. *Evolution* **59(6)**, 1194–1210.

Press, W. H., Teukolsky, S. A., Vetterling, W. T., & Flannery, B. P. 2007. *Numerical Recipies: The Art of Scientific Computing*. Cambridge: Cambridge University Press.

Rainey, P. B., & Travisano, M. 1998. Adaptive radiation in a heterogeneous environment. *Nature* **394(6688)**, 69–72.

Randerson, J. P., & Hurst, L. D. 2001a. A comparative test of a theory for the evolution of anisogamy. *Proceedings of the Royal Society of London Series B-Biological Sciences* **268(1469)**, 879–884.

Randerson, J. P., & Hurst, L. D. 2001b. The uncertain evolution of the sexes. *Trends in Ecology & Evolution* **16(10)**, 571–579.

Regoes, R. R., & Bonhoeffer, S. 2005. The HIV coreceptor switch: A population dynamical perspective. *Trends in Microbiology* **13(6)**, 269–277.

Regoes, R. R., & Bonhoeffer, S. 2006. Emergence of drug-resistant influenza virus: Population dynamical considerations. *Science* **312(5772)**, 389–391.

Renshaw, E. 1991. *Cambridge Studies in Mathematical Biology 11. Modelling Biological Populations in Space and Time*. Cambridge: Cambridge University Press.

Richerson, P. J., & Boyd, R. 2005. *Not by Genes Alone: How Culture Transformed Human Evolution*. Chicago: University of Chicago Press.

Ritchie, M. G. 2007. Sexual selection and speciation. *Annual Review of Ecology Evolution and Systematics* **38**, 79–102.

Rosenzweig, M. L. 1978. Competitive speciation. *Biological Journal of the Linnean Society* **10(3)**, 275–289.

Rosenzweig, R. F., Sharp, R. R., Treves, D. S., & Adams, J. 1994. Microbial evolution in a simple unstructured environment—Genetic differentiation in *Escherichia coli*. *Genetics* **137(4)**, 903–917.

Roughgarden, J. 1979. *Theory of Population Genetics and Evolutionary Ecology: An Introduction*. New York, New York, USA: Macmillan Publishing Co., Inc.

Roughgarden, J., & Pacala, S. 1989. In *Speciation and its Consequences* (D. Otte & J. A. Endler, eds.), 403–432. Sunderland, MA: Sinauer Associates, Inc.

Rozen, D. E., de Visser, J. A. G. M., & Gerrish, P. J. 2002. Fitness effects of fixed beneficial mutations in microbial populations. *Current Biology* **12(12)**, 1040–1045.

Rozen, D. E., & Lenski, R. E. 2000. Long-term experimental evolution in *Escherichia coli*. viii. Dynamics of a balanced polymorphism. *American Naturalist* **155(1)**, 24–35.

Rozen, D. E., Schneider, D., & Lenski, R. E. 2005. Long-term experimental evolution in *Escherichia coli*. xiii. Phylogenetic history of a balanced polymorphism. *Journal of Molecular Evolution* **61(2)**, 171–180.

Rueffler, C., Egas, M., & Metz, J. A. J. 2006a. Evolutionary predictions should be based on individual-level traits. *American Naturalist* **168(5)**, E148–E162.

Rueffler, C., Van Dooren, T. J. M., Leimar, O., & Abrams, P. A. 2006b. Disruptive selection and then what? *Trends in Ecology & Evolution* **21(5)**, 238–245.

Rummel, J. D., & Roughgarden, J. 1985. A theory of faunal buildup for competition communities. *Evolution* **39(5)**, 1009–1033.

Ruxton, G. D., & Doebeli, M. 1996. Spatial self-organization and persistence of transients in a metapopulation model. *Proceedings of the Royal Society of London Series B-Biological Sciences* **263(1374)**, 1153–1158.

Sasaki, A., & Ellner, S. 1995. The evolutionarily stable phenotype distribution in a random environment. *Evolution* **49**, 337–350.

Schluter, D. 2000. Ecological character displacement in adaptive radiation. *American Naturalist* **156**, S4–S16.

Schluter, D. 2009. Evidence for ecological speciation and its alternative. *Science* **323(5915)**, 737–741.

Schneider, D., & Lenski, R. E. 2004. Dynamics of insertion sequence elements during experimental evolution of bacteria. *Research in Microbiology* **155(5)**, 319–327.

Schneider, K. A. 2006. A multilocus-multiallele analysis of frequency-dependent selection induced by intraspecific competition. *Journal of Mathematical Biology* **52(4)**, 483–523.

Schneider, K. A. 2007. Long-term evolution of polygenic traits under frequency-dependent intraspecific competition. *Theoretical Population Biology* **71(3)**, 342–366.

Schneider, K. A., & Bürger, R. 2006. Does competitive divergence occur if assortative mating is costly? *Journal of Evolutionary Biology* **19(2)**, 570–588.

Schreiber, S. J., & Saltzman, E. 2009. Evolution of predator and prey movement into sink habitats. *American Naturalist* **174(1)**, 68–81.

Schuster, H. G., & Just, W. 2005. *Deterministic Chaos*. Weinheim, Germany: John Wiley and Sons Ltd.

Seger, J. 1985. In *Evolution: Essays in Honour of John Maynard Smith* (P. H. H. Greenwood, P. J. Harvey, & M. Slatkin, eds.), 43–54. New York, NY, USA; Cambridge, England: Cambridge University Press.

Sharkovskii, A. N. 1964. Co-existence of cycles of a continuous mapping of the line into itself. *Ukrainian Mathematical Journal* **16(1)**, 61–71.

Slatkin, M. 1980. Ecological character displacement. *Ecology* **61(1)**, 163–177.

Slatkin, M. 1984. Ecological causes of sexual dimorphism. *Evolution* **38(3)**, 622–630.

Spencer, C. C., Bertrand, M., Travisano, M., & Doebeli, M. 2007a. Adaptive diversification in genes that regulate resource use in *Escherichia coli. PLoS Genetics* **3(1)**, e15.

Spencer, C. C., Saxer, G., Travisano, M., & Doebeli, M. 2007b. Seasonal resource oscillations maintain diversity in bacterial microcosms. *Evolutionary Ecology Research* **9(5)**, 775–787.

Spencer, C. C., Tyerman, J., Bertrand, M., & Doebeli, M. 2008. Adaptation increases the likelihood of diversification in an experimental bacterial lineage. *Proceedings of the National Academy of Sciences of the United States of America* **105(5)**, 1585–1589.

Sperber, D. 1996. *Explaining culture: A naturalistic approach*. Oxford, UK and Cambridge, MA: Blackwell.

Spiers, A. J., Kahn, S. G., Bohannon, J., Travisano, M., & Rainey, P. B. 2002. Adaptive divergence in experimental populations of pseudomonas fluorescens. i. Genetic and phenotypic bases of wrinkly spreader fitness. *Genetics* **161(1)**, 33–46.

Stearns, S. C. 1992. *The Evolution of Life Histories*. Oxford: Oxford University Press.

Stewart, I., Elmhirst, T., & Cohen, J. 2003. In *Bifurcations, Symmetry and Patterns*, 3–54. Basel: Birkhauser.

Sugden, R. 1986. *The Economics of Rights, Co-operation and Welfare*. Oxford: Blackwell.

Svanback, R., Pineda-Krch, M., & Doebeli, M. 2009. Fluctuating population dynamics promotes the evolution of phenotypic plasticity. *American Naturalist* **174(2)**, 176–189.

Szathmáry, E., & Maynard Smith, J. 1995. The major evolutionary transitions. *Nature* **374(6519)**, 227–232.

Takens, F. 1996. In *Stochastic and Spatial Structures of Dynamical Systems* (S. J. van Strien, & S. M. Verduyn Lunel, eds.), KNAW, deel 45, 3–15. North Holland, Netherlands: Verhandelingen Afd. Natuurkunde.

Takimoto, G. 2002. Polygenic inheritance is not necessary for sympatric speciation by sexual selection. *Population Ecology* **44(2)**, 87–91.

Takimoto, G., Higashi, M., & Yamamura, N. 2000. A deterministic genetic model for sympatric speciation by sexual selection. *Evolution* **54(6)**, 1870–1881.

Taper, M. L., & Case, T. J. 1985. Quantitative genetic models for the coevolution of character displacement. *Ecology* **66(2)**, 355–371.

Taper, M. L., & Case, T. J. 1992. Models of character displacement and the theoretical robustness of taxon cycles. *Evolution* **46(2)**, 317–333.

ten Tusscher, K. H. W. J., & Hogeweg, P. 2009. The role of genome and gene regulatory network canalization in the evolution of multi-trait polymorphisms and sympatric speciation. *BMC Evolutionary Biology* **9(1)**, 159.

Toro-Roman, A., Mack, T. R., & Stock, A. M. 2005. Structural analysis and solution studies of the activated regulatory domain of the response regulator arca: A symmetric dimer mediated by the alpha 4-beta 5-alpha 5 face. *Journal of Molecular Biology* **349(1)**, 11–26.

Tregenza, T., & Butlin, R. K. 1999. Speciation without isolation. *Nature* **400(6742)**, 311–312.

Treves, D. S., Manning, S., & Adams, J. 1998. Repeated evolution of an acetate-crossfeeding polymorphism in long-term populations of *Escherichia coli*. *Molecular Biology and Evolution* **15(7)**, 789–797.

Troost, T. A., Kooi, B. W., & Kooijman, S. A. L. M. 2005a. Ecological specialization of mixotrophic plankton in a mixed water column. *American Naturalist* **166(3)**, E45–E61.

Troost, T. A., Kooi, B. W., & Kooijman, S. A. L. M. 2005b. When do mixotrophs specialize? Adaptive dynamics theory applied to a dynamic energy budget model. *Mathematical Biosciences* **193(2)**, 159–182.

Tuchman, B. 1985. *The March of Folly: From Troy to Vietnam*. New York: Ballantine Books.

Turchin, P. 2003. *Complex Population Dynamics: A Theoretical/Empirical Synthesis*. Princeton: Princeton University Press.

Turchin, P., & Taylor, A. D. 1992. Complex dynamics in ecological time-series. *Ecology* **73(1)**, 289–305.

Turner, G. F., & Burrows, M. T. 1995. A model of sympatric speciation by sexual selection. *Proceedings of the Royal Society of London Series B-Biological Sciences* **260(1359)**, 287–292.

Tyerman, J., Havard, N., Saxer, G., Travisano, M., & Doebeli, M. 2005. Unparallel diversification in bacterial microcosms. *Proceedings of the Royal Society B-Biological Sciences* **272(1570)**, 1393–1398.

Tyerman, J. G., Bertrand, M., Spencer, C. C., & Doebeli, M. 2008. Experimental demonstration of ecological character displacement. *BMC Evolutionary Biology* **8**, 9.

Udovic, D. 1980. Frequency-dependent selection, disruptive selection, and the evolution of reproductive isolation. *American Naturalist* **116(5)**, 621–641.

Vamosi, S. M. 2002. Predation sharpens the adaptive peaks: Survival trade-offs in sympatric sticklebacks. *Annales Zoologici Fennici* **39(3)**, 237–248.

Van Dooren, T. J. M., Durinx, M., & Demon, I. 2004. Sexual dimorphism or evolutionary branching? *Evolutionary Ecology Research* **6(6)**, 857–871.

Van Doorn, G. S., & Dieckmann, U. 2006. The long-term evolution of multilocus traits under frequency-dependent disruptive selection. *Evolution* **60(11)**, 2226–2238.

Van Doorn, G. S., Dieckmann, U., & Weissing, F. J. 2004. Sympatric speciation by sexual selection: A critical reevaluation. *American Naturalist* **163(5)**, 709–725.

Van Doorn, G. S., Edelaar, P., & Weissing, F. J. 2009. On the origin of species by natural and sexual selection. *Science* **326**, 1704–1707.

Van Doorn, G. S., Luttikhuizen, P. C., & Weissing, F. J. 2001. Sexual selection at the protein level drives the extraordinary divergence of sex-related genes during sympatric speciation. *Proceedings of the Royal Society of London Series B-Biological Sciences* **268(1481)**, 2155–2161.

Vandermeer, J. H., & Boucher, D. H. 1978. Varieties of mutualistic interaction in population models. *Journal of Theoretical Biology* **74(4)**, 549–558.

Vincent, T. L., & Brown, J. L. 2005. *Evolutionary Game Theory, Natural Selection, and Darwinian Dynamics*. Cambridge: Cambridge University Press.

White, D. 2000. *The Physiology and Biochemistry of Prokaryotes, Ed. 2*. Oxford: Oxford University Press.

Whitehouse, H. 1995. *Inside the cult*. Oxford, UK: Oxford University Press.

Wiese, L., Wiese, W., & Edwards, D. A. 1979. Inducible anisogamy and the evolution of oogamy from isogamy. *Annals of Botany* **44(2)**, 131–139.

Wolin, C. L., & Lawlor, L. R. 1984. Models of facultative mutualism—Density effects. *American Naturalist* **124(6)**, 843–862.

Index

A-loci, 75
Abrams, Peter, 279
acetate, 264
adaptive diversification, 4; due to mutualistic
  interactions, 148; due to predator-prey
  interactions, 113, 236; due to resource
  competition, 38, 74, 218; generic, 261; in
  cultural evolution, 195; in dispersal rates,
  178; in experimental evolution, 262; in
  microbes, 6, 265; in microbial
  metabolism, 265; model-independent,
  258; in plant-pathogen models, 243; in
  religions, 210; in small populations, 63; in
  socially structured cultures, 211; in
  spatially structured populations, 178, 242
adaptive dynamics, 279; one-dimensional,
  13; two-dimensional, 23, 117, 291;
  three-dimensional, 33; four-dimensional,
  56; attractor, 14; canonical equation, 13,
  43, 280; classification in one dimension,
  304; Continuous Snowdrift game, 142;
  degeneracy condition, 124; equilibrium
  points, 13; function-valued traits, 300;
  introduction to, 279; of anisogamy, 182;
  of asymmetric resource competition, 65;
  of cooperation, 142; of culture, 199; of
  dispersal, 172; of host-pathogen
  interactions, 136; of microbial
  metabolism, 270; of predator-
  prey interactions, 117; of resource
  competition, 43; of symmetric resource
  competition, 50; of trophic preference,
  193; singular coalition, 55; singular
  points, 4, 282; Tragedy of the Commons,
  147; vs. partial differential equations, 225
adaptive landscape, 279
adaptive radiation, 6, 39
adaptive speciation, 7, 74, 82, 100, 233; due
  to host-pathogen interactions, 247; due to
  mutualistic interactions, 158; due to
  postzygotic isolation, 243; due to
  predator-prey interactions, 131; due to
  resource competition, 85; in partial

differential equation models, 231; in
  spatially structured populations, 255;
  one-allele models, 82; two-allele models,
  100
adaptive tree, 287
agar plates, 265
Alemannian, 197
Allee effects, 88; dependence on initial
  conditions, 90, 235; evolution of
  assortative mating, 96; normalized mating
  functions, 96; one-allele models, 89;
  partial differential equation models, 234;
  two-allele models, 102
Alsatian, 197
anisogamy, 180; evolution of, 185
Arabidopsis thaliana, 243
arms race, 140, 241; cyclic, 240; partial
  differential equations, 241; predator-prey
  models, 114. *See also* coevolution
assortative mating, 7, 74; evolution of, 90;
  marker traits, 77; mutualism model, 158;
  normalized mating function, 94;
  one-allele model, 82; partial differential
  equation models, 229; predator-prey
  model, 130; preference function, 82;
  sexual selection, 79; spatially structured
  populations, 251; two-allele model, 104
asymmetric competition, 42, 64, 144
attack rate, 115, 236
autoimmune reactions, 243
autotrophy, 189

Bürger, Reinhard, 77
Basel, Switzerland, 197
batch culture, 264
bifurcation diagram, 174
bifurcation point, 175, 261
birth-death process, 280
blending inheritance, 81, 196

canonical equation, 13, 43, 280, 301
carbohydrate metabolism, 268

carrying capacity, 40, 47, 218; m-dimensional, 221; Gaussian, 52; in consumer-resource models, 42; in spatially structured populations, 249; quadratic, 58; quartic, 86, 225, 294

catabolite repression, 264

catch-22, 108

chaos, 166, 168; spatio-temporal, 166

chromosome inversion, 109

coevolution; host-pathogen, 132, 243; mutualism, 148; predator-prey, 113, 236

coexistence, 3, 21; anisogamy, 180; dispersal, 178; flowering schedules, 302; microbial metabolism, 268; mutualism, 146; predator-prey, 116; religions, 209; resource competition, 53; trophic levels, 194

cognitive constraints, 199

colony size, 265

competition, 38; asymmetric, 42, 64, 144; for mating partners, 111; for pollinators, 300; frequency-dependent, 44; Gaussian, 52; in sexual models, 74; partial differential equation models, 218; symmetric, 42, 51

competition kernel, 42; m-dimensional, 221; asymmetric, 65; consumer-resource dynamics, 46; Gaussian, 52; partial differential equations, 218; platykurtic, 86, 220; positive, 226; predator-prey model, 124; quartic, 90; symmetric, 42

complicated evolutionary dynamics, 234, 305

concave trade-off, 16

consumer-resource model, 45

Continuous Prisoner's Dilemma, 141

Continuous Snowdrift game, 140; asymmetric, 145

Continuous Tragedy of the Commons, 147

continuously stable strategy, 289, 303

convergence stability, 13; definition of, 284

convex trade-off, 16, 271

cooperative investments, 140

cost of dispersal, 170

coupled map lattices, 166

coupled partial differential equations, 236

covariance matrix, 34, 281

crossfeeding, 262

cultural diffusion, 205

cultural evolution, 148, 195; Catholic Church, 201; Eastern Orthodox Church, 215; epidemiologcal models, 197; group selection, 216; Judaism, 210; meme, 195; of language, 196; of religion, 200; Protestant Church, 201

cultural meme, 195

cultural transmission, 196

Darwin, Charles, 38

Dawkins, Richard, 195

delta function, 152, 220

Dennett, Daniel, 215

deterministic model, 39

diauxic growth, 265

Dieckmann, Ulf, 289

Diekmann, Odo, 280

difference equations, 149, 164

diffusion, 205

disassortative mating, 95, 101

dispersal, 163; coexistence, 178; cyclic evolutionary dynamics, 178; difference equations, 170; habitat patches, 164; spatial heterogeneity, 163; spatial predator-prey models, 164

diversification, 2, 279. *See* adaptive diversification

Dobzhansky-Muller incompatibility, 247

dominance, 78

dynamics of phenotype distributions, 210, 220

E-loci, 75

ecological character displacement, 39

effective density, 41, 218, 250

effector proteins, 243

Endler, John, 248

environmental gradient, 255

epidemiological model of religion, 201

epidemiological models for cultural evolution, 197

epidemiology of culture, 216

epistatic interactions, 243

equivariant bifurcation theory, 260

Escherichia coli, 262

Euler method, 219

evolution, 1

 —adaptive, 1

 —in sexual populations, 74

 —microbial, 262

 —neutral, 2

 —of anisogamy, 180

 —of assortative mating, 90; Allee effects, 88, 96, 105; one-allele models, 91; two-allele models, 104

—of competitive interactions, 50
—of cooperation, 139
—of culture, 195
—of dispersal, 163
—of diversity, 4
—of dominance, 78
—of isogamy, 186
—of language, 197
—of mutualism, 148
—of predator-prey interactions, 113, 236
—of religion, 200
—of trophic preference, 189
—of virulence, 132
evolutionary bifurcation, 305
evolutionary branching, 4, 15; asymmetric,
    136; in bacterial batch cultures, 271, 275;
    definition, 297; dimorphic divergence,
    291; due to convex tradeoffs, 15, 271; due
    to resource competition, 51; fast switcher,
    slow switcher, 275; function-valued traits,
    302; in cultural evolution, 206; in disperal
    rates, 169; in gamete size, 182; in
    host-pathogen models, 132; in microbial
    populations, 271; in mutualism models,
    157; in predator-prey models, 129; in
    religions, 209; in sexual populations, 74,
    85; in small populations, 64; in trophic
    preference, 192; introduction to, 292;
    multiple branching, 55, 63, 68, 209; niche
    position, 69; secondary, 30, 63, 130
evolutionary branching point, 4, 15;
    convergent stable, 15; definition, 297
evolutionary game theory, 3, 139
evolutionary repellor, 143, 304;
    anisogamy, 183
evolutionary stability, 14; three-dimensional
    trait space, 35; definition of, 288;
    introduction to, 287; multi-dimensional
    trait space, 221; predator-prey model, 119
evolutionary suicide, 68
experimental evolution, 262. See also
    microbial evolution
extinction, 56, 68, 178

facilitation of diversification through spatial
    structure, 256
fast switcher, 266
Felsenstein, Joseph, 75, 100
Fisherian runaway process, 110
fitness, 3, 279; cultural, 198; ecological
    definition of, 279; landscape, 3;

maximum, 14; minimum, 4, 14; temporal
    fluctuations, 36; frequency-dependent, 4
fixation, 19
flowering schedules, 300
Fourier transform, 226
frequency dependence in multi-dimensional
    phenotype spaces, 221
frequency-dependent selection, 3, 4
frequency-independent selection, 3
function-valued traits, 299; canonical
    equation, 301

game theory, 140, 197; Continuous
    Snowdrift game, 140; Continuous
    Tragedy of the Commons, 147;
    Hawk-Dove game, 3, 140; Prisoner's
    Dilemma game, 140; Snowdrift game, 140
gamete size, 180
Garden-of-Eden configurations, 289
Gaussian equilibrium distribution, 209, 219,
    228
Gaussian overcrowding function, 206
Gaussian transmission function, 204
Gaussian unstable equilibrium
    distribution, 231
Gaussian viability functions, 11
gene expression, 273
gene flow, 5
gene-culture coevolution, 216
generalist, 15, 26, 46, 277
generalist strategy. See generalist
genetic architecture, 77
genetic assimilation, 268
geographical isolation, 2
geographical range, 251
Geritz, Stefan, 9, 30
Gillespie algorithm, 60, 127, 207
glucose, 262
glucose-acetate environment, 264
glycolysis, 264
gradient dynamics, 279
group action, 259
group selection, 216
growth profile, 265

habitat asymmetry, 22
habitat choice, 31
habitat specialization, 10
habitat viability, 11
Hawk-Dove game, 3, 140
Hessian matrix, 34, 288
heterotrophy, 190

heterozygote advantage, 10
Hopf bifurcation, 305
host race formation, 9
host-pathogen model, 132; burst size, 133;
    epidemiological dynamics, 132; evolution
    of virulence, 132; evolutionary branching,
    136; pessimization principle, 133;
    superinfection, 134; trade-off, 137
human hosts, 197, 215
human society, 201
hybrid necrosis, 243
hypergeometric model, 76

immune system, 243
individual-based models, 6; demographic
    stochasticity, 175; for cultural
    epidemiology, 207; for dispersal, 173; for
    metapopulations, 173; for mutualistic
    interactions, 156; for predator-prey
    coevolution, 127; in discrete time, 174; of
    cultural evolution, 207; of resource
    competition in asexual populations, 60; of
    resource competition in sexual
    populations, 82
indulgences, 201
infinite-dimensional phenotype space,
    299
infinitesimal model, 81
insertion element, 273
interactions between phenotypic
    components, 222
invariant, 260
invasion dynamics, 18
invasion fitness, 12, 280; first order, 16;
    second order, 17
isogamy, 186
Ispolatov, Yaroslav, 219, 232

Jacobian matrix, 24, 284; determinant, 118;
    eigenvalues, 25; eigenvector, 186;
    trace, 118

kernel. *See* competition kernel; mutation
    kernel
Kisdi, Eva, 9, 119
Kondrashov, Alex, 76

Lande, Russell, 279
language meme, 198
Law, Richard, 279
Leimar, Olof, 287
Lenski, Richard, 262

Levene model, 9, 32
limiting resource, 38
linkage disequilibrium, 31, 77
local stability, 14, 165
logistic equation, 40
long-term evolution, 4, 74, 262, 279
long-term per capita growth rate, 279
loss of culture, 203
loss rate, 207
Lotka-Volterra model, 113
lottery model, 32

maintenance of phenotypic diversity, 221
mate choice function, 82, 229
Maynard Smith, John, 9, 188; model, 9, 76
Mayr, Ernst, 9
meme, 195
metabolic pathways, 264
metapopulation model, 170
Metz, Hans, 4, 279
Michaelis-Menten kinetics, 269
microarray, 273
microbial evolution, 262; ancestral strain,
    264; coexistence, 267; colony size
    dimorphism, 265; competition
    experiments, 267; crossfeeding, 270;
    derived strains, 273; ecological niche,
    277; epistasis, 277; fitness landscape, 276;
    fossil record, 275; frequency dependence,
    267; Gause's exclusion principle, 262;
    genomic background, 277; glucose
    specialist, 262, 271; most recent common
    ancestor, 275; point mutation, 271;
    resource generalist, 278; trade-off, 267,
    270
microbial metabolism, 268; *aceB* gene, 272;
    aerobic growth, 273; anaerobic growth,
    273; arcA gene, 273; diauxic growth, 265;
    downregulation, 273; global regulator,
    273; glycolysis, 264; glyoxylate shunt,
    272; *iclR* gene, 273; oxygen, 273;
    regulatory gene, 272; trade-off, 267, 270
migration, 30, 164, 250
mixotrophy, 189
multidimensional phenotype space, 221,
    297; diversification in, 220
multilocus models, 76
multimodal equilibrium, 86
multimodal phenotype distribution, 225; due
    to resource competition, 221; in asexual
    predator-prey models, 238; in sexual
    populations, 230; in sexual predator-prey

models, 240; in spatially structured populations, 254; plant-pathogen model, 248

multiple branching, 55, 63, 68, 209

mutant trait, 12, 18

mutation: one-dimensional adaptive dynamics, 13, 44; two-dimensional adaptive dynamics, 54, 117; covariance, 34, 281; in cultural evolution, 204; in individual-based models, 61; in partial differential equation models, 228; in plant-pathogen model, 246; in predator-prey models, 117; large, 105, 121; matrix, 281; rare, 280; small, 20, 62

mutation kernel, 204, 228

mutational matrix, 281

mutual invasibility, 53, 291; competition experiments, 267; predator-prey model, 121

mutualism, 148; asymmetric, 158; coevolution, 148; sexual populations, 158; symmetric, 155

neutral evolution, 1

niche width, 46, 79, 278

nonequilibrium dynamics, 72, 86, 169, 234

nonlinear functional response, 123

normal form, 259

Nowak, Martin, 196, 279

optimization, 1

out-of-phase two-cycles, 170

out-of-phase attractors, 171

overcrowding function, 203

pairwise invasibility plot, 18, 20, 27, 121, 187

parameter space for frequency dependence, 3

partial differential equation models, 217; Allee effect, 235; bimodal equilibrium solution, 260; coevolutionary, 236; cyclic arms race, 238; dispersal kernel, 250; equivariant bifurcation theory, 259; for asexual predator-prey interactions, 236; for asexual resource competition models, 218; for cultural evolution, 203; for plant-pathogen interactions, 247; for sexual predator-prey interactions, 235; for sexual resource competition models, 230; for spatially structured asexual populations, 254; for spatially structured sexual populations, 251; mutation kernel,

228; mutation-selection balance, 228; nonequilibrium dynamics, 234, 240; normal form, 259; pods, 259; predator attack rate, 236; segregation kernel, 229; temporal instability, 234; with assortative mating, 229

partial differential equations, 217; boundary conditions, 219, 249, 253; for Gaussian competition model, 219; Fourier transform, 226; pattern formation, 226

patch models, 170

pattern formation, 217, 226, 239; in spatially structured populations, 253

per capita birth rate, 39

per capita death rate, 39

period-doubling, 71, 168

permutations, 260

pessimization principle, 133

phenotype space, 4; two-dimensional, 113; multi-dimensional, 221, 300; spatial structure, 250

phenotypic bistability, 268

phenotypic plasticity, 6, 32, 78

plant-pathogen model, 243; *Arabidopsis thaliana*, 243; hybrid inviability, 243; hybrid necrosis, 243; immune system, 243; postzygotic isolation, 243

platykurtic, 86, 211

platykurtic competition kernels, 220, 229

pods, 259

point mutation in global regulator gene, 277

Poisson distribution, 73, 174

polymorphic populations, 30, 44, 63, 217, 236

polymorphism, 9, 31

positive feedback; linkage and selection, 105

postzygotic isolation, 243

power spectrum analysis, 128

predator-prey model, 113, 236; assortative mating, 130; attack optimization, 126; attack rate, 115; conversion efficiency, 115; cyclic arms race, 119, 240; saturating functional response, 123; segregation kernel, 130

prezygotic isolation, 7, 74, 243

Prisoner's Dilemma game, 140

promotor, 80

protected polymorphism, 291

Pseudomonas flourescens, 263
public good, 139

quadratic form, 221
quantitative genetics, 279
quantitative trait, 4, 10, 39, 76, 279;
    cooperation, 140; cultural, 197; dispersal,
    164; religion, 202

random mating, 31, 75
rare mutant, 279
rate-yield trade-off, 263
reaction norms, 300
recombination, 74; between E-loci and
    A-loci, 100
religion, 200. *See* cultural evolution
religious diversification, 206
religious meme, 201
religious variation, 202
replicator dynamics, 141
repressor, 80
reproductive isolation, 7, 74, 78, 243. *See
    also* prezygotic isolation; postzygotic
    isolation
resident attractor, 170
resident community, 280
resident trait, 12, 281
resource distribution, 45, 65
resource preference, 16, 42
resource utilization, 45
Rhagoletis pomonella, 9
Ricker model, 167
rituals, 202
Rosenzweig-MacArthur model, 123; stable
    equilibrium, 128
Runge-Kutta algorithm, 219

seasonal environment, 267
segregation kernel, 80; in partial differential
    equation models, 229; individual-based
    models, 83
selection
    —directional, 18, 69; in sexual
        populations, 79
    —disruptive, 4, 9, 32: in males and
        females, 110; in sexual populations, 75;
        nagging problem, 77
    —vanishing to second order, 70
    —for increased dispersal, 177
    —frequency-dependent, 3, 38
    —frequency-independent, 3, 44
    —hard selection, 29

    —sexual selection, 2, 78, 110
    —stabilizing, 44, 224
selection gradient, 12, 281; two-dimensional,
    22; three-dimensional, 33;
    four-dimensional, 56; adaptive dynamics,
    13; Jacobian matrix, 24; singular point, 13
sensitive dependence on initial conditions,
    168
sexual conflict, 111, 180
sexual dimorphism, 6, 78, 95
sexual selection, 2, 78, 110; speciation,
    110
Sigmund, Karl, 279
singular coalition, 55, 122, 154, 295
singular point, 4, 13; definition of, 282;
    fourth order, 294
slang, 198
slow switcher, 267
small populations, 64
Snowdrift game, 139
social status, 212
social structure, 212
spatial heterogeneity, 36, 173, 250
spatial pattern formation, 254
spatial structure, 163, 242
spatially localized competition, 253
specialist predators, 130
specialist strategy, 25
speciation, 2; allopatric, 4, 110; competitive,
    38; due to sexual selection, 110;
    ecological, 2; mutation order, 5;
    parapatric, 4; sympatric, 4, 75
stability: convergence stability, 13;
    definition of convergence stability, 284;
    definition of evolutionary stability, 288;
    evolutionary stability, 14; focus point,
    116, 128; local stability of population
    dynamics, 168; strong convergence
    stability, 34, 286
stage-structured populations, 211
stochastic equilibrium, 84
stochastic master equation, 279
stochastic models, 60, 279. *See also*
    individual-based models
strategy, 3, 26; generalist, 15, 26, 46, 277;
    specialist, 25, 130
strong convergence stability, 34, 285
superinfection, 134
symmetric competition, 42, 50
symmetry conditions, 260
symmetry group, 260
synchronous dynamics, 170

taxon cycles, 69
Taylor expansion, 168, 282
temporal heterogeneity, 36
trade-off, 10; anisogamy, 180; between size
    and number of gametes, 180; between
    virulence and transmission, 132; concave,
    16; convex, 16, 271; diauxic growth, 267;
    genetic causes, 268; in host-pathogen
    models, 132; in microbial metabolism,
    264; in model for microbial growth, 270;
    in mutualism, 154; in plant-pathogen
    model, 244; in predator-prey models, 126
    in trophic preference, 190; rate-yield, 263;
    seasonal environment, 270
Tragedy of the Commons, 147
transcription factors; promotors, 80;
    repressors, 80
transmission function, 134, 204
transmission rate, 132, 204
trophic levels, 189

trophic preference, 189
two-allele model, 75, 100; evolution to one
    allele mechanism, 108
type II functional response, 123

uniform distribution, 49, 61, 94, 224
unimodal phenotype distribution, 225
unit of selection, 196, 215
Universal Grammar, 197
unstable equilibrium distribution, 231
utilization curve, 45
utilization function, 45

variance-covariance structure, 301
virulence, 132

wrinkly spreader, 263

zygote size, 180

MONOGRAPHS IN POPULATION BIOLOGY

EDITED BY SIMON A. LEVIN AND HENRY S. HORN

1. *The Theory of Island Biogeography*, by Robert H. MacArthur and Edward O. Wilson
2. *Evolution in Changing Environments: Some Theoretical Explorations*, by Richard Levins
3. *Adaptive Geometry of Trees*, by Henry S. Horn
4. *Theoretical Aspects of Population Genetics*, by Motoo Kimura and Tomoko Ohta
5. *Populations in a Seasonal Environment*, by Steven D. Fretwell
6. *Stability and Complexity in Model Ecosystems*, by Robert M. May
7. *Competition and the Structure of Bird Communities*, by Martin L. Cody
8. *Sex and Evolution*, by George C. Williams
9. *Group Selection in Predator-Prey Communities*, by Michael E. Gilpin
10. *Geographic Variation, Speciation, and Clines*, by John A. Endler
11. *Food Webs and Niche Space*, by Joel E. Cohen
12. *Caste and Ecology in the Social Insects*, by George F. Oster and Edward O. Wilson
13. *The Dynamics of Arthropod Predator-Prey Systems*, by Michael P. Hassel
14. *Some Adaptations of Marsh-Nesting Blackbirds*, by Gordon H. Orians
15. *Evolutionary Biology of Parasites*, by Peter W. Price
16. *Cultural Transmission and Evolution: A Quantitative Approach*, by L. L. Cavalli-Sforza and M. W. Feldman
17. *Resource Competition and Community Structure*, by David Tilman
18. *The Theory of Sex Allocation*, by Eric L. Charnov
19. *Mate Choice in Plants: Tactics, Mechanisms, and Consequences*, by Nancy Burley and Mary F. Wilson
20. *The Florida Scrub Jay: Demography of a Cooperative-Breeding Bird*, by Glen E. Woolfenden and John W. Fitzpatrick
21. *Natural Selection in the Wild*, by John A. Endler
22. *Theoretical Studies on Sex Ratio Evolution*, by Samuel Karlin and Sabin Lessard
23. *A Hierarchical Concept of Ecosystems*, by R.V. O'Neill, D.L. DeAngelis, J.B. Waide, and T.F.H. Allen
24. *Population Ecology of the Cooperatively Breeding Acorn Woodpecker*, by Walter D. Koenig and Ronald L. Mumme
25. *Population Ecology of Individuals*, by Adam Lomnicki
26. *Plant Strategies and the Dynamics and Structure of Plant Communities*, by David Tilman
27. *Population Harvesting: Demographic Models of Fish, Forest, and Animal Resources*, by Wayne M. Getz and Robert G. Haight

28. *The Ecological Detective: Confronting Models with Data*, by Ray Hilborn and Marc Mangel

29. *Evolutionary Ecology across Three Trophic Levels: Goldenrods, Gallmakers, and Natural Enemies*, by Warren G. Abrahamson and Arthur E. Weis

30. *Spatial Ecology: The Role of Space in Population Dynamics and Interspecific Interactions*, edited by David Tilman and Peter Kareiva

31. *Stability in Model Populations*, by Laurence D. Mueller and Amitabh Joshi

32. *The Unified Neutral Theory of Biodiversity and Biogeography*, by Stephen P. Hubbell

33. *The Functional Consequences of Biodiversity: Empirical Progress and Theoretical Extensions*, edited by Ann P. Kinzig, Stephen J. Pacala, and David Tilman

34. *Communities and Ecosystems: Linking the Aboveground and Belowground Components*, by David Wardle

35. *Complex Population Dynamics: A Theoretical/Empirical Synthesis*, by Peter Turchin

36. *Consumer-Resource Dynamics*, by William W. Murdoch, Cheryl J. Briggs, and Roger M. Nisbet

37. *Niche Construction: The Neglected Process in Evolution*, by F. John Odling-Smee, Kevin N. Laland, and Marcus W. Feldman

38. *Geographical Genetics*, by Bryan K. Epperson

39. *Consanguinity, Inbreeding, and Genetic Drift in Italy*, by Luigi Luca Cavalli-Sforza, Antonio Moroni, and Gianna Zei

40. *Genetic Structure and Selection in Subdivided Populations*, by Franois Rousset

41. *Fitness Landscapes and the Origin of Species*, by Sergey Gavrilets

42. *Self-Organization in Complex Ecosystems*, by Ricard V. Solé and Jordi Bascompte

43. *Mechanistic Home Range Analysis*, by Paul R. Moorcroft and Mark A. Lewis

44. *Sex Allocation*, by Stuart West

45. *Scale, Heterogeneity, and the Structure of Diversity of Ecological Communities*, by Mark E. Ritchie

46. *From Populations to Ecosystems: Theoretical Foundations for a New Ecological Synthesis*, by Michel Loreau

47. *Resolving Ecosystem Complexity*, by Oswald J. Schmitz

48. *Adaptive Diversification*, by Michael Doebeli

Milton Keynes UK
Ingram Content Group UK Ltd.
UKHW020641290824
447545UK00007B/217